Philosophy of Pseudoscience

Reconsidering the
Demarcation Problem

Philosophy of Pseudoscience

Edited by
Massimo Pigliucci
and Maarten Boudry

The University of Chicago Press
Chicago and London

Massimo Pigliucci is professor of philosophy at the City University of New York.
Maarten Boudry is a postdoctoral fellow of the Flemish Fund for Scientific Research at
Ghent University.

The University of Chicago Press, Chicago 60637
The University of Chicago Press, Ltd., London
© 2013 by The University of Chicago
All rights reserved. Published 2013.
Printed in the United States of America

22 21 20 19 18 17 16 15 14 13   1 2 3 4 5

ISBN-13: 978-0-226-05179-6   (cloth)
ISBN-13: 978-0-226-05196-3   (paper)
ISBN-13: 978-0-226-05182-6   (e-book)

Library of Congress Cataloging-in-Publication Data

Philosophy of pseudoscience : reconsidering the demarcation problem / edited by
Massimo Pigliucci and Maarten Boudry.
     pages ; cm
   Includes bibliographical references and index.
   ISBN 978-0-226-05179-6 (cloth : alkaline paper) — ISBN 978-0-226-05196-3
(paperback : alkaline paper) — ISBN 978-0-226-05182-6 (e-book)   1. Pseudoscience.
2. Science.   I. Pigliucci, Massimo, 1964–, editor.   II. Boudry, Maarten, 1984–,
editor.
   Q172.5.P77P48 2013
   001.9—dc23

                                                                2013000805

CONTENTS

Introduction: Why the Demarcation Problem Matters   1

MASSIMO PIGLIUCCI AND MAARTEN BOUDRY

PART I
WHAT'S THE PROBLEM WITH THE
DEMARCATION PROBLEM?

1   The Demarcation Problem   9

A (Belated) Response to Laudan

MASSIMO PIGLIUCCI

2   Science and Pseudoscience   29

How to Demarcate after the (Alleged) Demise of the Demarcation Problem

MARTIN MAHNER

3   Toward a Demarcation of Science from Pseudoscience   45

JAMES LADYMAN

4   Defining Pseudoscience and Science   61

SVEN OVE HANSSON

5   Loki's Wager and Laudan's Error   79

On Genuine and Territorial Demarcation

MAARTEN BOUDRY

PART II
HISTORY AND SOCIOLOGY OF PSEUDOSCIENCE

6   The Problem of Demarcation   101

History and Future

THOMAS NICKLES

7   Science, Pseudoscience, and Science Falsely So-Called   121

DANIEL P. THURS AND RONALD L. NUMBERS

8   Paranormalism and Pseudoscience as Deviance   145

ERICH GOODE

9   Belief Buddies versus Critical Communities   165

The Social Organization of Pseudoscience

NORETTA KOERTGE

PART III
THE BORDERLANDS BETWEEN SCIENCE
AND PSEUDOSCIENCE

10   Science and the Messy, Uncontrollable World of Nature   183

CAROL E. CLELAND AND SHERALEE BRINDELL

11   Science and Pseudoscience   203

The Difference in Practice and the Difference It Makes

MICHAEL SHERMER

12   Evolution   225

*From Pseudoscience to Popular Science, from Popular Science
to Professional Science*

MICHAEL RUSE

PART IV
SCIENCE AND THE SUPERNATURAL

13 Is a Science of the Supernatural Possible? 247

EVAN FALES

14 Navigating the Landscape between Science and Religious Pseudoscience 263

Can Hume Help?

BARBARA FORREST

PART V
TRUE BELIEVERS AND THEIR TACTICS

15 Argumentation and Pseudoscience 287

The Case for an Ethics of Argumentation

JEAN PAUL VAN BENDEGEM

16 Why Alternative Medicine Can Be Scientifically Evaluated 305

Countering the Evasions of Pseudoscience

JESPER JERKERT

17 Pseudoscience 321

The Case of Freud's Sexual Etiology of the Neuroses

FRANK CIOFFI

18 The Holocaust Denier's Playbook and the Tobacco Smokescreen 341

Common Threads in the Thinking and Tactics of Denialists and Pseudoscientists

DONALD PROTHERO

PART VI
THE COGNITIVE ROOTS OF PSEUDOSCIENCE

19 Evolved to Be Irrational? 361

Evolutionary and Cognitive Foundations of Pseudosciences

STEFAAN BLANCKE AND JOHAN DE SMEDT

20 **Werewolves in Scientists' Clothing**   381

Understanding Pseudoscientific Cognition

KONRAD TALMONT-KAMINSKI

21 **The Salem Region**   397

Two Mindsets about Science

JOHN S. WILKINS

22 **Pseudoscience and Idiosyncratic Theories of Rational Belief**   417

NICHOLAS SHACKEL

23 **Agentive Thinking and Illusions of Understanding**   439

FILIP BUEKENS

Contributors   459

Index   461

Why the Demarcation Problem Matters

MASSIMO PIGLIUCCI AND MAARTEN BOUDRY

Ever since Socrates, philosophers have been in the business of asking questions of the type "What is X?" The point has not always been to actually find out what X is, but rather to explore how we think about X, to bring up to the surface wrong ways of thinking about it, and hopefully in the process to achieve an increasingly better understanding of the matter at hand. In the early part of the twentieth century one of the most ambitious philosophers of science, Karl Popper, asked that very question in the specific case in which X = science. Popper termed this the "demarcation problem," the quest for what distinguishes science from nonscience and pseudoscience (and, presumably, also the latter two from each other).

As the first chapters in this collection explain, Popper thought he had solved the demarcation problem by way of his criterion of falsifiability, a solution that seemed very convincing when he compared the eminently falsifiable theory of general relativity with the entirely unfalsifiable theory of psychoanalysis (Freudian or otherwise). Modern philosophers—made more wary by widespread appreciation of the issues raised in this context by the works of Pierre Duhem and W. V. O. Quine—have come to the conclusion that Popper was a bit too quick in declaring victory. They recognize that science is not a unified type of activity and that an ever-changing continuous landscape may connect it with nonscientific endeavors.

Nonetheless, the contributors to this volume also think that Larry Laudan's famous dismissal of the demarcation problem—almost three decades ago now—as an ill-conceived and even pernicious pseudoproblem, and of terms like "pseudoscience" as pieces of hollow rhetoric, was just as premature and misguided. Laudan may have forgotten Socrates' lesson: even if we do not arrive at a neat and exceptionless formal definition of some X, based on a small set of necessary and jointly sufficient conditions, we may still come to learn a lot in the process. If we raise the bar for the demarcation project too high, settling for nothing less than a timeless and essential definition, a death pronouncement such as Laudan's is all too easy to make. As Daniel Dennett put it in *Darwin's Dangerous Idea: Evolution and the Meanings of Life* (1995), "nothing complicated enough to be really interesting could have an essence."

Philosophers and scientists readily recognize a pseudoscience when they see one. Of course, certain interesting borderline cases are hotly disputed among scientists and philosophers, but even Popper's notorious critic Thomas Kuhn acknowledged that, despite their philosophical differences about demarcation, both of them were in remarkable agreement about paradigmatic cases, as were most of their colleagues. To argue that philosophers can neither spell out which criteria we implicitly rely on to tell science from pseudoscience, nor are able to evaluate and further refine those criteria, would be to relinquish one of the most foundational tasks of philosophy (what is knowledge? how do we attain it?). For too long, philosophers have been dwelling over technical problems and exceptions to formal demarcation criteria, only to rashly conclude that the demarcation problem is dead and that there is no such thing as "pseudoscience." We think this is mistaken.

This volume testifies to a lively and constructive discussion about demarcationism among philosophers, sociologists, historians, and professional skeptics. By proposing something of a new philosophical subdiscipline, the Philosophy of Pseudoscience, we hope to convince those who have followed in Laudan's footsteps that the term "pseudoscience" does single out something real that merits our attention. A ballpark demarcation of pseudoscience—with a lot of blanks to be filled in—is not difficult to come up with: if a theory strays from the epistemic desiderata of science by a sufficiently wide margin while being touted as scientific by its advocates, it is justifiably branded as pseudoscience.

The nature of science and the difference between science and pseudoscience are crucial topics for philosophers, historians, and sociologists of science for two fundamental reasons. First, science is having an ever-increasing impact in modern society. Science commands much public attention and prestige; it is funded at very high levels by governments and the private sector; its departments take more and more space and resources on university campuses; and its products may be beneficial to human welfare or bring about great destruction on a scale never before imaginable. It is therefore of compelling interest to all of us to understand the nature of science, its epistemic foundations, its limits, and even its power structure—which, of course, is precisely what philosophy, history, and sociology of science are set up to do.

Second, and in a complementary way, we also need a philosophical (and historical and sociological) understanding of the phenomenon of pseudoscience. The lack of interest for pseudoscience in some philosophical quarters derives from the tacit assumption that some ideas and theories are so *obviously* wrong that they are not even worth arguing about. Pseudoscience is still too often considered a harmless pastime indulged in by a relatively small number of people with an unusual penchant for mystery worship. This is far from the truth. In the form of creationism and its challenges to the study of evolution, pseudoscience has done great damage to public education in the United States and elsewhere; it has swindled people of billions of dollars in the form of "alternative" medicine like homeopathy; it has caused a lot of emotional distress, for example, to people who are told by mystics and assorted charlatans that they can talk with their dead loved ones. Conspiracy theories about AIDS, which are widespread in many African countries and even in the United States, have literally killed countless human beings throughout the world. Denialism about climate change, which seems to be ineradicable in conservative political circles, may even help to bring about a worldwide catastrophe. Dangerous cults and sects such as Scientology, which are based on pseudoscientific belief systems, continue to attract followers and wreak havoc in people's lives. Even apart from the very real consequences of pseudoscience, we should pause to consider the huge amount of intellectual resources that are wasted in shoring up discredited theories like creationism, homeopathy, and psychoanalysis, not to mention the never-ending quest for evidence of the paranormal and the indefatigable activism of conspiracy theorists.

Pseudoscience can cause so much trouble in part because the public does not appreciate the difference between real science and something that masquerades as science. Pseudoscientists seem to win converts because of a combination of science parroting and of distrust of academic authorities, both of which appear to be particularly palatable to so many people. In addition, pseudoscience thrives because we have not fully come to grips yet with the cognitive, sociological, and epistemological roots of this phenomenon. This is why the demarcation problem is not only an exciting intellectual puzzle for philosophers and other scholars, but is one of the things that makes philosophy actually relevant to society. Philosophers, accordingly, do not just have a scholarly duty in this area, but ethical and social ones as well. For all these reasons, we asked some of the most prominent and original thinkers on science and pseudoscience to contribute to this edited volume. The result is a collection of twenty-four essays, grouped under six thematic sections, to help bring some order to a large, complex, and inherently interdisciplinary field.

In the first part on "the problem with the demarcation problem," Massimo Pigliucci assesses in some detail Laudan's objections to the research program and goes on to propose an approach based on a quantifiable version of Wittgensteinian family resemblance. In a similar vein, Martin Mahner suggests a cluster approach to demarcationism, drawing inspiration from the taxonomy of biological species, which does not yield to essentialist definitions either. James Ladyman deploys Harry Frankfurt's famous analysis of "bullshit" to highlight the difference between pseudoscience and straightforward scientific fraud. Sven Hansson recasts the demarcation problem in terms of epistemic warrant and proposes an approach that views science as unified on an epistemological level, while still accounting for diversity in its methods. Maarten Boudry tries to clear up some confusion between what he calls genuine demarcation (the science/pseudoscience boundaries) and the "territorial" demarcation between science and other epistemic fields (philosophy, mathematics).

The second part deals with the history and sociology of pseudoscience. Thomas Nickles gets things started with a brief but comprehensive history of the demarcation problem, which leads into Daniel Thurs and Ronald Numbers's historical analysis of pseudoscience, which tracks down the coinage and currency of the term and explains its shifting meaning in tandem with the emerging historical identity of science. While we purposefully steered clear from the kind of sociology inspired by social constructivism and postmodernism—which we regard as a type of pseudodiscipline in its own right—sociologist Erich Goode provides an analysis of paranormalism as a "deviant discipline" violating the consensus of established science, and Noretta Koertge draws our attention to the characteristic social organization of pseudosciences as a means of highlighting the sociological dimension of the scientific endeavor.

The third part explores the territory marking the "borderlands" between science and pseudoscience. Carol Cleland and Sheralee Brindell deploy the idea of causal asymmetries in evidential reasoning to differentiate between what are sometime referred to as "hard" and "soft" sciences, and argue that misconceptions about this difference explain the higher incidence of pseudoscience and antiscience in the nonexperimental sciences. Professional skeptic of pseudoscience Michael Shermer looks at the demographics of pseudoscientific belief and examines how the demarcation problem is treated in legal cases. In a surprising twist, Michael Ruse tells us of a time when the concept of evolution was in fact treated as pseudoscience and then popular science, before blossoming into a professional science, thus challenging a conception of demarcation in terms of timeless and purely formal principles.

Part 4, on science and the supernatural, begins with Evan Fales arguing that, contrary to recent philosophical discussions, the appeal to the supernatural should not be ruled out as science for methodological reasons, but rather because the notion of supernatural intervention probably suffers from fatal flaws. Meanwhile, Barbara Forrest enlists David Hume to help navigating the treacherous territory between science and religious pseudoscience and to assess the epistemic credentials of supernaturalism.

The fifth part of the volume focuses on the tactics deployed by "true believers" in pseudoscience, beginning with Jean Paul Van Bendegem's discussion of the ethics of argumentation about pseudoscience, followed by Jesper Jerkert's contention that alternative medicine can be evaluated scientifically—contra the immunizing strategies deployed by some of its most vocal supporters. Frank Cioffi, whose 2012 passing we mourn, summarizes his misgivings about Freudian psychoanalysis and argues that we should move beyond assessments of the testability and other logical properties of a theory, focusing instead on spurious claims of validation and other recurrent misdemeanors on the part of pseudoscientists. Donald Prothero describes the different strategies used by climate change "skeptics" and other denialists, outlining the links between new and "traditional" pseudosciences.

Finally, we close with a section examining the complex cognitive roots of pseudoscience. Stefaan Blancke and Johan De Smedt ask whether we actually evolved to be irrational, describing a number of evolved heuristics that are rational in ecologically relevant domains, but lead us astray in other contexts. Konrad Talmont-Kaminski explores the noncognitive functions of superempirical beliefs and analyzes the different attitudes of science and pseudoscience toward intuitive beliefs. John Wilkins distinguishes between two mindsets about science and explores the cognitive styles relating to authority and tradition in both science and pseudoscience. Nicholas Shackel proposes that belief in pseudoscience may be partly explained in terms of idiosyncratic theories about the ethics of belief, and Filip Buekens ends the volume with a chapter on pseudohermeneutics and the illusion of understanding, drawing inspiration from the cognitive psychology and philosophy of intentional thinking.

This collection will certainly not represent the final word on the issue of demarcation. On the contrary, it is meant to renew and stimulate discussion in an area of philosophy of science that is both intrinsically interesting from an intellectual point of view and that, for once, can actually make philosophy directly relevant to people's lives.

Acknowledgments

We gratefully acknowledge Séverine Lahousse, An Ravelingien, Griet Vandermassen, Alexis De Tiège, and Stefaan Blancke for generous help with the index; Christie Henry as our wonderful editor; Amy Krynak for her invaluable assistance; and Lisa Wehrle as our stupendous copy editor.

PART I

What's the Problem with the Demarcation Problem?

1

The Demarcation Problem

A (Belated) Response to Laudan

MASSIMO PIGLIUCCI

The Premature Obituary of the Demarcation Problem

The "demarcation problem," the issue of how to separate science from pseudoscience, has been around since fall 1919—at least according to Karl Popper's (1957) recollection of when he first started thinking about it. In Popper's mind, the demarcation problem was intimately linked with one of the most vexing issues in philosophy of science, David Hume's problem of induction (Vickers 2010) and, in particular, Hume's contention that induction cannot be logically justified by appealing to the fact that "it works," as that in itself is an inductive argument, thereby potentially plunging the philosopher straight into the abyss of a viciously circular argument.

Popper famously thought he had solved both the demarcation and induction problems in one fell swoop, by invoking falsification as the criterion that separates science from pseudoscience. Not only, according to Popper, do scientific hypotheses have to be falsifiable (while pseudoscientific ones are not), but since falsification is an application of *modus tollens*, and hence a type of deductive thinking, we can get rid of induction altogether as the basis for scientific reasoning and set Hume's ghost to rest once and for all.

As it turns out, however, although Popper did indeed have several important things to say about both demarcation and induction, philosophers are still very much debating both issues as live ones (see, e.g., Okasha 2001 on induction, and Hansson 2009 on demarcation). The fact that we continue to discuss the issue of demarcation may seem peculiar, though, considering that Laudan (1983) allegedly laid to rest the problem once and for all. In a much referenced paper quite definitively entitled "The Demise of the Demarcation Problem," Laudan concluded that "the [demarcation] question is both uninteresting and, judging by its checkered past, intractable. If we would stand up and be counted on the side of reason, we ought to drop terms like 'pseudoscience' and 'unscientific' from our vocabulary" (Laudan 1983, 125).

At the risk of being counted on the side of unreason, in this chapter I argue that Laudan's requiem for the demarcation problem was much too premature. First, I quickly review Popper's original arguments concerning demarcation and falsification (but not those relating to induction, which is beyond the scope of this contribution); second, I comment on Laudan's brief history of the demarcation problem as presented in parts 2 and 4 of his paper; third, I argue against Laudan's "metaphilosophical interlude" (part 3 of his paper), where he sets out the demarcation problem as he understands it; and last, I propose to rethink the problem itself, building on an observation made by Kuhn (1974, 803) and a suggestion contributed by Dupré (1993, 242). (Also see in this volume, Boudry, chapter 5; Hansson, chapter 4; Koertge, chapter 9; and Nickles, chapter 6.)

Popper's Attack

Popper (1957) wanted to distinguish scientific theories or hypotheses from nonscientific and pseudoscientific ones, and was unhappy with what he took to be the standard answer to the question of demarcation: science, unlike pseudoscience (or "metaphysics"), works on the basis of the empirical method, which consists of an inductive progression from observation to theories. If that were the case, Popper reckoned, astrology would have to rank as a science, albeit as a spectacularly unsuccessful one (Carlson 1985). Popper then set out to compare what in his mind were clear examples of good science (e.g., Albert Einstein's general theory of relativity) and pseudoscience (e.g., Marxist theories of history, Freudian psychoanalysis, and Alfred Adler's "individual psychology") to figure out what exactly distinguishes the first from the second group. I use a much broadened version of the same comparative approach toward the end of this essay to arrive at my own proposal for the problem raised by Popper.

Popper was positively impressed by the then recent spectacular confirmation of Einstein's theory after the 1919 total solar eclipse. Photographs taken by Arthur Eddington during the eclipse confirmed a daring and precise prediction made by Einstein, concerning the slight degree by which light coming from behind the sun would be bent by the latter's gravitational field. By the same token, however, Popper was highly unimpressed by Marxism, Freudianism, and Adlerianism. For instance, here is how he recalls his personal encounter with Adler and his theories:

> Once, in 1919, I reported to [Adler] a case which to me did not seem particularly Adlerian, but which he found no difficulty in analysing in terms of his theory of inferiority feelings, although he had not even seen the child. Slightly shocked, I asked him how he could be so sure. "Because of my thousandfold experience," he replied; whereupon I could not help saying: "And with this new case, I suppose, your experience has become thousand-and-one-fold." (Popper 1957, sec. 1)

Regardless of whether one agrees with Popper's analysis of demarcation, there is something profoundly right about the contrasts he sets up between relativity theory and psychoanalysis or Marxist history: anyone who has had even a passing acquaintance with both science and pseudoscience cannot but be compelled to recognize the same clear difference that struck Popper as obvious. I maintain in this essay that, as long as we agree that there *is* indeed a recognizable difference between, say, evolutionary biology on the one hand and creationism on the other, *then* we must also agree that there are demarcation criteria—however elusive they may be at first glance.

Popper's analysis led him to a set of seven conclusions that summarize his take on demarcation (Popper 1957, sec. 1):

1. Theory confirmation is too easy.
2. The only exception to statement 1 is when confirmation results from risky predictions made by a theory.
3. Better theories make more "prohibitions" (i.e., predict things that should *not* be observed).
4. Irrefutability of a theory is a vice, not a virtue.
5. Testability is the same as falsifiability, and it comes in degrees.
6. Confirming evidence counts only when it is the result of a serious attempt at falsification (this is, it should be noted, somewhat redundant with statement 2 above).

7. A falsified theory can be rescued by employing ad hoc hypotheses, but this comes at the cost of a reduced scientific status for the theory in question.

The problems with Popper's solution are well known, and we do not need to dwell too much on them. Briefly, as even Popper acknowledged, falsificationism is faced with (and, most would argue, undermined by) the daunting problem set out by Pierre Duhem (see Needham 2000). The history of science clearly shows that scientists do *not* throw a theory out as soon as it appears to be falsified by data, as long as they think the theory is promising or has been fruitful in the past and can be rescued by *reasonable* adjustments of ancillary conditions and hypotheses. It is what Johannes Kepler did to Nicolaus Copernicus's early insight, as well as the reason astronomers retained Newtonian mechanics in the face of its apparent inability to account for the orbit of Uranus (a move that quickly led to the discovery of Neptune), to mention but two examples.[1] Yet, as Kuhn (1974, 803) aptly noticed, even though his and Popper's criteria of demarcation differed profoundly (and he obviously thought Popper's to be mistaken), they did seem to agree on where the fault lines run between science and pseudoscience: which brings me to an examination and critique of Laudan's brief survey of the history of demarcation.

Laudan's Brief History of Demarcation

Two sections of Laudan's (1983, secs. 2, 4) critique of demarcation are devoted to a brief critical history of the subject, divided into "old demarcationist tradition" and "new demarcationist tradition" (and separated by the "metaphilosophical interlude" in section 3, to which I come next). Though much is right in Laudan's analysis, I disagree with his fundamental take on what the history of the demarcation problem tells us: for him, the rational conclusion is that philosophers have failed at the task, probably because the task itself is hopeless. For me, the same history is a nice example of how philosophy makes progress: by considering first the obvious moves or solutions, then criticizing them to arrive at more sophisticated moves, which are in turn criticized, and so on. The process is really not entirely disanalogous with that of science, except that philosophy proceeds in logical space rather than by empirical evidence.

For instance, Laudan is correct that Aristotle's goal of scientific analysis as proceeding by logical demonstrations and arriving at universals is simply not attainable. But Laudan is too quick, I think, in rejecting Parmenides' distinction between *episteme* (knowledge) and *doxa* (opinion), a rejection that he traces to the success of fallibilism in epistemology during the nineteenth century (more on this in a moment). But the dividing line between knowledge and opinion does not have (and in fact *cannot be*) sharp, just like the dividing line between science and pseudoscience cannot be sharp, so that fallibilism does not, in fact, undermine the possibility of separating knowledge from mere opinion. Fuzzy lines and gradual distinctions—as I argue later—still make for useful separations.

Laudan then proceeds with rejecting Aristotle's other criterion for demarcation, the difference between "know-how" (typical of craftsmen) and "know-why" (what the scientists are aiming at), on the ground that this would make pre-Copernican astronomy a matter of craftsmanship, not science, since pre-Copernicans simply knew how to calculate the positions of the planets and did not really have any scientific idea of what was actually causing planetary motions. Well, I will bite the bullet here and agree that protoscience, such as pre-Copernican astronomy, does indeed share some aspects with craftsmanship. Even Popper (1957, sec. 2) agreed that science develops from protoscientific myths: "I realized that such myths may be developed, and become testable; and that a myth may contain important anticipations of scientific theories."

Laudan makes much of Galileo Galilei's and Isaac Newton's contentions that they were not after causes, *hypothesis non fingo* to use Newton's famous remark about gravity, and yet they were surely doing science. Again, true enough, but both of those great thinkers stood at the brink of the historical period where physics was transitioning from protoscience to mature science, so that it was clearly way too early to search for causal explanations. But no physicist worth her salt today (or, indeed, shortly after Newton) would agree that one can be happy with a science that ignores the search for causal explanations. Indeed, historical transitions away from pseudoscience, when they occur (think of the difference between alchemy and chemistry), involve intermediate stages similar to those that characterized astronomy in the sixteenth and seventeenth centuries and physics in the seventeenth and eighteenth centuries. But had astronomers and physicists not eventually abandoned Galileo's and Newton's initial caution about hypotheses, we would have had two aborted sciences instead of the highly developed disciplines that we so admire today.

Laudan then steps into what is arguably one of the most erroneous claims of his paper: the above mentioned contention that the onset of fallibilism in

epistemology during the nineteenth century meant the end of any meaningful distinction between knowledge and opinion. If so, I wager that scientists
themselves have not noticed. Laudan does point out that "several nineteenth
century philosophers of science tried to take some of the sting out of this
volte-face [i.e., the acknowledgment that absolute truth is not within the grasp
of science] by suggesting that scientific opinions were more probable or more
reliable than non-scientific ones" (Laudan 1983, 115), leaving his readers to
wonder why exactly such a move did not succeed. Surely Laudan is not arguing that scientific "opinion" is *not* more probable than "mere" opinion. If
he were, we should count him amongst postmodern epistemic relativists, a
company that I am quite sure he would eschew.

Laudan proceeds to build his case against demarcation by claiming that,
once fallibilism was accepted, philosophers reoriented their focus to investigate and epistemically justify science as a *method* rather than as a body of
knowledge (of course, the two are deeply interconnected, but we will leave
that aside for the present discussion). The history of that attempt naturally
passes through John Stuart Mill's and William Whewell's discussions about the
nature of inductive reasoning. Again, Laudan reads this history in an entirely
negative fashion, while I—perhaps out of a naturally optimistic tendency—
see it as yet another example of progress in philosophy. Mill's ([1843] 2002)
five methods of induction and Whewell's (1840) concept of inference to the
best explanation represent marked improvements on Francis Bacon's (1620)
analysis, based as it was largely on enumerative induction. These are mile-
stones in our understanding of inductive reasoning and the workings of science, and to dismiss them as "ambiguous" and "embarrassing" is both pre-
sumptuous and a disservice to philosophy as well as to science.

Laudan then moves on to twentieth-century attempts at demarcation, beginning with the logical positivists. It has become a fashionable sport among
philosophers to dismiss logical positivism out of hand, and I am certainly not
about to mount a defense of it here (or anywhere else, for that matter). But,
again, it strikes me as bizarre to argue that the exploration of another corner of
the logical space of possibilities for demarcation—the positivists' emphasis on
theories of meaning—was a waste of time. It is *because* the positivists and their
critics explored and eventually rejected that possibility that we have made
further progress in understanding the problem. This is the general method of
philosophical inquiry, and for a philosopher to use these "failures" as a reason
to reject an entire project is akin to a scientist pointing out that because New-

tonian mechanics turned out to be wrong, we have made no progress in our understanding of physics.

After dismissing the positivists, Laudan turns his guns on Popper, another preferred target amongst philosophers of science. Here, however, Laudan comes close to admitting what a more sensible answer to the issue of demarcation may turn out to be, one that was tentatively probed by Popper himself: "One might respond to such criticisms [of falsificationism] by saying that scientific status is a matter of degree rather than kind" (Laudan 1983, 121). One might indeed do so, but instead of pursuing that possibility, Laudan quickly declares it a dead end on the grounds that "acute technical difficulties confront this suggestion." That may be the case, but it is nonetheless true that within the sciences themselves there has been quite a bit of work done (admittedly, much of it since Laudan's paper) to make the notion of quantitative comparisons of alternative theories more rigorous. These days this is done by way of either Bayesian reasoning (Henderson et al. 2010) or some sort of model selection approach like the Akaike criterion (Sakamoto and Kitagawa 1987). It is beyond me why this sort of approach could not be one way to pursue Popper's eminently sensible intuition that scientificity is a matter of degrees. Indeed, I argue below that something along these lines is actually a much more promising way to recast the demarcation problem, following an early suggestion by Dupré (1993). For now, though, suffice it to say that even scientists would agree that some hypotheses are more testable than others, not just when comparing science with proto- or pseudoscience, but within established scientific disciplines themselves, even if this judgment is not exactly quantifiable. For instance, evolutionary psychology's claims are notoriously far more difficult to test than similarly structured hypotheses from mainstream evolutionary biology, for the simple reason that human behavioral traits happen to be awful subjects of historical investigation (Kaplan 2002; Pigliucci and Kaplan 2006, chap. 7). Or consider the ongoing discussion about the (lack of) testability of superstring and allied family of theories in fundamental physics (Voit 2006; Smolin 2007).

Laudan eventually gets to what really seems to be bothering him: "Unwilling to link scientific status to any evidential warrant, twentieth century demarcationists have been forced into characterizing the ideologies they oppose (whether Marxism, psychoanalysis or creationism) as untestable in principle. Very occasionally, that label is appropriate" (Laudan 1983, 122). I am not sure why ideology needs to be brought in. I am certainly not naive enough to suggest that anyone—scientists, philosophers, or pseudoscientists—do not subscribe to ideological positions that influence their claims. But surely we can constructively do philosophy nonetheless, and do not have to confine ourselves to politics and psychology. Popper actually wrote that "the Marxist theory of history, in spite of the serious efforts of some of its founders and followers, ultimately adopted this soothsaying practice [making its predictions so vague that they become irrefutable]. In some of its earlier formulations (for example in Marx's analysis of the character of the 'coming social revolution') their predictions were testable, and in fact falsified" (Popper 1957, sec. 2). In other words, Popper saw Marxist theories of history as analogous to the modern case of cold fusion (Huizenga 1992), an initially legitimate scientific claim that was eventually falsified but that degenerated into a pseudoscience in the hands of a small cadre of people who simply refuse to give up the idea regardless of the evidence.

As far as Freudian and Adlerian theories are concerned, again they are no longer taken seriously as scientific ideas by the practicing cognitive science community, as much as they were important (particularly Freud's) in the historical development of the field (see Cioffi, this volume). When it comes to creationism, things are a bit more complicated: very few scientists, and possibly philosophers, would maintain that *specific* creationist claims are not testable. Just as in the case of claims from, say, astrology or parapsychology, one *can* easily test young creationists' contention that the earth is only a few thousand years old. But these tests do not make a science out of creationism for the simple reason that either one must accept that the contention has been conclusively falsified, or one must resort to the inscrutable and untestable actions, means, and motives of a creator god. When a young-earth creationist is faced with geological evidence of an old earth, he has several retorts that seem completely logical to him, even though they actually represent the very reasons why creationism is a pseudoscience: the methods used to date rocks are flawed (for reasons that remain unexplained); the laws of physics have changed over time (without any evidence to support the suggestion); or God simply created a world that *looks* like it is old so that He could test our faith (called "last Thursday" defense, which deserves no additional commentary). So, *pace* Laudan, there are perfectly good, principled, not ideological reasons to label Marxism, Freudianism, and creationism as pseudosciences—even though the details of these reasons vary from case to case.

The rest of Laudan's critique boils down to the argument that no demarcation criterion proposed so far can provide a set of necessary and sufficient conditions to define an activity as scientific, and that the "epistemic hetero-geneity of the activities and beliefs customarily regarded as scientific" means that demarcation is a futile quest. I agree with the former point, but I argue below that it represents a problem only for a too narrowly constructed de-marcation project; the second point has some truth to it, but its extent and consequences are grossly exaggerated by Laudan within the context of this discussion.

Laudan's "Metaphilosophy"

Laudan maintains that the debate about demarcation hinges on three con-siderations that he labels as "metaphilosophical" (though it is not clear to this reader, at least, why the "meta" prefix is necessary). Briefly, these are: "(1) What conditions of adequacy should a proposed demarcation criterion satisfy? (2) Is the criterion under consideration offering necessary or suffi-cient conditions, or both, for scientific status? (3) What actions or judgments are implied by the claim that a certain belief or activity is 'scientific' or 'unsci-entific'?" (Laudan 1983, 117). As we shall see, I agree with Laudan's answer to question 1, I think that question 2 is too simplistic as formulated, and I force-fully reject his answer to question 3.

Laudan correctly argues (question 1) that modern philosophers think-ing about demarcation ought to take seriously what most people, particularly most scientists, actually agree to count as science and pseudoscience. That is, it would be futile to pursue the question in a Platonic way, attempting to arrive at a priori conclusions regardless of whether and to what extent they match scientists' (and most philosophers') intuitions about what science is and is not. Indeed, I think of the target of demarcation studies along the lines sketched in figure 1.1: some activities (and the theories that characterize them) represent established science (e.g., particle physics, climate science, evolutionary biology, molecular biology); others are often treated as "soft" sciences (e.g., economics, psychology, sociology; Pigliucci 2002), character-ized by some of that "epistemic heterogeneity" referred to above; yet more efforts are best thought of as proto- or quasi-scientific (e.g., the Search for Extraterrestrial Intelligence, superstring physics, at least some evolution-ary psychology, and scientific approaches to history); finally, a number of activities unquestionably represent what most scientists and philosophers would regard as pseudoscience (Intelligent Design "theory," astrology, HIV denialism, etc.). Figure 1.1 is obviously far from exhaustive, but it captures

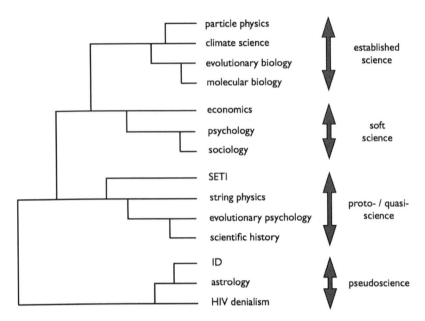

Figure 1.1

Laudan's idea that—no matter how we philosophize about it—demarcation analyses should come up with something that looks like the cluster diagram I sketched, or we would have reasonable doubts that the analysis was not on the right track. To some this might seem like an undue concession to empirical evidence based on common practice and intuition, and one could argue that philosophical analysis is most interesting when it does *not* support common sense. That may be, but our task here is to understand what differentiates a number of actual human practices, so empirical constraints are justified, within limits.

I also agree with Laudan (1983, 118) that "minimally, we expect a demarcation criterion to identify the *epistemic* or *methodological* features which mark off scientific beliefs from unscientific ones," though these criteria (necessarily plural, I think) would have to include much more than Laudan was likely thinking about, for instance, considerations of science as a social activity of a particular kind, with a number of structures in place (e.g., peer review) and desiderata (e.g., cultural diversity) that contribute indirectly to its epistemic and methodological features (Longino 1990).

My first major departure from Laudan's "metaphilosophy" is with respect to his answer to question 2 above: "Ideally, [a demarcation criterion] would specify a set of individually necessary and jointly sufficient conditions for deciding whether an activity or set of statements is scientific or unscientific" (Laudan 1983, 118). He goes on to clarify that a set of necessary but not sufficient conditions would permit us to point to activities that are not scientific (those lacking the necessary conditions) but could not specify which activities are indeed scientific. Conversely, a set of sufficient (but not necessary) conditions would tell us what counts as science, but not what is pseudoscientific. Hence the need for necessary and sufficient conditions (though no single set of criteria needs to be both at the same time).

This strikes me as somewhat old-fashioned, particularly for someone who has been telling his readers that many of philosophy's classic pursuits—such as a priori truths and the search for logical demonstrations—went out the window with the advent of more nuanced philosophical analyses in the modern era. It seems like the search for sets of necessary and sufficient conditions to sharply circumscribe concepts that are clearly not sharp in themselves ought to give pause at least since Ludwig Wittgenstein's talk of family resemblance concepts—which inspired the above mentioned suggestion by Dupré (1993).

As is well known, Wittgenstein (1958) discussed the nature of complex concepts that do not admit of sharp boundaries—or of sets of necessary and sufficient conditions—such as the concept of game. He suggested that the way we learn about these concepts is by example, not through logical definitions: "How should we explain to someone what a game is? I imagine that we should describe games to him, and we might add: 'This and similar things are called games.' and do we know any more about it ourselves? Is it only other people whom we cannot tell exactly what a game is? . . . But this is not ignorance. We do not know the boundaries because none have been drawn. . . . We can draw a boundary for a special purpose. Does it take that to make the concept usable? Not at all!" (Ibid., 69).

Figure 1.2 is my graphic rendition of Wittgenstein's basic insight: games make up a family resemblance concept (also known as a "cluster," in analogy to the type of diagram in figure 1.1) that cannot be captured by a set of necessary and sufficient conditions. Any such set will necessarily leave out some activities that ought to be considered as legitimate games while letting in activities that equally clearly do not belong there. But Wittgenstein correctly argued that this is neither the result of our epistemic limitations nor of some intrinsic incoherence in the concept itself. It is the way in which "language games" work, and philosophy of science is no exception to the general idea

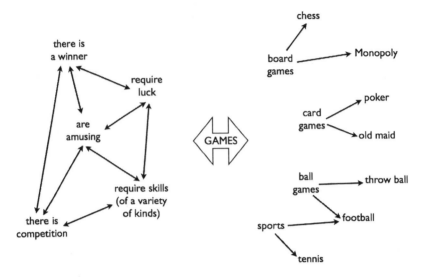

Figure 1.2

of a language game. I return to the possibility of understanding science as a Wittgenstein-type cluster concept below to make it a bit more precise.

I also markedly disagree with Laudan in answer to his question 3 above, where he says:

> Precisely because a demarcation criterion will typically assert the epistemic superiority of science over non-science, the formulation of such a criterion will result in the sorting of beliefs into such categories as "sound" and "unsound," "respectable" and "cranky," or "reasonable" and "unreasonable." Philosophers should not shirk from the formulation of a demarcation criterion merely because it has these judgmental implications associated with it. Quite the reverse, philosophy at its best should tell us what is reasonable to believe and what is not. But the value-loaded character of the term "science" (and its cognates) in our culture should make us realize that the labeling of a certain activity as "scientific" or "unscientific" has social and political ramifications which go well beyond the taxonomic task of sorting beliefs into two piles. (Laudan 1983, 119–20)

Seems to me that Laudan here wants to have his cake and eat it too. To begin with, the "value-loaded" character of science is not exactly an unqualified social positive for all things labeled as "scientific." We regularly see large sections of the public, especially in the United States, who flatly reject all sorts of scientific findings when said public finds them ideologically inconvenient or simply contrary to pet notions of one sort or another. Just think about the number of Americans who deny the very notion of human-caused climate change or who believe that vaccines cause autism—both positions held despite an overwhelming consensus to the contrary on the part of the relevant scientific communities. Obviously, labeling something "scientific" does not guarantee acceptance in society at large.

More important, Laudan simply cannot coherently argue that "philosophy at its best should tell us what is reasonable to believe and what is not" and then admonish us that "[the] social and political ramifications . . . go well beyond the taxonomic task of sorting beliefs into two piles." Of course there are political and social implications. Indeed, I would argue that if the distinction between science and pseudoscience did *not* have political and social implications, then it would merely be an academic matter of little import outside of a small cadre of philosophers of science. There simply is no way, nor *should* there be, for the philosopher to make arguments to the rest of the world concerning what is or is not reasonable to believe without not just having, but *wanting* political and social consequences. This is a serious game, which ought to be played seriously.

Rethinking Demarcation

As Bacon (1620) rightly admonished us, it is not good enough to engage in criticism (*pars destruens*); one also ought to come up with positive suggestions on how to move ahead (*pars construens*). So far I have built an argument against Laudan's premature death certificate for the demarcation problem, but I have also hinted at the directions in which progress can reasonably be expected. I now briefly expand on those directions.

The starting point is provided by Dupré's (1993) suggestion to treat science (and therefore pseudoscience) as a Wittgensteinian family resemblance, or cluster concept, along the lines sketched in figure 1.1. As is well known—and as illustrated for the concept of game in figure 1.2—family resemblance concepts are characterized by a number of threads connecting instantiations of the concept, with some threads more relevant than others to specific instantiations, and indeed sometimes with individual threads entirely absent from individual instantiations. For example, while a common thread for the

concept of games is that there is a winner, this is not required in all instantiations of the concept (think of solitaire).

Several useful concepts within science itself are best thought of as Wittgensteinian in nature, for instance, the idea of biological species (Pigliucci
2003). The debate on how exactly to define species has been going on for a
long time in biology, beginning with Aristotle's essentialism and continu-
ing through Ernst Mayr's (1996) "biological" species concept (based on reproductive isolation) to a number of phylogenetic concepts (i.e., based on
ancestry-descendant relations, see De Queiroz 1992). The problem can also
be seen as one generated by the same sort of "metaphilosophy" adopted by
Laudan: the search for a small set of jointly necessary and sufficient conditions
adequate to determine whether a given individual belongs to a particular species or not. My suggestion in that case—following up on an original remark
by Hull (1965) and in agreement with Templeton's (1992) "cohesion" species
concept—was that species should be treated as cluster concepts, with only
a few threads connecting very different instantiations like those represented
by, say, bacterial and mammalian species, and a larger number of threads connecting more similarly circumscribed types of species, like vertebrates and
invertebrates, for instance.

Clearly, a concept like science is at least as complex as one like "biological
species," which means that the number of threads underlying the concept, as
well as their relative importance for any given instantiation of the concept,
are matters for in-depth discussions that are beyond the scope of this chapter.
However, I am going to provide two complementary sketches of how I see the
demarcation problem, which I hope will move the discussion forward.

At a very minimum, two "threads" run throughout any meaningful treatment of the differences between science and pseudoscience, as well as of further distinctions within science itself: what I label "theoretical understanding" and "empirical knowledge" in figure 1.3. Presumably if there is anything
we can all agree on about science, it is that science attempts to give an empirically based theoretical understanding of the world, so that a scientific theory
has to have both empirical support (vertical axis in figure 1.3) and internal coherence and logic (horizontal axis in figure 1.3). I am certainly not suggesting
that these are the *only* criteria by which to evaluate the soundness of a science
(or pseudoscience), but we need to start somewhere. And of course, both
these variables in turn are likely decomposable into several factors related in
complex, possibly nonlinear ways. But again, one needs to start somewhere.

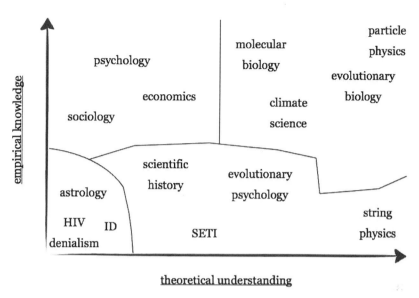

Figure 1.3

Figure 1.3, then, represents my reconstruction of how theoretical and empirical strengths begin to divide the space identified by the cluster diagram in figure 1.1: at the upper right corner of the empirical/theoretical plane we find well-established sciences (and the scientific notions that characterize them), like particle physics, evolutionary biology, and so forth. We can then move down vertically, encountering disciplines (and notions) that are theoretically sound but have decreasing empirical content, all the way down to superstring physics, based on a very sophisticated mathematical theory that—so far at least—makes no contact at all with (new) empirical evidence. Moving from the upper left to the upper right of the diagram brings us to fields and notions that are rich in evidence, but for which the theory is incomplete or entirely lacking, as in many of the social (sometimes referred to as "soft") sciences.

So far I doubt I have said anything particularly controversial about the empirical/theoretical plane so identified. More interesting things happen when one moves diagonally from the upper right to the lower left corner. For instance, the "proto-/quasi-science" cluster in figure 1.1 is found in the middle and middle-lower part of figure 1.3, where theoretical sophistication is intermediate and empirical content is low. Here belong controversial disciplines like evolutionary psychology, the Search for Extraterrestrial Intelligence (SETI), and "scientific" approaches to the study of history. Evolutionary psychology is theoretically sound in the sense that it is grounded on the general theory of evolution. But as I mention above, there are serious doubts about the testability of a number of specific claims made by evolutionary psychologists (e.g., that a certain waist-to-hip ratio in human females is universally attractive), simply because of the peculiar difficulties represented by the human species when it comes to testing historical hypotheses about traits that do not leave a fossil record (Kaplan 2002). In the case of SETI, despite the occasional ingenious defense of that research program (Cirkovic and Bradbury 2006), the fact remains that not only has it (so far) absolutely no empirical content, but its theoretical foundations are sketchy at best and have not advanced much since the onset of the effort in the 1960s (Kukla 2001). As for scientific approaches to the study of history (e.g., Diamond 1999, 2011; Turchin 2003, 2007), their general applicability remains to be established, and their degree of theoretical soundness is far from being a settled matter.

We finally get to the lower left corner of figure 1.3, where actual pseudoscience resides, represented in the diagram by astrology, Intelligent Design (ID) creationism, and HIV denialism. While we could zoom further into this corner and begin to make interesting distinctions among pseudosciences themselves (e.g., among those that pretend to be based on scientific principles versus those that invoke completely mysterious phenomena versus those that resort to supernatural notions), they all occupy an area of the diagram that is extremely low both in terms of empirical content and when it comes to theoretical sophistication. This most certainly does *not* mean that no empirical data bears on pseudosciences or that—at least in some cases—no theoretical foundation supports them. Take the case of astrology as paradigmatic: plenty of empirical tests of astrological claims have been carried out, and the properly controlled ones have all failed (e.g., Carlson 1985). Moreover, astrologers certainly can produce "theoretical" foundations for their claims, but these quickly turn out to be both internally incoherent and, more damning, entirely detached from or in contradiction with very established notions from a variety of other sciences (particularly physics and astronomy, but also biology). Following a Quinean conception of the web of knowledge (Quine 1951), one would then be forced to either throw out astrology (and, for similar reasons, creationism) or reject close to the entirety of the established sciences occupying the upper right corner of figure 1.3. The choice is obvious.

Could the notions captured in figures 1.1 and 1.3 be made a bit more precise than simply invoking the Wittgensteinian notion of family resemblance? I believe this can be done in a variety of ways, one of which is to dip into the resources offered by symbolic nonclassical logics like fuzzy logic (Hajek 2010). Fuzzy logic, as is well known, was developed out of fuzzy set theory to deal with situations that contain degrees of membership or degrees of truth, as in the standard problems posed by notions like being "old" versus "young," and generally related to Sorites paradox.

Fuzzy logic as a type of many-valued logic using *modus ponens* as its deductive rule is well equipped, then, to deal with the degree of "scientificity" of a notion or field, itself broken down in degrees of empirical support and theoretical sophistication as outlined above. For this to actually work, one would have to develop quantitative metrics of the relevant variables. While such development is certainly possible, the details would hardly be uncontroversial. But this does not undermine the general suggestion that one can make sense of science/pseudoscience as cluster concepts, which in turn can be treated—at least potentially—in rigorous logical fashion through the aid of fuzzy logic.

Here, then, is what I think are reasonable answers to Laudan's three "metaphilosophical" questions concerning demarcation:

(1) What conditions of adequacy should a proposed demarcation criterion satisfy?

A viable demarcation criterion should recover much (though not necessarily all) of the intuitive classification of sciences and pseudosciences generally accepted by practicing scientists and many philosophers of science, as illustrated in figure 1.1.

(2) Is the criterion under consideration offering necessary or sufficient conditions, or both, for scientific status?

Demarcation should *not* be attempted on the basis of a small set of individually necessary and jointly sufficient conditions because "science" and "pseudoscience" are inherently Wittgensteinian family resemblance concepts (fig. 1.2). A better approach is to understand them via a multidimensional continuous classification based on degrees of theoretical soundness and empirical support (fig. 1.3), an approach that, in principle, can be made rigorous by the use of fuzzy logic and similar instruments.

(3) What actions or judgments are implied by the claim that a certain belief or activity is "scientific" or "unscientific"?

> Philosophers *ought* to get into the political and social fray raised by discussions about the value (or lack thereof) of *both* science and pseudoscience. This is what renders philosophy of science not just an (interesting) intellectual exercise, but a vital contribution to critical thinking and evaluative judgment in the broader society.

Laudan (1983, 125) concluded his essay by stating that "pseudo-science" and "unscientific" are "just hollow phrases which do only emotive work for us. As such, they are more suited to the rhetoric of politicians and Scottish sociologists of knowledge than to that of empirical researchers." On the contrary, those phrases are rich with meaning and consequences precisely because both science and pseudoscience play important roles in the dealings of modern society. And it is high time that philosophers get their hands dirty and join the fray to make their own distinctive contributions to the all-important—sometimes even vital—distinction between sense and nonsense.

NOTE

1. As several authors have pointed out (e.g., Needham 2000), Duhem's thesis needs to be distinguished from Quine's (1951), even though often the two are jointly known as the Duhem-Quine thesis. While Duhem's "adjustments" to rescue a theory are local (i.e., within the circumscribed domain of the theory itself), Quine's are global, referring to changes that can be made to the entire web of knowledge—up to and including the laws of logic themselves. Accordingly, Duhem's thesis properly belongs to discussions of falsification and demarcation, while Quine's is better understood as a general critique of empiricism (in accordance to its appearance as an aside in his famous paper "Two Dogmas of Empiricism").

REFERENCES

Bacon, Francis. 1620. *Novum Organum.* http://archive.org/stream/baconsnovumorgan00bacoiala/baconsnovumorgan00bacoiala_djvu.txt.

Carlson, Shawn 1985. "A Double-Blind Test of Astrology." *Nature* 318:419–25.

Cirkovic, Milan M., and Robert J. Bradbury. 2006. "Galactic Gradients, Postbiological Evolution and the Apparent Failure of SETI." *New Astronomy* 11:628–39.

De Queiroz, Kevin 1992. "Phylogenetic Definitions and Taxonomic Philosophy." *Biology and Philosophy* 7:295–313.

Diamond, Jared 1999. *Guns, Germs, and Steel: The Fates of Human Societies.* New York: Norton.

————. 2011. *Collapse: How Societies Choose to Fail or Succeed.* London: Penguin.

Dupré, John. 1993. *The Disorder of Things: Metaphysical Foundations of the Disunity of Science.* Cambridge, MA: Harvard University Press.

Hajek, Peter 2010. "Fuzzy Logic." *Stanford Encyclopedia of Philosophy.* http://plato.stanford.edu/entries/logic-fuzzy/.

Hansson, Sven O. 2009. "Cutting the Gordian Knot of Demarcation." *International Studies in the Philosophy of Science* 23:237–43.

Henderson, Leah, Noah D. Goodman, Joshua B. Tenenbaum, and James F. Woodward. 2010. "The Structure and Dynamics of Scientific Theories: A Hierarchical Bayesian Perspective." *Philosophy of Science* 77:172–200.

Huizenga, John R. 1992. *Cold Fusion: The Scientific Fiasco of the Century.* Rochester, NY: University of Rochester Press.

Hull, David 1965. "The Effect of Essentialism on Taxonomy—Two Thousand Years of Stasis." *British Journal for the Philosophy of Science* 16:314–26.

Kaplan, Jonathan 2002. "Historical Evidence and Human Adaptation." *Philosophy of Science* 69:S294–S304.

Kuhn, Thomas S. 1974. "Logic of Discovery or Psychology of Research?" In *The Philosophy of Karl Popper*, edited by P. A. Schilpp, 798–819. Chicago: Open Court.

Kukla, André 2001. "SETI: On the Prospects and Pursuitworthiness of the Search for Extraterrestrial Intelligence." *Studies in the History and Philosophy of Science* 32:31–67.

Laudan, Larry. 1983. "The Demise of the Demarcation Problem." In *Physics, Philosophy and Psychoanalysis*, edited by R. S. Cohen and L. Laudan, 111–27. Dordrecht: D. Reidel.

Longino, Helen E. 1990. *Science as Social Knowledge: Values and Objectivity in Scientific Inquiry.* Princeton, NJ: Princeton University Press.

Mayr, Ernst. 1996. "What Is a Species, and What Is Not?" *Philosophy of Science* 63:262–77.

Mill, John Stuart. (1843) 2002. *A System of Logic, Ratiocinative and Inductive.* Honolulu: University Press of the Pacific.

Needham, Pual 2000. "Duhem and Quine." *Dialectica* 54:109–32.

Okasha, Samir 2001. "What Did Hume Really Show about Induction?" *Philosophical Quarterly* 51:307–27.

Pigliucci, Massimo. 2002. "Are Ecology and Evolutionary Biology 'Soft' Sciences?" *Annales Zoologici Fennici* 39:87–98.

————. 2003. "Species as Family Resemblance Concepts: The (Dis-)solution of the Species Problem?" *BioEssays* 25:596–602.

Pigliucci, Massimo, and J. Kaplan. 2006. *Making Sense of Evolution: The Conceptual Foundations of Evolutionary Biology.* Chicago: University of Chicago Press.

Popper, Karl. 1957. "Philosophy of Science: A Personal Report." In *British Philosophy in Mid-Century*, edited by C. A. Mace, 155–91. Crows Nest, New South Wales: Allen and Unwin.

Quine, Willard V. O. 1951. "Two Dogmas of Empiricism." *Philosophical Review* 60:20–43.

Sakamoto, Yosiyuki, Ishiguro, Makio, and Genshiro Kitagawa. 1987. *Akaike Information Criterion Statistics.* Alphen aan den Rijn, Netherlands: Kluwer.

Smolin, Lee 2007. *The Trouble with Physics: The Rise of String Theory, the Fall of a Science, and What Comes Next.* Boston: Houghton Mifflin Harcourt.

Templeton, Alan R. 1992. "The Meaning of Species and Speciation: A Genetic Perspective." In *The Units of Evolution: Essays on the Nature of Species*, edited by M. Ereshefsky, 159–83. Cambridge, MA: MIT Press.

Turchin, Peter 2003. *Historical Dynamics: Why States Rise and Fall*. Princeton, NJ: Princeton University Press.

———. 2007. *War and Peace and War: The Rise and Fall of Empires*. London: Penguin.

Vickers, John 2010. "The Problem of Induction." *Stanford Encyclopedia of Philosophy*. http:// plato.stanford.edu/entries/induction-problem/.

Voit, Peter 2006. *Not Even Wrong: The Failure of String Theory and the Search for Unity in Physical Law*. New York: Basic Books.

Whewell, William. 1840. *The Philosophy of the Inductive Sciences*. http://books.google .com/books?id=um85AAAAcAAJ.

Wittgenstein, Ludwig 1958. *Philosophical Investigations*. Hoboken, NJ: Blackwell.

2

Science and Pseudoscience

How to Demarcate after the (Alleged) Demise

of the Demarcation Problem

MARTIN MAHNER

Naturally, we would expect that the philosophy of science is able to tell us what science is and how it is to be distinguished from nonscience in general and from pseudoscience in particular. Curiously, however, we find that the topic of demarcation has long gone out of fashion. In an influential paper published already back in 1983, the philosopher Larry Laudan has even announced the demise of the demarcation problem (Laudan 1983)—a view that seems to have made its way into the philosophical mainstream (see, e.g., the review by Nickles 2006, as well as in this volume). In stark contrast to this lack of interest on the part of many philosophers, the general public and in particular science educators are faced with the advocates of pseudoscientific theories and practices, who work hard to either keep up their status in society or gain more influence. Think of alternative medicine, astrology, or creationism. If demarcation is dead, it seems that those who attempt to defend a scientific outlook against the proliferation of pseudoscience and esotericism can no longer count on the help of philosophers of science. Worse, they must be prepared to hear that their efforts are unfounded and hence misguided (e.g., by Laudan and his followers, see "Is Demarcation Really Dead?" below).

Indeed, for a long time now the demarcation problem has made only sporadic appearances in the philosophy of science, as is evidenced by the scarcity of academic publications on this subject. The main reason for this lack of interest probably is that demarcation has proven to be quite difficult. Practically all demarcation criteria proposed by philosophers of science have turned out to be either too narrow or too wide. Thus, the majority of contemporary philosophers of science believe that there simply is no set of individually necessary and jointly sufficient demarcation criteria. Indeed, already about fifteen years ago a survey conducted with 176 members of the Philosophy of Science Association in the United States showed that 89 percent of the respondents believed that no universal demarcation criteria had been found yet (Alters 1997).

To illustrate this negative answer, let us take a brief look at the most famous demarcation criterion: Popper's falsifiability condition (Popper 1963). It says that a statement is (logically) *falsifiable* if, and only if, there is at least one conceivable observation (statement) that is incompatible with it. Alternatively, if a statement is compatible with every possible state of affairs, it is *unfalsifiable*. The problem with the falsifiability criterion, however, is that many pseudosciences do contain falsifiable statements and therefore would count as sciences. For example, the central claim of astrology, that there is a clear connection between zodiac signs and human character traits, is testable—and it has been statistically tested and refuted many times (Dean 1977; Carlson 1985). Thus, being falsifiable, astrology fulfills Popper's demarcation criterion and would have to be accepted as a science. Yet there are many other problems with astrology, not captured by Popper's criterion, that give us good reasons to regard it as a pseudoscience. For example, as Kanitscheider (1991) shows, the "theory" of astrology is so defective that it would be incapable of explaining its own data without resorting to sheer magic, even if the empirical situation did show a significant correlation between the various constellations and human behavior. Another example is creationism. Its central doctrine, that a supernatural being created the world, is indeed unfalsifiable: it is compatible with every possible observation because any state of affairs can be seen as exactly what the omnipotent creator chose to do. However, other creationist claims, such as that our planet is only six thousand to ten thousand years old, are falsifiable and have been falsified.[1] Thus, the falsifiability criterion allows us to recognize *some* claims as pseudoscientific, but it fails us in the many cases of refutable pseudoscientific claims. In a similar vein, other demarcation criteria can be analyzed and rejected as not being necessary or sufficient or both (for a review, see Mahner 2007; Hansson 2008).

Despite the lack of generally accepted demarcation criteria, we find remarkable agreement among virtually all philosophers and scientists that fields like astrology, creationism, homeopathy, dowsing, psychokinesis, faith healing, clairvoyance, or ufology are either pseudosciences or at least lack the epistemic warrant to be taken seriously.[2] As Hansson (2008, 2009) observes, we are thus faced with the paradoxical situation that most of us seem to recognize a pseudoscience when we encounter one, yet when it comes to formulating criteria for the characterization of science and pseudoscience, respectively, we are told that no such demarcation is possible.[3]

Problems of Demarcation

Demarcation has proven so difficult for a number of reasons. The first is that there is not just the distinction between science and pseudoscience, but also the distinction between science and nonscience in general. Not everything nonscientific is pseudoscientific. Ordinary knowledge as well as the arts and humanities are not sciences, yet they are not pseudosciences. Second, there is the distinction between good and bad science (Nickles 2006). A scientist who follows a sloppy and careless experimental protocol, or who even omits a few data from his report to obtain "smoother" graphs and results (which borders on scientific fraud), is a bad scientist but not (yet) a pseudoscientist. Third is the problem of protoscience and heterodoxy. Under what conditions is a young field of research a protoscience rather than a pseudoscience? By definition, a protoscience does not possess all the features of a full-blown science, so how do we evaluate its status? For example, are the young and controversial fields of evolutionary psychology and memetics protosciences or else pseudosciences, as their critics maintain (see Pigliucci 2010 for various examples of such borderline cases)? When exactly is an alternative theory a piece of pseudoscience and when just a heterodox view? This distinction is important because heterodoxy should be welcomed as stimulating critical debate and research, whereas pseudoscience is just a waste of time.

The fourth problem is the debate about the unity or disunity of science (Cat 2006). Some philosophers have argued that the subject matter, the methods, and the approaches of the various scientific fields are so different that it is mistaken to uphold the old idea of a unity of science (Dupré 1993). Indeed, the neopositivists had argued for the unity of science in the sense that all scientific statements could and should be reduced to physical statements.[4] Although physical reductionism has still wide currency, nonreductionist approaches have succeeded in claiming a considerable territory, so the idea that the unity of science lies in a reduction to physics is no longer a majority view. Yet there is another conception of the unity of science: the fact that the sciences provide a consistent and unified view of the world. Its consilience, its network character makes for the unity of science (Reisch 1998; Wilson 1998). Generally, philosophers who believe in the unity of science are more inclined toward demarcation than those who argue for its disunity.

The fifth problem concerns the units of demarcation. The various attempts at demarcation have referred to quite different aspects and levels of science: statements, problems, methods, theories, practices, historical sequences of theories and/or practices (i.e., research programs in the sense of Lakatos), and fields of knowledge. For example, Popper's falsificationist approach concerns statements, as it essentially consists in the application of the *modus tollens* rule; Lakatos (1970) refers to theories and research programs; Kuhn (1970) focuses on problems and the problem-solving capacity of theories; Kitcher (1982) and Lugg (1987) examine theories and practices; Bunge (1983, 1984) and Thagard (1988) refer to entire fields of knowledge; and Wilson (2000) analyzes the differences in the reasoning of scientists and pseudoscientists, that is, their different logic and methodology. Plausibly, all these different aspects may be scientific or pseudoscientific, respectively, yet there is no unanimity as to an optimal unit of demarcation, if any.

The sixth problem is related to the units of demarcation: many demarcation attempts propose only a single criterion or a small number of criteria. For this reason they are bound to be unsatisfactory considering the variety of possible units of demarcation. For example, falsificationism is a single criterion approach, as is Kuhn's problem-solving criterion as well as Lakatos's progressiveness condition for research programs. Other authors admit further conditions like fertility, independent testability of auxiliary hypotheses, concern with empirical confirmation or disconfirmation, the application of the scientific method, and so on. Yet it seems that the diversity of the sciences cannot be captured by a small number of scientificity criteria only.

The seventh problem is the question of whether demarcation criteria should be ahistorical or time dependent. For example, as a logical criterion, falsifiability is ahistorical; that is, we can pick any statement from any time, without knowing anything about the history or development of the given field, and simply apply the criterion. By contrast, Lakatos's progressiveness condition is a time-dependent criterion; that is, to apply it, we need to examine the past history and development of the given theory or field over several years, if not decades, as to whether it progressed, stagnated, or regressed during that time. In the case of a new theory or field, we cannot pass judgment right away, but we need to wait and observe the future development of the field for, probably, several years. Thus, unlike ahistorical criteria, time-dependent ones pose additional practical problems.

The final problem to be considered here is the very logic of demarcation, that is, whether it makes sense to search for criteria that are individually necessary and jointly sufficient for a proper demarcation of a given unit of science (statement, theory, research program, field) as either scientific or pseudo-scientific. The standard logic of classification requires individually necessary and jointly sufficient conditions for the definition of a proper class, so it comes as no surprise that most philosophers of science have complied with this logical requirement. Yet the question is whether this requirement can be upheld, or whether we must settle for a less strict way of demarcation—or else for none at all, and give up the idea of demarcation altogether.

Is Demarcation Really Dead?

The view that we better give up the idea of demarcation altogether has been espoused by Laudan (1983). He claims that the problem of demarcation has turned out to be a pseudoproblem that should rightfully be put to rest. What matters, by contrast, would be "the empirical and conceptual credentials for claims about the world" (125). In other words, according to Laudan, the important distinction is not between scientific and nonscientific claims, but between reliable and unreliable knowledge or, more generally, between epistemically warranted and unwarranted beliefs. Thus, he recommends dropping "terms like 'pseudo-science' and 'unscientific' from our vocabulary" because "they are just hollow phrases which do only emotive work for us" (125). Consequently, identifying and fighting "pseudosciences" is a misguided endeavor.

Although Laudan's famous paper contributed to the topic of demarcation going out of fashion, not much has changed with respect to the institutional status of the philosophy of science. If there were really nothing special about science, and if all that matters were the distinction between epistemically warranted and unwarranted beliefs, wouldn't we assume that, as a consequence, the philosophy of science would by now (that is, almost thirty years after the publication of Laudan's article) have dissolved as a discipline? Wouldn't general epistemology have jumped in as a replacement because it can handle all the relevant problems of epistemic justification? Yet, as Pennock (2009) rightly points out, universities still offer courses in the philosophy of science. They still hire philosophers of science instead of replacing these positions with general epistemologists. Journals like *Philosophy of Science* still exist. In science education, the program of how to teach the nature of science has not been abandoned, although it would obviously be unfounded and misguided if there were no such thing as the nature of science. While Pennock's argument is not decisive because all this could just be due to historical and institutional inertia, it illustrates at least that Laudan's pronouncement that demarcation is dead has not been accepted with all its natural consequences. It seems, therefore, that science involves something special that we do not want to give up. If so, however, why are philosophers of science so reluctant to delineate their very subject matter? Do they really know, then, what they are talking about?

In my view, the major problem with Laudan's proposal is that the distinction between epistemically warranted and unwarranted beliefs just extends the demarcation line to a wider circle: the problem of how to distinguish between warranted and unwarranted beliefs remains unsolved. Although widely accepted standards of epistemic justification exist, the fact is (and Laudan is probably aware of it) that even in this case there is no unanimity. Just think of theology. Theologians and theistic philosophers claim that their beliefs are warranted and thus rational, whereas naturalist philosophers maintain that they are not. So how do we draw a line at this level? Are we faced with another pseudoproblem because we once again fail to find a clear-cut demarcation?

Another problem of Laudan's criticism is that it is based on the traditional approach of requiring a set of not only individually necessary and jointly sufficient but, as it seems, also ahistorical criteria; for he speaks about science throughout history, from Aristotle to modern science, as though ancient pre-science were on a par with the maturity of contemporary science. What if demarcation can be achieved only with time-dependent criteria? After all, modern science began only about four hundred years ago, and it has developed quite a bit since the times of Galileo and Newton. Thus, it may well be that some belief was perfectly scientific back in 1680, while it no longer would be if held today. (For further criticism, see Pigliucci and Boudry in this volume.)

Finally, even if the distinction between reliable and unreliable knowledge were ultimately the most important one, it would still remain legitimate to try to delineate a more restricted way of knowledge production, namely the scientific one, from other ways of gaining knowledge. After all, science and technology are still considered to be epistemically privileged due to their systematic and rigorous approach, as a result of which they produce the most reliable knowledge. For this reason, they are very important parts of our culture.

Why Demarcation Is Desirable

Indeed, science and technology are not just important for economic reasons, but also because citizens of a civilized and educated society ought to be able to make scientifically informed decisions in their personal lives and in their roles in society, politics, and culture. We are all faced, for example, with questions like these (Mahner 2007):

- Should our own as well as other peoples' health and even lives be entrusted to diagnostic or therapeutic methods that are not scientifically validated?
- Should public health insurances cover magical cures like homeopathy or Therapeutic Touch?
- Should dowsers be employed to search for people buried by avalanches or collapsed buildings?
- Should the police ask clairvoyants to search for missing children or to help solve crimes?
- Should evidence presented in court trials include astrological character analysis or testimony of mediums?
- Should taxpayers' money be spent on funding pseudoscientific "research," or is it a better investment to fund only scientific research?
- Should people living in a modern democratic society base their political decisions on scientific knowledge rather than on superstition or ideology?[5]

The preceding questions are not just questions of public policy, but also ethical and legal ones, as they may involve fraud or even negligent homicide, for example, if a patient dies because he was treated with a quack remedy. Thus, a case can be made for the need to distinguish science from pseudoscience.

The apparent inability of the philosophy of science to characterize its own subject matter also poses an obvious problem for science education. Why teach astronomy rather than astrology? Why evolutionary biology instead of creationism? Why physics rather than the pseudophysics of free-energy machines or antigravitation devices? Why standard history rather than von Däniken's ancient astronaut archeology? In general, why should we teach sciences rather than pseudosciences? Even more generally, how can we inform students about the nature of science—one of the central topics in science education (e.g., Alters 1997; Eflin et al. 1999; Matthews 2009)—when even the philosophy of science has given up on characterizing the nature of science?

As a consequence, the very aim of science education is called into question by Laudan and his followers (Martin 1994; Afonso and Gilbert 2009).

How Demarcation Is Possible

Obviously, a new approach to demarcation must avoid the mistakes of the past. The first step toward a feasible demarcation is to choose the most comprehensive unit of analysis: entire fields of knowledge, or epistemic fields (Bunge 1983; Thagard 1988, 2011). Roughly speaking, an epistemic field is a group of people and their theories and practices, aimed at gaining knowledge of some sort. Thus, both astronomy and astrology or both physics and theology are epistemic fields. Likewise, both biology in general and ecology in particular are epistemic fields. The first examples show that the knowledge acquired in an epistemic field needs neither be factual nor true: we may acquire knowledge about purely fictional rather than real entities, and our knowledge may be false or illusory.[6] The second example illustrates that epistemic fields often are more or less inclusive: they form hierarchies.

Choosing fields of knowledge as a starting point allows us to consider the many facets of science, namely that it is at the same time a body of knowledge and a social system of people including their collective activities. It also takes into account that science is something that did not come into existence ready-made, but rather has developed over several centuries from a mixed bag of ordinary knowledge, metaphysics, and nonscientific inquiry. In other words, it allows us to consider not just the philosophy of science, but also the history, sociology, and psychology of science.[7] Moreover, the study of epistemic fields requires that we also analyze their components. In this way, the smaller or lower-level units, such as statements, theories, methods, and so on, can be included as necessary components in our account. In view of this analysis, it is implausible that we can do with a single criterion or a few conditions for determining the scientific status of a field of knowledge (Bunge 1984). This may work in *some* cases, but it is likely to fail us in many others. For this reason, it is recommendable to have a checklist of demarcation criteria that is as comprehensive as possible: a whole battery of science indicators.

To illustrate this point, let us look at a recent definition of "pseudoscience" by Hansson (2009, 240). He defines a statement as pseudoscientific if and only if it satisfies the following three conditions: (1) it pertains to an issue within the domain of science; (2) it is not epistemically warranted, and (3) it is part of a doctrine whose major proponents try to create the impression that it is epistemically warranted. As useful as this definition certainly is for quick reference, it refers only to the lowest-level components of a field, namely statements, and it leaves open the question of how exactly epistemic warrant is achieved (see Hansson, this volume). Thus, "epistemic warrant" can only be a shorthand term for a more extensive list of criteria determining what such justification consists in. Referring to entire fields of knowledge and to a more comprehensive list of science indicators has the advantage of being able to cover all these aspects and their respective weights. Its disadvantage, however, is that it no longer allows us to formulate short and handy definitions of either science or pseudoscience.

If we have to rely on a whole battery of science indicators, the question arises whether these indicators are descriptive or normative or both. Of course, we can just factually describe what scientists do and what the representatives of other epistemic fields do. And any demarcation will have to rely on such descriptions. However, a demarcation proper involves a judgment as to the epistemic status of a given field: it is expected to tell us why the theories and practices in a given field produce at least reliable or even true knowledge, whereas those in some others fail to do so. To justify such an evaluation, we need normative criteria. For example, whereas falsifiability is considered to be a normative (logico-methodological) criterion, the fact that the people involved in some epistemic field do or do not form a research community seems to be just a descriptive social feature. It appears, therefore, that only normative criteria are relevant for demarcation purposes, whereas descriptive ones are not. And if epistemic justification is what matters in the end, then only normative, in particular methodological, criteria appear to be relevant for demarcation. So much, at least, for the traditional view.

However, the distinction between descriptive and normative indicators is not so straightforward. Is Lakatos's progressiveness criterion descriptive or normative? That some field progresses or stagnates over a certain time is first of all a factual description. Yet we attribute a positive value to progressiveness because it indicates that the field keeps solving problems, so that there is a growth of knowledge; this, in turn, indicates that the given field or its theories, respectively, most likely achieve a deeper and deeper as well as a reasonably correct representation of its subject matter, presumably in the sense of an approximation to the truth. So progressiveness is generally accepted as a normative rather than a descriptive criterion.

Now what about a social feature like being organized in a research community? At first sight this looks like a purely descriptive feature, which is not really relevant to the epistemic warrant of the theories and practices in question. But a closer look shows that the collective activity of a research community is an important feature of science (see Koertge, this volume). Let us take a look at dowsing, a field that has no research community. Dowsers share some information, but most have their own private theories about the alleged laws and mechanisms of dowsing, of what can and cannot be accomplished by dowsing. Despite some minor and mostly linguistic rather than conceptual overlaps, these theories are mutually incompatible. That is, there is no general theory shared by a community of dowsers, no mutual evaluation of methods and theories, no collective mechanism of error correction, and so forth. So the lack of a research community is a clear indicator that what these people do is not a science.[8] Consequently, it may turn out that a social feature like this has a normative component after all. Yet this remains to be seen.

The upshot of all this is that a comprehensive list of science indicators should not a priori be restricted to normative criteria only. What might such a checklist look like? Let me give some examples (following Bunge 1983, 1984; Mahner 2007). As just mentioned, we may first take a look at the people involved in the given field:

- Do they form a research community, or are they just a loose collection of individuals doing their own thing?
- Is there an extensive mutual exchange of information, or is there just an authority figure passing on his doctrines to his followers?
- Is the given group of people free to research and publish whatever they want, or are they censored by the reigning ideology of the society they live in (e.g., Aryan physics, Lysenkoism)?
- Does the domain of study consist of concrete objects, or does it contain fuzzy "energies" or "vibrations," if not ghosts or other spiritual entities?
- What are the philosophical background assumptions of the given field?
- Does its ontology presuppose a natural, causal, and lawful world only, or does it also admit supernatural entities or events?

Traditionally, a large number of indicators are found in the logic and methodology adopted by any given field:

- Does it accept the canons of valid and rational reasoning?
- Do the principles of noncircularity and noncontradiction matter?

- Does it admit fallibilism or endorse dogmatism? Occam's razor or unfettered speculation?
- How important are testability and criticism?
- How important is evidential support?
- Can the reliability of its methods or techniques be independently tested?
- Do the theories have genuine explanatory or predictive power, or both?
- Are the theories fruitful?
- Are the data reproducible?
- Are there mechanisms of error elimination? Are its claims compatible with well-established knowledge, in particular law statements?
- Does the field borrow knowledge and methods from adjacent fields?
- Does it in turn inform and enrich neighboring fields, or is it isolated?
- Do the problems tackled in the field spring naturally from research or theory construction, or are the problems pulled out of a hat?[9]
- Is the corpus of knowledge of the given field up to date and well confirmed, or is it obsolete, if not anachronistic? Is it growing or stagnating?

This list could be extended further, but let these examples suffice. The sheer amount of possible indicators, both descriptive and normative, shows that it is unlikely that each of them is fulfilled in every case of demarcation. For example, does the reproducibility condition hold for historical sciences like phylogenetics or cosmology? And modern parapsychology does a lot of research, some of which even uses up-to-date statistical methods, but is this sufficient to regard it as a science? The aforementioned indicators, therefore, fail to constitute a set of individually necessary and jointly sufficient conditions. The rationality condition, for instance, is certainly necessary for a field to count as a science, but it is not sufficient because there are other rational human enterprises. If we have to give up the goal of finding a set of necessary and sufficient criteria of scientificity, how can we perform a reasonable demarcation then?

To see how, it is helpful to take a look at biological systematics, which is plagued by a similar problem: it is very hard to characterize biological species and often even higher taxonomic units by a set of jointly necessary and sufficient features. Due to the high variation of traits, some organisms of a particular species lack certain features, so they do not fulfill the definition of the given species. For this reason, extensive debate has waged about essentialism versus anti-essentialism in the philosophy of systematics (Mahner and Bunge 1997). The compromise position suggests what is called "moderate species essentialism." This is the idea that biological species can be characterized by a variable cluster of features instead of a strict set of individually necessary and jointly sufficient properties (Boyd 1999; Wilson 1999; Pigliucci 2003).[10] Thus, whereas no particular property need be present in all the members of the given species, "enough" shared properties always make these organisms belong to the given kind. If we graphically represent the features of different species in a trait space diagram, they often form rather distinct clusters, despite the occasional overlap.

Applying this approach to the demarcation problem may allow us to characterize a science or a pseudoscience, respectively, by a variable cluster of properties too. For example, if we had ten conditions of scientificity (all of equal weight), we might require that, say, an epistemic field fulfill at a minimum seven out of these ten conditions to be regarded as scientific. However, it would not matter which of these ten conditions are actually met. According to the formula $N!/n!(N-n)!$, where $N = 10$ and $n = 7$, and adding the permutations for $n = 8$, $n = 9$, and $n = 10$, there would in this case be a total of 176 possible ways of fulfilling the conditions of scientificity (Mahner 2007).[11] Of course, it remains to be shown how many criteria there actually are (my guess is at least thirty to fifty), so that the numbers above can be adapted to a realistic demarcation scenario. Moreover, as the indicators are in fact not equally important, a weighting factor may have to be introduced to make such a cluster analysis more realistic. In particular, the distinction between descriptive and normative traits would have to be considered in this weighting procedure. Finally, to calibrate such a list of science indicators, and in particular to get an idea about the number of required positive checks out of any such full list, several uncontroversial cases of pseudosciences would have to be carefully analyzed and compared to uncontroversial sciences. An ideal task for a PhD dissertation!

A final question may come to mind: does such a cluster approach amount to a demarcation proper? After all, one could contend that if we have to give up demarcation by means of jointly necessary and sufficient criteria, the result of our analysis is no longer a demarcation proper. At least two possible answers emerge. We may concede the case and henceforth avoid the term "demarcation" in favor of weaker terms like "delimitation," "delineation," or even just "distinction"; or we may simply redefine the concept of demarcation and accept this as an example of conceptual change. Which way we choose is probably just a matter of taste. In this chapter, I opt for the term "demarcation" in its new, weaker sense.

Conclusion

The consequence of a cluster approach to demarcation is that, as some authors have suggested earlier (e.g., Thagard 1988; Eflin et al. 1999), we must do with a reasonable *profile* of any given field rather than with a clear-cut assessment. Such a profile would be based on a comprehensive checklist of science/pseudoscience indicators and hence a thorough analysis. A cluster demarcation also entails that the reasons we give for classifying a given field as a pseudoscience may be quite different from field to field. For example, the criteria by which we reject the theory and practice of dowsing as pseudoscientific may be different from the criteria by which we reject creationism. We must indeed say goodbye to the idea that a small set of demarcation criteria applies to all fields of knowledge, allowing us to clearly partition them into scientific and nonscientific ones. The actual situation is more complicated than that, but it is not hopeless either. Although we lose the benefit of unambiguous categorization, the cluster approach should enable us to come to a well-reasoned and hence rational conclusion concerning the scientific or pseudoscientific status of any epistemic field including its components.

NOTES

1. By adding protective ad hoc hypotheses, creationists often immunize their claims from falsification. For example, they say that while the earth is in fact only ten thousand years old, God made it *appear* to be much older to test our faith (Pennock 2009). This would be irrefutable also if the latter were not an ad hoc addition, but an essential part of the original claim, for example, from a theological viewpoint that allows for a deceptive God.

2. For a comprehensive review of pseudoscientific fields, see Hines (2003). Pigliucci (2010) also examines many interesting borderline cases, such as evolutionary psychology and string theory.

3. This situation is not unique: we find a similar situation in the field of religious studies, where there are claims that a definition of "religion" is impossible due to the immense variation of religious views and practices (see, e.g., Platvoet and Molendijk 1999), though apparently this is not stopping anyone from making meaningful statements about religions.

4. More precisely, the original neopositivist goal was to reduce all scientific statements to physical *observation* statements. This goal had to be given up soon, so that we can ignore this phenomenalist aspect and focus on physical reductionism.

5. A topical example is the debate about global warming (Pigliucci 2010).

6. Speaking of reliable knowledge or even false knowledge presupposes a Popperian approach to knowledge, according to which all knowledge is fallible. In the traditional con-

ception of knowledge as true justified belief, "false knowledge" would be an oxymoron; and speaking of "reliable knowledge" presupposes that there is such a thing as an approximation to the truth.

7. For example, Kitcher (1993) has argued that pseudoscience is actually a problem of psychology rather than the philosophy of science: it is mainly the mindset of certain people that makes them pseudoscientists.

8. Note that some pseudosciences mimic research communities. For example, creationists organize congresses and publish their own peer-reviewed journals. However, this community is isolated as it makes virtually no contact with other scientific research communities, that is, it is not a proper subcommunity of the international scientific community.

9. For example, Erich von Däniken's pseudoarcheology does not solve genuine problems, but fabricates them by mystifying normal archeological facts to solve these mysteries in terms of extraterrestrial interventions (Mahner 2007).

10. A forerunner is Wittgenstein's fuzzy concept of family resemblance (Dupré 1993; Pigliucci 2003), as well as Beckner's idea of defining species disjunctively (Beckner 1959). After the manuscript of this chapter had already been submitted, Irzik and Nola (2011) published a paper exploring a family resemblance approach to the nature of science in science education.

11. Note that a cluster construal like this does not require nonstandard forms of logic such as fuzzy logic as, for example, suggested by Pigliucci, chapter 1, in this volume.

REFERENCES

Afonso, Ana S., and John K. Gilbert. 2009. "Pseudo-science: A Meaningful Context for Assessing Nature of Science." *International Journal of Science Education* 32 (3): 329–48. doi:10.1080/09500690903055758.

Alters, Brian J. 1997. "Whose Nature of Science?" *Journal of Research in Science Teaching* 34:39–55.

Beckner, Morton. 1959. *The Biological Way of Thought*. New York: Columbia University Press.

Boyd, Richard. 1999. "Homeostasis, Species, and Higher Taxa." In *Species: New Interdisciplinary Essays*, edited by R. A. Wilson, 141–85. Cambridge, MA: MIT Press.

Bunge, Mario. 1983. *Treatise on Basic Philosophy*. Vol. 6, *Epistemology and Methodology II*. Dordrecht: Reidel.

———. 1984. "What Is Pseudoscience?" *Skeptical Inquirer* 9 (1): 36–46.

Carlson, S. 1985. A Double-Blind Test of Astrology. *Nature* 318:419–25.

Cat, Jordi. 2006. "Unity and Disunity of Science." In *The Philosophy of Science: An Encyclopedia*, edited by Sahotra Sarkar and Jessica Pfeifer, 2:842–47. New York: Routledge.

Dean, Geoffrey. 1977. *Recent Advances in Natal Astrology*. Subiaco: Analogic.

Dupré, John. 1993. *The Disorder of Things: Metaphysical Foundations of the Disunity of Science*. Cambridge, MA: Harvard University Press.

Eflin, Judy T., Stuart Glennan, and George Reisch. 1999. "The Nature of Science: A Perspective from the Philosophy of Science." *Journal of Research in Science Teaching* 36:107–16.

Hansson, Sven O. 2008. "Science and Pseudo-science." *Stanford Encyclopedia of Philosophy*. http://plato.stanford.edu/entries/pseudo-science.

———. 2009. "Cutting the Gordian Knot of Demarcation." *International Studies in the Philosophy of Science* 23:237–43.

Hines, Terence. 2003. *Pseudoscience and the Paranormal*. Amherst, NY: Prometheus Books.

Irzik, Gürol, and Robert Nola. 2011. A Family Resemblance Approach to the Nature of Science in Science Education. *Science & Education* 20:591–607.

Kanitscheider, Bernulf. 1991. "A Philosopher Looks at Astrology." *Interdisciplinary Science Reviews* 16:258–66.

Kitcher, Philip. 1982. *Abusing Science: The Case Against Creationism*. Cambridge, MA: MIT Press.

———. 1993. *The Advancement of Science*. New York: Oxford University Press.

Kuhn, Thomas S. 1970. "Logic of Discovery or Psychology of Research?" In *Criticism and the Growth of Knowledge*, edited by Imre Lakatos and Alan Musgrave, 1–24. New York: Cambridge University Press.

Lakatos, Imre. 1970. "Falsification and the Methodology of Research Programmes." In *Criticism and the Growth of Knowledge*, edited by Imre Lakatos and Alan Musgrave, 91–197. New York: Cambridge University Press.

Laudan, Larry. 1983. "The Demise of the Demarcation Problem." In *Physics, Philosophy, and Psychoanalysis*, edited by Robert S. Cohen and Larry Laudan, 111–27. Dordrecht: D. Reidel.

Lugg, Andrew 1987. "Bunkum, Flim-Flam and Quackery: Pseudoscience as a Philosophical Problem." *Dialectica* 41:221–30.

Mahner, Martin. 2007. "Demarcating Science from Non-science." In *Handbook of the Philosophy of Science*. Vol. 1, *General Philosophy of Science—Focal Issues*, edited by Theo A.F. Kuipers, 515–75. Amsterdam: North Holland.

Mahner, Martin, and Mario Bunge. 1997. *Foundations of Biophilosophy*. Heidelberg: Springer.

Martin, Michael. 1994. "Pseudoscience, the Paranormal and Science Education." *Science & Education* 3:357–71.

Matthews, Michael. 2009. Teaching the Philosophical and Worldview Components of Science. *Science & Education* 18:697–728.

Nickles, Thomas. 2006. "The Problem of Demarcation." In *The Philosophy of Science: An Encyclopedia*. Vol. 1, edited by Sahotra Sarkar and Jessica Pfeifer, 188–97. New York: Routledge.

Pennock, Robert T. 2009. "Can't Philosophers Tell the Difference between Science and Religion? Demarcation Revisited." *Synthese* 178 (2): 177–206. doi:10.1007/s11229-009-9547-3.

Pigliucci, Massimo. 2003. Species as Family Resemblance Concepts: The (Dis-)solution of the Species Problem? *BioEssays* 25:596–602.

———. 2010. *Nonsense on Stilts: How to Tell Science from Bunk*. Chicago: University of Chicago Press.

Platvoet, Jan G., and Arie L. Molendijk, eds. 1999. *The Pragmatics of Defining Religion*. Leiden: Brill.

Popper, Karl R. 1963. *Conjectures and Refutations*. New York: Basic Books.

Reisch, George A. 1998. "Pluralism, Logical Empiricism, and the Problem of Pseudoscience." *Philosophy of Science* 65:333–48.

Thagard, Paul. 1988. *Computational Philosophy of Science*. Cambridge, MA: MIT Press.

———. 2011. "Evolution, Creation, and the Philosophy of Science." In *Epistemology and Science Education: Understanding the Evolution vs. Intelligent Design Controversy*, edited by Roger S. Taylor and Michael Ferrari, 20–37. New York: Routledge.

Wilson, Edward O. 1998. *Consilience: The Unity of Knowledge*. New York: Vintage Books.

Wilson, Fred. 2000. *The Logic and Methodology of Science and Pseudoscience*. Toronto: Canadian Scholars' Press.

Wilson, Robert A. 1999. "Realism, Essence, and Kind: Resuscitating Species Essentialism?" In *Species: New Interdisciplinary Essays*, edited by Robert A. Wilson, 187–207. Cambridge, MA: MIT Press.

3

Toward a Demarcation of Science
from Pseudoscience

JAMES LADYMAN

One of the most salient features of our culture is that there is so much bullshit.
—Harry Frankfurt (2005, 1)

Pseudoscience is a complex phenomenon just as science is, and like bullshit it may be sophisticated and artfully crafted. It is socially, politically, and epistemically important correctly to taxonomize these phenomena, and this chapter offers a modest contribution to that project. I argue, first, that the concept of pseudoscience is distinct from that of nonscience, bad science, and science fraud; second, that the concept of pseudoscience is a useful and important one in need of theoretical elaboration; and third, that progress can be made in this regard by learning from Harry Frankfurt's celebrated account of bullshit. Bullshitting, according to Frankfurt, is very different from lying. Pseudoscience is similarly different from science fraud. The pseudoscientist, like the bullshitter, is less in touch with the truth and less concerned with it than either the fraudster or the liar. I consider the difference between accounts of science and pseudoscience that focus on the product and those that focus on the producer, and I sketch an account in terms of the social organization and relations of the producers, their relationship to the product, as well the reliability of the production process.

How Pseudoscience Differs from Nonscience,
Bad Science, and Science Fraud

Science often errs, and . . . pseudoscience may happen to stumble on the truth.
—Karl Popper (1963, 33)

For the moment, let's take the concept of science for granted. Physics and biology are very different in many ways, but both are undoubtedly sciences. Clearly, there is a lot of intellectual activity that is not scientific, such as political philosophy or literary criticism (though both may draw on science, especially the former). Some such activity may have as its goal the acquisition of knowledge, and it may even be based on the gathering of empirical evidence, as with history, for example. The concept of nonscience implies no value judgments concerning its extension, and in particular it is not pejorative to describe something as nonscientific. By contrast, since according to the *Oxford English Dictionary*, "pseudo" means "false, pretended, counterfeit, spurious, sham; apparently but not really, falsely or erroneously called or represented, falsely, spuriously," it is pretty clear that the term "pseudoscience" is normatively loaded. However, an important distinction is drawn between two uses in the *Oxford English Dictionary*: the first, a count noun, involves either a sense derivative of the second, or what is mistakenly taken to be science or based on the scientific method. The second, a mass noun, is what is pretended to be science. Below it is argued that it is the second sense, or the first sense derivative of it, that philosophers of science usually have in mind when they use the term, and that mistaken science or something being mistaken for science does not amount to pseudoscience in an interesting or important sense.

The history of science is full of mistakes and falsehoods, even if we count it as not starting until the Scientific Revolution. For example, light is not composed of corpuscles as Isaac Newton believed, flammable substances do not contain phlogiston, and the rate of the expansion of the universe is not decreasing as was orthodoxy in cosmology until the 1990s. None of the scientists responsible for promulgating these false beliefs seems to deserve to be called a pseudoscientist, and it would not be appropriate to call every erroneous scientific theory pseudoscience. It seems clear that the connotation of either fraudulence or some kind of pretense is essential to contemporary uses of the term "pseudoscience," or at least that it should be part of any regimentation of the concept that is proposed. Even very bad science that is advocated as good science is not necessarily aptly described as pseudoscience. For example, Lemarckian inheritance may have been somewhat vindicated recently, but the basic idea that acquired phenotypic characteristics are not inherited is correct. Professional tennis players develop much heavier bones and larger muscles in one arm and shoulder, but their children do not have any such variation. In the 1920s, William McDougall claimed that the offspring of rats that had learned the layout of a particular maze were able to run it faster than the offspring of rats that had not learned the maze. Oscar Werner Tiegs and Wilfred Eade Agar and their collaborators showed that McDougall's work was based on poor experimental controls, which made it bad science but not fraudulent or based on any kind of pretense. More prosaically, an incompetent undergraduate physicist who gets the wrong answer when experimentally determining the acceleration due to gravity is not considered a pseudoscientist, nor is his laboratory report considered pseudoscience.

So pseudoscience is not just nonscience, nor is it simply bad science. Perhaps the idea of fraud or pretense is the only missing ingredient of pseudoscience as the dictionary suggests. After all, pseudoscientists often pretend that certain beliefs are supported by scientific evidence or theorizing when they are not, just as fraudsters do. Clearly, not all nonscience or bad science is science fraud, so maybe the latter is the only additional concept we need. Science fraud certainly exists and can be extremely damaging, and since false results and very bad science do not amount to fraud without the falsification of data or the intention to deceive about how results have been arrived at, we would seem to have made the connection to our second dictionary definition of pseudoscience.

However, this will not do for at least two reasons. First, the deliberate intention to deceive about explicitly expressed facts about the world (usually including experimental data) is a necessary condition for science fraud, but not for pseudoscience. For example, Karl Popper (1963) famously argued that although Freudian and Alderian psychology and Marxism were claimed to be scientific by many of their respective adherents, all are in fact pseudoscientific. Yet it is not at all clear that any, or even most, of them were in any way insincere. It is preposterous to suggest that Sigmund Freud's lifelong and obsessive researches in psychology did not amount to a genuine attempt to grapple with the profound problems of understanding the human mind and personality, motivation and the bizarre forms of pathological and self-destructive behavior that came to his attention.[1] Similarly, Frederick Engels surely believed his famous claim that just as Charles Darwin had understood biological evolution, so Karl Marx had discovered the laws of the evolution of human societies. So, pseudoscientists need not be disingenuous about their avowed beliefs that form the subject matter of the pseudoscience, even if they are deceiving or self-deceiving in other ways. Not all pseudoscience is science fraud. On the other hand, science fraudsters intend to mislead others about the truth (or what they take it to be). Hence, not all pseudoscience is science fraud, though some of the former may involve the latter.

It also seems wrong to call most examples of scientific fraud pseudoscience since it misdiagnoses the problem. In science fraud, it is not the avowed methodology, the nature of the subject, and the kind of theories that are in question, or the basic principles on which the discipline is based that are problematic. The fakery of pseudoscience is more profound than the mere faking of results; it is the nature of the enterprise and its methods that are falsely pretended to be scientific. Furthermore, one may of course use science fraud to establish a theory that is continuous with established science, expected and no threat to orthodoxy, and in fact true. Consider a scientist who rushes to publish preliminary results and claims that extensive checking with more data has been carried out when it has not. The results may be correct and more data may have supported them had it been gathered, but this would still be a case of scientific fraud. Hence, science fraud is not pseudoscientific as such, though as mentioned above, the two may overlap in some cases.

Why We Need the Concept of Pseudoscience

> The demarcation between science and pseudoscience is not merely a problem of armchair philosophy: it is of vital social and political relevance.
>
> —Imre Lakatos (1977, 1)

Pseudoscience is not the same thing as nonscience, bad science, or science fraud, though they may all overlap, and in particular science fraud is in general very different from pseudoscience. When the theory of Intelligent Design (ID) is described as pseudoscientific, there may be the implication of dishonesty since some advocates of ID were previously advocates of young-earth creationism; hence it appears that they promote ID not because it represents their beliefs, but because they think it will undermine the hegemony of evolutionary biology in science education and cede ground to the religious interests that are their ultimate concern. However, this kind of deception is not necessary as we saw above. Not all (nor probably most, nor even possibly any) advocates of homeopathy promote it without believing in its efficacy. Yet homeopathy is a paradigmatic example of pseudoscience. It is neither simply bad science nor science fraud, but rather profoundly departs from scientific methods and theories while being described as scientific by some of its adherents (often sincerely).[2]

So a complete taxonomy requires a concept that is distinct from non-science, bad science, and pseudoscience. But is the concept of pseudoscience worth having in practice? Is it socially, politically, and/or epistemologically important as Imre Lakatos says it is? Certainly judging by the amount the word "pseudoscience" gets used, lots of scientists and science writers think the concept is important. However, the danger with words that are used to express strong value judgments is that often the motivation to condemn something causes people to reach for the word, deploying it as a catchall for expressing disapprobation, even though it does not pick out a true kind in the world. For example, terms like "conservative" and "liberal" are often deployed to criticize in public political discourse, even though they do not necessarily pick out genuine kinds since the types of views and individuals they refer to are incredibly diverse and often contradictory. Clearly, terms like "Jewish science" are also spurious. They refer only by ostension or stipulation—as in "Special Relativity is Jewish science"—and not because there is such a kind in the world. Scientists are an important sector of society and they have social interests and agendas like everyone else, so the mere fact that the term "pseudoscience" seems indispensable to them and their cultural allies is not sufficient to establish that it is worth using. The previous section established that there is logical space for a concept distinct from the others discussed that fits certain paradigmatic examples of pseudoscience, but not that using the concept of pseudoscience is theoretically or practically worthwhile.

Larry Laudan (1982) argues that it is a mistake to engage with pseudoscience in the abstract, and to characterize it and how it differs from science in terms of general criteria to do with the kinds of theories it uses or methods it employs. Instead, he argues, we should concentrate on evaluating first-order claims about the world piecemeal and considering whether the evidence supports them. According to this view, beliefs concerning the efficacy of heterodox medical treatments or the age and geology of the earth that are labeled pseudoscientific are not to be mistrusted for that reason but because there is no evidence for them. It is then argued that to combat what we call "pseudoscience," we do not need any such notion but simply the idea of beliefs that are not in accordance with the facts and the evidence. But confronting pseudoscience in this way is problematic: it consumes too much time and too many resources, is not useful when engaging in public debates that operate at a general level, and is too detailed for scientifically nonliterate audiences.

There are pragmatic considerations, but note also that Laudan does not show that a genuine kind is not picked out by the term "pseudoscience." As pointed out above, the fact that a word is used by a group of people to label a kind does not imply that a genuine kind exists. However, correspondingly, the fact that a group sometimes misuses a term does not imply that it is meaningless or useless, as Laudan seems to assume.[3] However, perhaps his skepticism about the worth of the term "pseudoscience" is just the correlate of his similar attitude toward the term "science." Much effort spent seeking demarcation criteria for the latter has not produced an agreed-on definition.

In Laudan's case, his extensive study of the history of science and its methods convinced him that it has changed so much that no core set of stable characteristics can be identified. Statistical methods, for example, are essential to science, and extensive training in them is part of the education of scientists in very diverse fields; and yet prior to the twentieth century, they barely existed. The techniques of measurement and the criteria for good explanations that we have now are very different from those of the eighteenth century. However, even Thomas Kuhn emphasizes five core criteria that all scientific theories may be judged against:

1. Accuracy—empirical adequacy with experimentation and observation
2. Consistency—both internally and externally with other theories
3. Scope—broad implications for phenomena beyond those the theory was initially designed to explain
4. Simplicity—the simplest explanation is to be preferred
5. Fecundity—new phenomena or new relationships among phenomena should result

Unfortunately, Kuhn does not think every scientist will agree about how to weigh these criteria against one another when theories score well on some and poorly on others. Furthermore, he does not think they will agree even about how theories score on a single criterion since that involves judgment, background beliefs, and epistemic values. According to Kuhn, then, there is neither a unique measure on each criterion nor a unique function to weigh them and produce a ranking of theories. This is similar to Pierre Duhem's (1954) approach to the problem of theory choice, which was to deny that there was a rule that would determine which of a set of empirically equivalent theories should be chosen, based on the virtues of theories that go beyond their mere conformity with the data. Duhem was quite certain that there were such virtues (e.g., simplicity) and that they do get weighed in judgments about which theory to choose. However, he thought that the right choice was a matter of irreducible "good sense" that cannot be formalized. Duhem differed from Kuhn in being convinced that all such judgments were temporary, lasting only until more empirical evidence became available.[4]

In the light of these arguments about the demarcation of science from nonscience and the contested nature of the scientific method, it is not immediately obvious that even the term "science" is useful since science is so heterogeneous and it is contested whether certain parts of it are really scientific; think, for example, of skepticism about social science or string theory. Maybe it is better to disambiguate between, for example, medical science, physical science, life science, and so on, and not use the term "science" at all.

However, the fact that science evolves and that it is hard to capture its nature in a definition may be outweighed by the theoretical simplicity and utility of the concept of science. There is continuity over time in science too, and we have no trouble at all understanding Newton's theories and models, and the problems he set out to solve, and phenomenal laws like Boyle's and Kepler's. We can be reasonably confident that the great scientists of the past would regard our current theories and experimental knowledge as the fulfillment of their ambitions. Robert Boyle would recognize modern chemistry for its empirical success, even if he found much of it baffling; and we would presumably find it equally easy to convince Robert Hooke of our knowledge of microbiology. Our understanding of the rainbow and the formation of the solar system are still fundamentally related to the explanations of those phenomena given by Descartes. All, or almost all, the established empirical knowledge of past science is retained in otherwise radical theory change. Newtonian gravitation is a low-energy limit of general relativity, for example. A family resemblance certainly exists between the sciences, and the success of fields such as thermodynamics and biophysics shows that science as a whole has a great deal of continuity and unity.

Like the concept of science, the concept of pseudoscience is useful because it taxonomizes the phenomena reasonably precisely and without too much error. This is all that can be asked of any such concept. For instance, many branches of quackery and snake oil marketing have in common that they emulate scientific theories and explanations and often employ scientific terms, or terms that sound like scientific terms, as if they were connected to scientific knowledge. Being able to deploy the concept of pseudoscience is important in advancing the public understanding of science and in ensuring that public policymaking and public health are informed by genuine science.

It is a familiar if frustrating fact to philosophers that important and even fundamental concepts such as that of knowledge resist analysis into necessary and sufficient conditions. Had we concluded our preliminary investigations into the concept of pseudoscience by identifying it with the concepts of bad science or science fraud, we would have merely postponed a full definition until the concept of science itself had been explicated since both bad science and science fraud are defined negatively in relation to science. Pseudoscience must surely also involve some kind of emulation of science or some of its characteristics or appearance. What is distinctive about pseudoscience may also be illuminated by considering another case of a concept that is closely related to the propagation of falsehoods, but that turns out to be interestingly sui generis and similarly undermining of the genuine quest for truth.

On Pseudoscience and Bullshit

> Bullshit is a greater enemy of the truth than lies are.
>
> —Harry Frankfurt (2005, 61)

Frankfurt's celebrated investigation of bullshit seems motivated by his sense of its significance for our lives, not just by its intellectual interest. The above quotation may be aptly applied to pseudoscience: pseudoscience is a greater enemy of knowledge than science fraud. Frankfurt makes a very important point about how bullshit differs from lies, namely that the latter are designed to mislead us about the truth whereas the former is not concerned with the truth at all. Plainly, this distinction is analogous to that between pseudoscience and science fraud. As a first approximation, we may say that pseudoscience is to science fraud as bullshit is to lies.

This is only a first approximation because we usually assume that bullshitters know what they are doing whereas, as pointed out above, many pseudoscientists are apparently genuinely seeking truth. However, one can bullshit unwittingly, and pseudoscience is often akin to that. Just because one's first-order self-representations are that one is sincerely seeking the truth, it may be argued that, in a deeper sense, one does not care about it because one does not heed to the evidence. A certain amount of self-deception on the part of its advocates explains how pseudoscience is often disconnected from a search for the truth, even though its adherents think otherwise. This is important because it means that what makes an activity connected or disconnected to the truth depends on more than the individual intentions of its practitioners. We shall return to this below.

Note that the analogy between lies and science fraud is notwithstanding the fact that, as pointed out above, science fraud may involve the propagation of claims that are true. After all, lies may turn out to be true. Roughly speaking, someone tells a lie when he or she says something believed to be false with the intention of bringing about in the audience what he or she thinks is a false belief concerning both the matter of fact, and what he or she believes about the matter of fact. If the liar is mistaken about the matter of fact, and so inadvertently tells the truth, he or she still lies. Science fraud always involves lying, even when it supports true claims, because the fraud consists in the data or methodology cited to support those claims being falsified.

The analogy between bullshit and pseudoscience is quite apt. Both seem to capture something important that makes them distinct from and, in some ways, more dangerous than the peddling of false claims. The reason pseudoscience is so dangerous is analogous to the reason Frankfurt thinks that bullshit is more dangerous than lies, namely that lies, being straightforward claims about reality, can be proved lies with sufficient scrutiny, whereas bullshit and pseudoscience resist refutation by not making definite claims at all. They progressively disconnect us from the truth in a way that is more insidious than lying, for we may end up not just with false beliefs but with no beliefs at all. Indeed, they undermine the habit of making sure that our thoughts are determinate and make contact with reality. Negative accounts of pseudoscience define it in terms of it not being science, but the analogy with bullshit shows that pseudoscience has a positive nature akin to that of bullshit, consisting in the production of epistemic noise that resembles propositions and beliefs only superficially.

However, one important difference between bullshit and pseudoscience is that the latter, but not the former, often asserts factual claims of some kind. Consider, for example, pseudoscientific accounts of novel medical treatments, many exotic varieties of which exist as well as relatively mainstream ones such as homeopathy: no matter how much blather and bluster surrounds them, they clearly claim that the treatments are efficacious. However, it is arguable that these false (or at least dubious) factual claims are not what makes pseudoscience pseudoscience. Let's introduce a distinction that also applies to theories of bullshit and of science and pseudoscience, namely, that between the producer and the product. Clearly, we may give accounts that focus on the texts or theories that are produced or on the mental states and attitudes of the people who produce them.

Many people have sought to demarcate science from pseudoscience in terms of the product. Most influentially, Popper argued that genuine scientific claims must be testable in the sense of being falsifiable, and that the core claims of many putative pseudosciences are unfalsifiable. This condition is very popular in contemporary discussions of pseudoscience. The problem with it is that "testability, revisability and falsifiability are exceedingly weak requirements" (Laudan 1982, 18)—at least, that is, if all that is required is in-principle testability. For example, imagine a pseudoscientific cult that builds an elaborate theory around a predicted alien intervention in the year 3000. This is testable but nonetheless pseudoscientific. Of course, it is not testable for a long time, and so may be argued to be testable only in principle and not in practice. However, drawing the line between when exactly we need to be able to perform a test is not possible, and we do not want to rule out proposed tests in some parts of science that are not at the moment testable in practice, as there have been many such cases in the past that have subsequently turned out to be tested by means that were thought impossible.[5] Furthermore, the requirement of testability or falsifiability is too strong, at least when applied to individual propositions, because high-level scientific hypotheses have no direct empirical consequences. Hence, there are many scientific statements that are not falsifiable, or at least not directly. For example, the principle of the conservation of energy does not imply anything until we add hypotheses about the kinds of energy they are. Indeed, the idea of testability in principle is very unstable since if we do not hold fixed the rest of science and technology, then the propositions that are testable change.

There is something right about the criterion of testability and falsifiability, but it is not to be found solely in its application to theories or propositions. Before turning to that, it is important to note that the notion of testability may also be given an inductivist reading that is compelling about certain cases. Rudolf Carnap (1966) argued against vital forces theories in biology for not making definite and precise predictions, and many people claim that general relativity and/or quantum electrodynamics are the most successful scientific theories ever, not just because they make very precise predictions but because these have been confirmed by experiment. Novel and precise predictive success is a key difference between science and pseudoscience, but it is often neglected for emphasis on falsifiability. Pseudoscience can be characterized negatively in so far as it does not make precise and accurate predictions, while science in general does. The point about this criterion is that it raises problems for some areas of social science and theoretical physical science.

It is important to distinguish sentences that make factual claims about the world from those that do not. Practitioners of bullshit and pseudoscience both produce sentences intended to convince their audience that some factual claim is being made, when it is not. For example, a politician, asked how he will achieve some goal in light of specific criticism, may respond: "The important thing is to ensure that going forward we put in place robust processes that deliver the services that people rightly expect to be of the highest quality, and that is why I have taken steps to ensure that our policies will be responsive to the needs on the ground." He succeeds in taking up the time available, uses his tone of voice and facial expressions to convey a possibly false impression of his affective states and values, and says nothing (or at least nothing other than expected platitudes). The function of this kind of bullshit, as Frankfurt says, is not to bring about beliefs in the audience, or at least not beliefs about the subject matter, but about the politician and his good offices.

Similarly, a pseudoscientific explanation of a bizarre medical treatment may refer to the rebalancing of the energy matrix as if some determinate concept of physics were involved when, in fact, the description has no scientific meaning. Pseudoscientific words often combine genuine scientific and non-scientific terms. A speaker who uses the made-up term "photon torpedoes," for example, suggests that she has a clever and sensible theory and that her audience can assume her causal claims are true: viz., the crystal will help a bad back or the spaceship will get its engines disabled. The point is that for many people, scientific terms are indistinguishable from those of pseudoscience. The layperson has no way of knowing that a term such as "magnetic flux density" is genuine whereas "morphic energy field resonance" is not.

In any case, Popper knew falsifiability of its product is not sufficient to distinguish science from pseudoscience, for he characterizes pseudoscience in terms of its producers as well as its products. Whereas he argues that the theories of psychoanalysis are unfalsifiable, he accepts that Marxist theories make predictions about the phenomena, but insists that Marxism is a pseudoscience because Marxists keep modifying the product to make it compatible with the new data and refuse to accept the falsification of their core commitments. However, Popper has been extensively criticized for making unreasonable demands on the producers of scientific knowledge, demands that are not met in the history of science. Individual scientists may well cling tenaciously to their theories and keep meeting failure with modification of peripheral components and working hard to make their framework fit the facts. This is necessary because theories require painstaking effort and the exploration of many blind alleys, and persistence and commitment are required.

Nonetheless, there is something right about the stress on empirical testability. Just as collective endorsement requires a generally accepted proof in mathematics, so in science, collective endorsement requires that theories pass stringent tests of empirical adequacy that even their opponents in the scientific community set for them. Atomism triumphed in the late nineteenth century because it succeeded on multiple fronts in successfully predicting quantities that its critics said it would never successfully predict. Pseudoscience has no analogue of this in its history. The lesson from this failure of accounts of the scientific method that focus on theories, for example, by proposing criteria of testability, is that they leave out the producers' attitudes to those theories. But the problem with prescribing instead the mental states and attitudes of individual producers is that it neglects that science is a collective enterprise; the way individual scientists think and behave is conditioned to a great extent by their interactions with their peers and their work. The reliability of science as a means of producing knowledge of the world is not to be found in the content of its theories, nor in a model of the ideal scientific mind, but rather in the emergent properties of the scientific community and the interactions among its members, as well as between them and their products.

In science, theories and propositions form hierarchies and webs of relations that through the use of mathematics give many concrete applications and specific predictions. Scientists of all stripes routinely collaborate in highly productive ways in engineering, medical or technological applications, or simply in gathering data. This unity of scientific practice and its theories is missing in pseudoscience. We have common measurement units, the conservation of energy, and the second law of thermodynamics, and dimensional analysis reduces all quantities in the physical sciences to the same fundamental basis. The atomic theory of matter and the periodic table is deployed in every science from the study of stars and galaxies to the study of the climate, living beings, and geology. Everywhere in science we find people working to connect its boundaries between levels of organization, between different domains and between different regimes. Science is hugely collaborative and involves rich relationships among experimenters, theorists, engineers, statisticians, and so on. Pseudoscience is largely about cult figures and networks whose relational structure involves a lot of chat, but lacks the integration with rich mathematics, material interventions, and technology that characterizes science.

The scientific method is reliably connected to the truth of the theories produced. At the very least, reliability signals genuine knowledge, if it does not completely define it. Reliability is of two kinds in epistemology: believing what is true and not believing what is false. These two goals are clearly in tension with each other since it is easy to achieve the second by being utterly skeptical, but one will not then achieve the first. Likewise, credulity engenders many true beliefs but also false ones. Reliable knowledge needs to avoid both kinds of error, which are analogous to Type I and Type II errors in statistics or false positives and false negatives. Suppose, for example, someone is undergoing a test for a medical condition: a false-positive test result indicates that the patient has the condition when he does not, and a false-negative test result indicates that the patient does not have the condition when he does.

Reliability in knowledge means what epistemologists call "sensitivity" and "safety." The former means if it were not true we would not believe it, the latter that in different but similar circumstances in which it was true we would still believe it. Similarly, if the test is positive, "sensitivity" means it would have been negative when the person did not have the condition, and "safety" means it would have still been positive even if it had been tested a few minutes earlier or later, or by a different technician, and so on. Knowledge is different from mere belief because it is not accidental or random that we believe the truth when we know. The mathematician does not just happen to believe Pythagoras's theorem; she believes it in a way that is intimately connected to it being true. Similarly, many of our basic perceptual beliefs about the world count as knowledge because we are endowed to reliably arrive at them in the kinds of circumstances in which they are formed, and they are reasonably judged to be safe and sensitive in the above senses. Bird (2007) has argued that the progress of science should be understood in terms of the growth in knowledge, not merely in terms of growth in true beliefs. Without explaining or endorsing his account, there is no doubt that he is on to something important, and pseudoscience insofar as it involves beliefs about the world, pretends not just at true belief but at knowledge.

I hope to have shown that Frankfurt's seminal investigation of bullshit applies to pseudoscience because (a) like bullshit, pseudoscience is largely characterized not by a desire to mislead about how things are (as with science fraud) but by failing to say anything much at all about how things are; and

(b) it offers a helpful distinction between defining bullshit/pseudoscience in terms of the producer or the product or both.

Conclusion

Pseudoscience is attractive to people for two reasons. First is the general mistrust some laypeople feel for scientists and science as an institution. Trust in science has always been partial and contested, and abuses of scientific knowledge and the power of science make this an understandable reaction in many cases. Mistrust has been and continues to be engendered by pseudoscience and science fraud within mainstream science, leading some to the conclusion that the distinction between science and pseudoscience is like the distinction between orthodox and heterodox in being purely a matter of power and authority. Certain key examples of bogus mainstream science have reflected and thus indirectly condoned harmful societal ideologies of sex or race. Appalling medical science abuses have given the public reason to doubt that the medical establishment always has their best interests in mind.[6] In psychiatry not that long ago, women in the UK were incarcerated for what is now considered nothing more than a healthy interest in sex. Until 1973, the American Psychiatric Association regarded homosexuality as a mental disorder. And surely some medical researchers are corrupted by corporate interests and either exaggerate the efficacy of potential lucrative treatments or downplay or deny their negative effects.

The second reason for pseudoscience's continued influence is that many people suffer from physical, mental, and emotional ailments and afflictions for which medical science can do little, if anything, or for which the appropriate treatment would require a lot of resources. People may even have a lot to gain by believing in pseudoscientific solutions to their problems. Work on the placebo effect shows that they may be right that pseudoscience "helps" them, although of course which bogus therapy they choose is more or less irrelevant, though some may prefer a pseudoscientific rather than supernatural wrapper.

Science fraud, scientific corruption, and ideologically biased science are the greatest friends of pseudoscience, for they all help create the epistemic climate of skepticism and mistrust of epistemic authority in which it can flourish. We need epistemic authority because nobody can check everything for him- or herself and because many of us lack the knowledge and/or intellectual powers to follow the reasoning in science, mathematics, and medicine. Unreliable pseudoscience may seem authoritative, but it is full of bullshit.

NOTES

1. This is not to rule out that Freud did commit some science fraud in so far as he fabricated claims of clinical success and studies that did not exist (see Cioffi 1999).

2. It is worth noting that there may be a fair amount of science fraud associated with pseudoscience since the temptation to fabricate results to substantiate what pseudoscientists may sincerely believe to be the truth can be very great. In this respect, however, it is no different from science.

3. See chapter 2 of the present volume in which Martin Mahner argues that the pseudoscience concept lacks necessary and sufficient conditions but can be individuated as a cluster concept like species concepts in biology. Also see chapter 1 by Massimo Pigliucci for a similar argument.

4. See Ivanova (2010).

5. The Michaelson-Morley experiment is a good example since it was previously believed that such accuracy was unattainable.

6. I have in mind the Tuskegee experiment and others in which subjects were actually given syphilis.

REFERENCES

Bird, A. 2007. "What Is Scientific Progress?" *Noûs* 41:64–89.

Carnap, Rudolf. 1966. "The Value of Laws: Explanation and Prediction." In *Philosophical Foundations of Physics*, edited by Martin Gardner, 12–16. New York: Basic Books.

Cioffi, Frank. 1999. *Freud and the Question of Pseudoscience*. Peru, IL: Open Court.

Duhem, Pierre Maurice Marie. 1954. *The Aim and Structure of Physical Theory*, translated by Philip P. Wiener. Princeton, NJ: Princeton University Press.

Frankfurt, Harry G. 2005. *On Bullshit*. Princeton, NJ: Princeton University Press.

Ivanova, Milena. (2010). "Pierre Duhem's Good Sense as a Guide to Theory Choice." *Studies in History and Philosophy of Science* pt. A41 (1): 58–64.

Lakatos, Imre. 1977. "Science and Pseudoscience." In *Philosophical Papers*, vol. 1. Cambridge: Cambridge University Press.

Laudan, Larry. 1982. "Commentary: Science at the Bar—Causes for Concern." *Science, Technology and Human Values* 7 (41): 16–19.

Popper, Karl. 1963. *Conjectures and Refutations*. London: Routledge and Kegan Paul.

4

Defining Pseudoscience and Science

SVEN OVE HANSSON

For a scientist, distinguishing between science and pseudoscience is much like riding a bicycle. Most people can ride a bicycle, but only a few can account for how they do it. Somehow we are able to keep balance, and we all seem to do it in about the same way, but how do we do it?

Scientists have no difficulty in distinguishing between science and pseudoscience. We all know that astronomy is science and astrology not, that evolution theory is science and creationism not, and so on. A few borderline cases remain (psychoanalysis may be one, see Cioffi, chapter 17, in this volume), but the general picture is one of striking unanimity. Scientists can draw the line between science and pseudoscience, and with few exceptions they draw the line in the same place. But ask them by what general principles they do it. Many of them find it hard to answer that question, and the answers are far from unanimous.

Just like keeping balance on a bicycle, distinguishing between science and pseudoscience seems to be a case of tacit knowledge, knowledge that we cannot make fully explicit in verbal terms so that others can understand and replicate what we do (Polanyi 1967). In the modern discipline of knowledge management, tacit knowledge is as far as possible articulated (i.e., turned into explicit, communicable knowledge). When tacit knowledge becomes articulated, it is more easily taught and learned, and more accessible to criticism and systematic improvements. These are all good reasons to articulate many forms of tacit knowledge, including that of the science/pseudoscience demarcation.

But whereas the tacit knowledge of bicycle riding has been successfully articulated (Jones 1970), the same cannot be said of the science/pseudoscience demarcation. Philosophers of science have developed criteria for the demarcation, but no consensus has been reached on these criteria. To the contrary, the lack of philosophical agreement in this area stands in stark contrast to the virtual agreement among scientists in the more specific issues of demarcation (Kuhn 1974, 803). In my view, the reason for this divergence is that philosophers have searched for a criterion of demarcation on the wrong level of epistemological specificity. I intend to show here how a change in that respect makes it possible to formulate a demarcation criterion that avoids the problems encountered by previous proposals.

A science/pseudoscience demarcation with sufficient generality has to be based on general epistemological criteria. But the subject area of science, in the common sense of the word, is not delimited exclusively according to epistemological criteria. In the next section, I discuss how the subject area of science (and consequently that of pseudoscience) should be delimited to make an epistemically cogent demarcation possible.

Science has limits not only against pseudoscience but also against other types of nonscience. "Unscientific" is a wider concept than "pseudoscientific," and "nonscientific" an even wider one. Therefore, it is inadequate (though unfortunately not uncommon) to define pseudoscience as that which is not science. Attention must be paid to the specific ways in which pseudoscience violates the inclusion criteria of science and to what the term "pseudoscientific" means in addition to "nonscientific." This is discussed in the two sections that follow. Based on these considerations, a definition of pseudoscience is proposed. It differs from most previous proposals by operating on a higher level of epistemic generality. Finally, I defend that feature of the definition and explain how it contributes to avoiding some of the problems besetting previously proposed definitions.

The Subject Area of Science

The term "science" itself is vague. Its delimitation depends not only on epistemological principles but also on historical contingencies. Originally, the word "science" denoted any form of systematic knowledge, practical or theoretical. In the nineteenth century, its meaning was restricted to certain forms of academic knowledge, mainly those based on the study of nature (Layton 1976). Today, "science" refers to the disciplines investigating natural phenomena and individual human behavior and to some of the disciplines studying human societies. Other disciplines concerned with human societies and culture are instead called humanities. Hence, according to the conventions of the English language, political economy is a science (one of the social sciences) but classical philology and art history are not.

But the sciences and the humanities have something important in common: their very raison d'être is to provide us with the most epistemically warranted statements that can be made, at the time being, on the subject matter within their respective domains. Together they form a community of knowledge disciplines characterized by mutual respect for each other's results and methods (Hansson 2007). An archaeologist or a historian will have to accept the outcome of a state-of-the art chemical analysis of an archaeological artifact. In the same way, a zoologist will have to accept the historians' judgments on the reliability of an ancient text describing extinct animals. To understand ancient descriptions of diseases, we need cooperations between classical scholars and medical scientists—not between classical scholars and homeopaths or between medical scientists and bibliomancers.

The interconnections among the knowledge disciplines have increased for a long time and continue to do so. Two hundred years ago, physics and chemistry were two independent sciences with only few connections. Today they are closely knit together not least by integrative subdisciplines such as physical chemistry, quantum chemistry, and surface science. The interdependencies between natural sciences and the humanities are also rapidly growing. Although the comparison is difficult to make, archaeologists seem to depend more on chemistry and physics today than what biologists did two hundred years ago. These and many other bonds between the natural sciences and the humanities have increased dramatically in the half century that has passed since C. P. Snow's ([1959] 2008) pessimistic prediction of a widening gap between natural science and the humanities. As one of many examples of this, methods and concepts from studies of biological evolution (such as the serial founder effect) have recently been successfully applied to throw light on the development of human societies and even on the development of languages tens of thousands of years before written evidence (Henrich 2004; Pagel et al. 2007; Lycett and von Cramon-Taubadel 2008; Atkinson 2011).

Unfortunately, neither "science" nor any other established term in the English language covers all the disciplines that are parts of this community of knowledge disciplines. For lack of a better term, I will call them "science(s) in a broad sense." (The German word *Wissenschaft*, the closest translation of "science" into that language, has this wider meaning; that is, it includes all the academic specialties, including the humanities. So does the Latin *scientia*.) Science in a broad sense seeks knowledge about nature (natural science), about ourselves (psychology and medicine), about our societies (social science and history), about our physical constructions (technological science), and about our thought constructions (linguistics, literary studies, mathematics, and philosophy). (Philosophy, of course, is a science in this broad sense of the word; cf. Hansson 2003.)

Two side remarks should be made about the community of knowledge disciplines. First, some branches of learning have not received academic status. This applies, for instance, to philately and to the history of conjuring, both of which are pursued by devoted amateurs rather than professional scholars. Philately is a particularly illuminating example since the closely related subject area of numismatics has a strong academic standing. A major reason for the difference is the usefulness of numismatics in the dating of archaeological sites where coins have been found. But in the few cases when historians need the help of philatelists to date an undated but stamped letter, they will have to rely on the expertise of amateur philatelists in the same way that they rely on numismatists in other contexts. This is a good reason for including philately in the community of knowledge disciplines. It is not the academic status but the methodology and the type of knowledge that should determine whether a discipline is scientific (in the broad sense).

The second minor issue concerns scientists and scholars who opt out of the community and decide not to respect other disciplines. Examples of this are scholars who disbelieve carbon dating or choose to disregard archaeological evidence of the functions of ancient artifacts (as discussed in Nickell 2007 and Krupp 1984, respectively). Judging by experience, such disregard of other disciplines is a sure sign of low scientific quality. One of the most remarkable examples is what has been self-consciously announced as the "strong programme" in the sociology of knowledge (Bloor 1976). Proponents of this approach programmatically disregard what is known about the truth or falsity of scientific theories in fields other than their own. This is an unsuccessful methodology for the obvious reason that the successes and failures of scientific theories are important factors that need to be taken into account to understand science and its role in society.

A principled demarcation of pseudoscience cannot be based on the

standard concept of science that excludes the humanities. As already mentioned, a sizable number of promoters tout severely flawed theories in history and literature studies—Holocaust deniers; ancient astronaut theorists;
fabricators of Atlantis myths; sindonologists (investigators of the shroud of
Turin), practitioners of scripture-bound, "biblical" archaeology; proponents
of fringe theories on Shakespearian authorship; promotors of the Bible code;
and many others (Stiebing 1984; Thomas 1997; Shermer and Grobman 2000).
What places them outside the community of knowledge disciplines is primarily their neglect of historical and literary scholarship. In many of these
cases, neglect or falsification of natural science adds to the unreliability of the
teachings.

We have a terminological problem here. On one hand, it may seem strange
to use the term "pseudoscience" for instance about an Atlantis myth that has
nothing to do with science in the ordinary (narrow) sense of the word. On the
other hand, the creation of a new category for the "pseudohumanities" is unwarranted since the phenomenon overlaps and largely coincides with that of
pseudoscience. I follow the former option and use the term "pseudoscience"
to cover not only failed versions of science in the traditional sense but also
failed versions of science in the broad sense (including the humanities). In
this way, we can obtain a principled and epistemologically unified account
of the demarcation issue that is not available with the traditional, too narrow
notion of science in the English language.

How Pseudoscience Violates the Quality Criteria of Science

The phrases "demarcation of science" and "demarcation of science from pseudoscience" are often taken to be synonyms. This reduces the demarcation issue to a binary classification: for a given theory or statement, we have to determine whether it is a piece of science or a piece of pseudoscience. No other
options are considered.

This, of course, is a much too simplified picture. Science has nontrivial
borders to nonscientific phenomena other than pseudoscience, such as religion, ethics, and various forms of practical knowledge. Consider, for instance,
the (somewhat unsharp) border between the science of musicology and
practical musicianship. Practical musicianship is not science, but neither is it
pseudoscience. Borders also exist between religious studies and confessional
theology, between political economics and economic policy, between gender
studies and gender politics, and so on.

Pseudoscience is characterized not only by not being science but also, importantly, by deviating substantially from the quality criteria of science. To find the defining characteristics of pseudoscience, we therefore need to have a close look at the quality criteria of science. There are three major types of scientific quality criteria. The first and most basic of these is reliability: a scientific statement should be correct, or rather, as close to correctness as can currently be achieved. If a pharmacologist tells us that a certain substance reduces bleeding, then it should do so. If an anthropologist tells us that shamans in Amazonia have given leaves containing the substance to wounded tribesmen, then that should be so. The requirement of reliability is fundamental in all the knowledge disciplines.

The second criterion is scientific fruitfulness. Consider two scientists investigating birdsong. The first scientist records and analyzes the song of one hundred male birds of the same species. The outcome is an analysis identifying the different elements of song and the ways in which they are combined by different individuals. The second scientist also records and analyzes the song of one hundred birds of the same species, but she selects the individuals so that she can compare the song of birds with neighboring territories. Her analysis provides valuable information on the capacity of adult members of this species to learn new song patterns (cf. Doupe and Kuhl 1999). Therefore, even though the two investigations do not differ in reliability, the second represents better science since, against the background of other available information, the information it provides is scientifically more valuable.

The third criterion is practical usefulness. Consider two scientists both investigating the synthesis of serotonin in the neural system. One of them provides knowledge on these processes, but there is no foreseeable practical use of the new information. The other discovers a precursor that can be used as an antidepressant drug. Assuming that the two investigations provide equally reliable information, the latter is better science as judged by the criterion of practical usefulness.

The justification of scientific studies depends on their scientific fruitfulness, practical usefulness, or both. Science is often classified as either "basic" or "applied," under the assumption that basic research aims exclusively at scientific fruitfulness and applied research at practical usefulness. But in many areas of science, such as biochemistry and materials science, it is common for investigations to combine scientific fruitfulness with practical usefulness. It is also important to recognize that the three types of scientific quality are interconnected, not least since reliability is a necessary prerequisite for the other two. If the birdsong researcher confuses the different elements of birdsong, then her research cannot score high on scientific fruitfulness. If the neuroscientist misidentifies the neurotransmitters, then the practical usefulness of her research is essentially nil.

How does this relate to pseudoscience? My proposal is that only one of the three types of scientific quality, namely reliability, is involved in the distinction between science and pseudoscience. Consider the following examples of (in the broad sense) scientific work that satisfies the reliability criterion but neither of the other two:

- A chemist performs meticulous spectroscopic measurements on a large number of sulfosalt minerals. Some new spectral lines are found, but the new data does not lead to any new knowledge of the structure or properties of these minerals, and neither does it have any practical applications.
- A researcher in structural mechanics investigates the behavior of variously shaped aluminum bars under different mechanical loads. The results unsurprisingly confirm what was already known, and nothing new is learnt.
- A historian spends five years examining previously well-studied sources from the reign of Queen Mary I of England. The outcome is essentially a confirmation of what was already known; no new conclusions of any importance are drawn.

Although none of this is important science, it would seem wrong to call such endeavors pseudoscientific (or unscientific). An investigation does not qualify as pseudoscience merely by lacking in scientific fruitfulness and practical usefulness. It has to fail in terms of reliability (epistemic warrant), the most basic of the three quality criteria.

We can summarize this by saying that pseudoscience is characterized by suffering from such a severe lack of reliability that it cannot at all be trusted. This is the criterion of unreliability. It can be taken as a necessary condition in the definition of pseudoscience.

The "Pseudo" of Pseudoscience

A friend of mine who works in a chemistry laboratory once had problems with a measuring instrument. A whole series of measurements had to be repeated after the instrument had been properly repaired and recalibrated. The faulty measurements satisfied our criterion of unreliability, that is, they suffered from such a severe lack of reliability that they could not at all be trusted. However, it would be strange to call these measurements "pseudoscientific." They were just faulty, nothing else. As this example shows, the criterion of unreliability is not sufficient to define pseudoscience. Something more must be said about the use and function of unreliable claims in pseudoscience.

An obvious starting point for this discussion is the prefix "pseudo-" (ψευδο-) of "pseudoscience." Etymologically, it means "false." Many writers on pseudoscience have emphasized that pseudoscience is nonscience posing as science. The foremost modern classic on pseudoscience bears the title *Fads and Fallacies in the Name of Science* (Gardner 1957). According to Brian Baigrie (1988, 438), "what is objectionable about these beliefs is that they masquerade as genuinely scientific ones." These authors characterize pseudoscience as nonscientific teachings or statements that pose as science. Somewhat more precisely, it is a common characteristic of the pseudosciences that their main proponents try to create the impression that they are scientific (the criterion of scientific pretense).

However, it is easy to show that this criterion is too wide. Consider the following two examples:

1. A biologist studying moths on the Faroe Islands tries hard to identify the individuals she collects, but in spite of her best intentions she makes several classification mistakes. Therefore, colleagues refrain from drawing any conclusions from her report of this study.
2. A biochemist fabricates experimental data, purportedly confirming a recent hypothesis on the biosynthesis of spider silk. Although his report is a fake, the hypothesis is soon afterward confirmed in legitimate experiments.

The first is an example of a scientist trying honestly, yet failing, to provide reliable results. It would be inordinately harsh to call her research pseudoscientific. The second is an example of fraud in science. Like other such cases, it clearly operates within the domains of science, and it certainly satisfies the unreliability criterion. It also satisfies the criterion of scientific pretense; scientific fraud is surely "false science." Nevertheless, we tend to treat fraud and pseudoscience as different categories. Fraud in otherwise legitimate branches of science is seldom, if ever, called "pseudoscience" (but it can certainly be called "unscientific").

The crucial element that is missing in these cases is a deviant doctrine. All the typical cases of pseudoscience are cases in which a deviant doctrine plays a crucial role (Hansson 1996). Pseudoscience, as it is commonly conceived, involves a sustained effort to promote teachings that do not have scientific legitimacy at the time. We therefore can specify the criterion of scientific pretense and characterize a pseudoscientific statement as one that is part of a doctrine whose major proponents try to create the impression that it is scientific (the criterion of an allegedly scientific doctrine).

This criterion explains why mistakes such as the biologist's and scientific fraud such as the biochemist's in the above examples are not regarded as pseudoscientific. Successful perpetrators of scientific fraud tend not to associate themselves with an unorthodox doctrine. Their chances to avoid disclosure are much higher when the data they fabricate conform to the predictions of established scientific theories.

The criterion of an allegedly scientific doctrine has significantly improved the definition. But we are not finished. Consider the following three examples:

3. A biochemist performs a long series of experiments with questionable quality. She consistently interprets them as showing that a particular protein has a role in muscle contraction, a conclusion not accepted by other scientists.

4. A homeopath claims that his remedies (consisting chemically of nothing else than water) are better than those of conventional medicine. He maintains that his therapeutic claims are supported by science and makes attempts to show that this is so.

5. A homeopath claims that his remedies (consisting chemically of nothing else than water) are better than those of conventional, science-based medicine. However, he does not claim that homeopathy is scientific. Instead, he claims that it is based on another form of knowledge that is more reliable than science.

Number four is a paradigm of a pseudoscience: a demonstrably false doctrine is put forward and claimed to be scientific. Number three answers the same description. In number five, an unscientific claim within the subject area of science is announced as reliable knowledge, but its promoters do not call it "science." Writers on pseudoscience commonly use the term "pseudoscience" in cases like this, and some have explicitly stated that it is right to do so: "there are many pseudoscientific doctrines which seek public legitimation and support by claiming to be scientific; others purport to offer alternative accounts to those of science or claim to explain what science cannot explain" (Grove 1985, 219). A good reason can be given for this extended use of the notion of pseudoscience: science is not just one of several competing approaches to knowledge. For an account of a subject area to qualify as science, it has to be the most reliable, epistemically most warranted account of that area that is accessible to us (at the point in time in question). Therefore, the definition of pseudoscience should not refer to usage of the word "science" but to claims that correspond to the meaning of that word. This should lead us to widen the criterion of an allegedly scientific doctrine and characterize a pseudoscience as part of a doctrine whose major proponents try to create the impression that it represents the most reliable knowledge on its subject matter (the criterion of deviant doctrine).

Definitions and Demarcations

Let us now summarize these deliberations in the form of definitions of science and pseudoscience. The outcome of our search for an appropriate definition of science can be summarized as follows:

> Science (in the broad sense) is the practice that provides us with the most reliable (i.e., epistemically most warranted) statements that can be made, at the time being, on subject matter covered by the community of knowledge disciplines (i.e., on nature, ourselves as human beings, our societies, our physical constructions, and our thought constructions).

The above discussion on pseudoscience can be condensed into the following definition:

> A statement is pseudoscientific if and only if it satisfies the following three criteria:
>
> 1. It pertains to an issue within the domains of science in the broad sense (the criterion of scientific domain).
> 2. It suffers from such a severe lack of reliability that it cannot at all be trusted (the criterion of unreliability).
> 3. It is part of a doctrine whose major proponents try to create the

impression that it represents the most reliable knowledge on its subject matter (the criterion of deviant doctrine).

This is an improved version of a previously proposed definition (Hansson 2009). It differs from most other definitions of pseudoscience in having its focus on pseudoscience itself. Rather than using the demarcation of pseudoscience from science as a vehicle to define science, I have suggested that we first need to clarify what science is. Based on that we can determine which of the many forms of deviation from science should be called "pseudoscience." This definitional structure has the advantage of treating the notion of pseudoscience as secondary to that of science, which seems to be right in terms of conceptual priority.

However, due to this structure, the definition is not operational on its own for the demarcation of pseudoscience. For that purpose it needs to be complemented with a specification of the reliability criterion. Obviously, various such specifications can be added, giving rise to different (e.g., falsificationist or verificationist) demarcations between science and pseudoscience. Let us now finally turn to such specifications. They are the subject matter of most traditional accounts of the demarcation between science and pseudoscience.

Levels of Epistemic Generality

The literature on the science/pseudoscience demarcation contains two major types of demarcation proposals. The first type provides an exhaustive definition, that is, a set of necessary and sufficient criteria that will allegedly tell us in each specific case whether a statement is scientific or pseudoscientific. The best known of these proposals is Karl Popper's falsifiability criterion, according to which "statements or systems of statements, in order to be ranked as scientific, must be capable of conflicting with possible, or conceivable observations" (Popper 1962, 39). This is often contrasted with the logical positivists' verificationist criterion, according to which a scientific statement can be distinguished from a metaphysical one by being at least in principle possible to verify. However, this is not a historically accurate comparison. The verificationists were mainly interested in meaningfulness, and their discussions focused on the difference in terms of meaningfulness between scientific and metaphysical statements. Originally, Popper had the same focus and presented falsifiability as a criterion for the distinction between empirical science and metaphysics (Popper 1932, 1935), but later he shifted to a focus on the distinction between science and pseudoscience (e.g., Popper 1962). It is the latter formulation of falsificationism that has become influential in philosophy and science (cf. Bartley 1968).

Lakatos (1970, 1974a, 1974b, 1981) proposed that the demarcation should not be applied to an isolated hypothesis or theory, but rather to a whole research program that is characterized by a series of theories successively replacing each other. A new theory that is developed in such a program is in his view scientific if it has a larger empirical content than its predecessor; otherwise it is degenerative. Thagard (1978) and Rothbart (1990) have further developed this criterion. Thomas Kuhn (1974) distinguished between science and pseudoscience in terms of the former's ability to solve puzzles. George Reisch (1998) maintained that a scientific discipline is characterized by being adequately integrated into the other sciences. All of these proposals have been subject to severe criticism, and none of them has gained anything approaching general acceptance.

The second type of demarcation proposal follows a multicriteria approach. Each of these proposals provides of a list of mistakes committed in the pseudosciences. Usually, the assumption is that if a statement or a theory fails according to one of these criteria, then it is pseudoscientific. However, no claim of exhaustiveness is made; in other words, it is left open whether a statement or theory can be pseudoscientific without violating any of the listed criteria (presumably because it violates some other, unlisted criterion). A large number of such lists have been published (usually with five to ten criteria), for instance, by Langmuir ([1953] 1989), Gruenberger (1964), Dutch (1982), Bunge (1982), Radner and Radner (1982), Kitcher (1982, 30–54), Hansson (1983), Grove (1985), Thagard (1988), Glymour and Stalker (1990), Derkson (1993, 2001), Vollmer (1993), Ruse (1996, 300–306), and Mahner (2007). One such list reads as follows:

1. Belief in authority: it is contended that some person or persons have a special ability to determine what is true or false. Others have to accept their judgments.

2. Unrepeatable experiments: reliance is put on experiments that cannot be repeated by others with the same outcome.

3. Handpicked examples: handpicked examples are used although they are not representative of the general category that the investigation refers to.

4. Unwillingness to test: a theory is not tested although it is possible to do so.

5. Disregard of refuting information: observations or experiments that conflict with a theory are neglected.

6. Built-in subterfuge: the testing of a theory is so arranged that the theory
can only be confirmed, never disconfirmed, by the outcome.

7. Explanations are abandoned without replacement: tenable explanations
are given up without being replaced, so that the new theory leaves much
more unexplained than the previous one. (Hansson 1983)

The two types of demarcation proposals have in common that they operate with concrete and directly applicable criteria. If we wish to determine
whether Freudian psychoanalysis is a pseudoscience, we can directly apply,
for instance, Popper's criterion of falsifiability, Kuhn's criterion of puzzle-
solving ability, or Reisch's criterion of integration into the other sciences. We
can also apply, for the same purpose, the criteria found on the multicriteria
lists, such as unrepeatable experiments, handpicked examples, deference to
authority, and so forth. However, the two types of demarcation proposals differ in another respect: the first type provides a criterion that is intended to be
sufficient to determine in each particular case whether a statement, practice,
or doctrine is scientific or pseudoscientific. The second type has more modest
claims, and no attempt is made to show that the list of criteria that it provides
is exhaustive.

The pseudoscience definition proposed in the previous section does not
belong to either of these two categories. Like the first-mentioned group (Popper, Kuhn, etc.), it intends to provide a necessary and sufficient criterion,
valid in all cases where a distinction between science and pseudoscience has
to be made. However, it differs from both the above-mentioned types in not
providing concrete and directly applicable criteria. If we want to determine
whether Freudian psychoanalysis is a science or a pseudoscience, a reformu-
lation of the question in terms of reliable knowledge or epistemic warrant
leaves most of the work undone. We are not told what type of data or investigations to look for, or what types of strengths and weaknesses we should look
for in the psychoanalytical literature.

The rationale for choosing a criterion that is not directly applicable to
concrete issues of demarcation is that such direct applicability comes at a high
price: it is incompatible with the desired exhaustiveness of the definition. The
reason for this incompatibility is that the unity of the different branches of science that was referred to above does not include methodological uniformity.
What unites the sciences, across disciplines and over time, is the basic com-

mitment to finding the most reliable knowledge in various disciplinary areas. However, the precise means to achieve this differ among subject areas, and the chosen methods are also in constant development. It is not only the detailed methods that change but also general methodological approaches, such as methods of hypothesis testing, experimental principles like randomization and blinding, and basic assumptions about what types of explanations can be used in science (such as action over distance). The capacity that science has for self-improvement applies not least to its methodologies.

Most of the demarcation literature has had its focus on methodological demands on experimental studies in the natural sciences. However, the requirements on experimental studies, such as repeatability, randomization, blinding, and so on are not relevant in most historical studies. It would also be pointless to apply them to experiments performed in the seventeenth century, before modern experimental methodology had been developed. We would then be bound to dismiss some of the best science of those days as pseudoscientific, which would certainly be misleading.

Popper's criterion of falsifiability can serve as an example of these problems. It is a criterion for hypothesis-testing science, but far from all of science is hypothesis-testing. Studies in the humanities are mostly driven by open-ended questions rather than by hypotheses, and the same applies to important parts of the experimental natural sciences. Many experiments are conducted to determine the veracity of a predecided hypothesis, but many other experiments are explorative. Such experiments aim at answering an open question such as "what is the structure of this protein?" rather than a yes-or-no (hypothesis-testing) question such as "does this protein have the structure X?" A small statistical study of articles in *Nature* indicates that explorative studies may well outnumber hypothesis-testing ones in modern natural science (Hansson 2006). Science progresses through the combined use of explorative and hypothesis-testing investigations. Successful explorative studies tend to result in precise hypotheses that are suitable for further testing. Hypotheses that have survived testing often give rise to new research questions that are most adequately attacked, in the initial phase, with explorative studies.

We can choose between two types of science/pseudoscience demarcations. We can have a demarcation that is general and timeless. It cannot then provide us with concrete criteria for the evaluation of specific investigations, statements, or theories. Such criteria will have to refer to methodological particulars that differ between subject areas and change with the passage of time.

Alternatively, we can have demarcation criteria that are specific enough to tell us what is required in a particular context, such as contemporary experimental science. We can use both these types of demarcations, of course, for different purposes. However, one and the same demarcation criterion cannot both be general and timeless and also be sufficiently precise to tell us how to evaluate the scientific status of specific investigations.

Many of the philosophical demarcation proposals have tried to do the impossible in this respect. They have thereby neglected what is perhaps the most fundamental strength of the scientific tradition, namely its remarkable ability of self-improvement, not only in details but also in basic methodology. The unity of science operates primarily on another, more fundamental level than that of concrete scientific methodology.

Acknowledgments

I would like to thank Maarten Boudry, Massimo Pigliucci, and Martin Rundkvist for valuable comments on an earlier version.

REFERENCES

Atkinson, Quentin D. 2011. "Phonemic Diversity Supports a Serial Founder Effect Model of Language Expansion from Africa." *Science* 332:346–49.

Baigrie, B. S. 1988. "Siegel on the Rationality of Science." *Philosophy of Science* 55:435–41.

Bartley, W. W., III. 1968. "Theories of Demarcation between Science and Metaphysics." In *Problems in the Philosophy of Science, Proceedings of the International Colloquium in the Philosophy of Science* (London 1965), vol. 3, edited by Imre Lakatos and Alan Musgrave, 40–64. Amsterdam: North-Holland.

Bloor, David. 1976. *Knowledge and Social Imagery*. London: Routledge & Kegan Paul.

Bunge, Mario. 1982. "Demarcating Science from Pseudoscience." *Fundamenta Scientiae* 3:369–88.

Derksen, A. A. 1993. "The Seven Sins of Pseudoscience." *Journal for General Philosophy of Science* 24:17–42.

———. 2001. "The Seven Strategies of the Sophisticated Pseudoscience: A Look into Freud's Rhetorical Tool Box." *Journal for General Philosophy of Science* 32:329–50.

Doupe, Alison J., and Patricia K. Kuhl. 1999. "Birdsong and Human Speech: Common Themes and Mechanisms." *Annual Review of Neuroscience* 22:567–631.

Dutch, Steven I. 1982. "Notes on the Nature of Fringe Science." *Journal of Geological Education* 30:6–13.

Gardner, Martin. 1957. *Fads and Fallacies in the Name of Science*. New York: Dover. (Expanded version of his *In the Name of Science*, 1952.)

Glymour, Clark, and Stalker, Douglas. 1990. "Winning through Pseudoscience." In *Philosophy of Science and the Occult*, 2nd ed., edited by Patrick Grim, 92–103. Albany: State University of New York Press.

Grove , J. W. 1985. "Rationality at Risk: Science against Pseudoscience." *Minerva* 23:216–40.

Gruenberger, Fred J. 1964. "A Measure for Crackpots." *Science* 145:1413–15.

Hansson, Sven Ove. 1983. *Vetenskap och ovetenskap*. Stockholm: Tiden.

———. 1996. "Defining Pseudoscience." *Philosophia Naturalis* 33:169–76.

———. 2003. "Is Philosophy Science?" *Theoria* 69:153–56.

———. 2006. "Falsificationism Falsified." *Foundations of Science* 11:275–86.

———. 2007. "Values in Pure and Applied Science." *Foundations of Science* 12:257–68.

———. 2009. "Cutting the Gordian Knot of Demarcation." *International Studies in the Philosophy of Science* 23:237–43.

Henrich, Joseph. 2004. "Demography and Cultural Evolution: How Adaptive Cultural Processes Can Produce Maladaptive Losses: The Tasmanian Case." *American Antiquity* 69:197–214.

Jones, David E. H. 1970. "The Stability of the Bicycle." *Physics Today* 23 (4): 34–40.

Kitcher, Philip. 1982. *Abusing Science: The Case Against Creationism*. Cambridge, MA: MIT Press.

Krupp, Edwin C., ed. 1984. *In Search of Ancient Astronomies*. Harmondsworth: Penguin Books.

Kuhn, Thomas S. 1974. "Logic of Discovery or Psychology of Research?" In *The Philosophy of Karl Popper, The Library of Living Philosophers*, vol. 14, bk. 2, edited by P. A. Schilpp, 798–819. La Salle: Open Court.

Lakatos, Imre. 1970. "Falsification and the Methodology of Research program." In *Criticism and the Growth of Knowledge*, edited by Imre Lakatos and Alan Musgrave, 91–197. Cambridge: Cambridge University Press.

———. 1974a. "Popper on Demarcation and Induction." In *The Philosophy of Karl Popper, The Library of Living Philosophers*, vol. 14, bk. 1, edited by P. A. Schilpp, 241–73. La Salle: Open Court.

———. 1974b. "Science and Pseudoscience." *Conceptus* 8:5–9.

———. 1981. "Science and Pseudoscience." In *Conceptions of Inquiry: A Reader*, edited by Stuart Brown, 114–21. London: Methuen.

Langmuir, Irving. (1953) 1989. "Pathological Science." *Physics Today* 42 (10): 36–48.

Layton, Edwin. 1976. "American Ideologies of Science and Engineering." *Technology and Culture* 17:688–701.

Lycett, Stephen J., and Noreen von Cramon-Taubadel. 2008. "Acheulean Variability and Hominin Dispersals: A Model-Bound Approach." *Journal of Archaeological Science* 35:553–62.

Mahner, Martin. 2007. "Demarcating Science from Non-Science." In *Handbook of the Philosophy of Science: General Philosophy of Science—Focal Issues*, edited by Theo Kuipers, 515–75. Amsterdam: Elsevier.

Nickell, Joe. 2007. *Relics of the Christ*. Lexington: University Press of Kentucky.

Pagel, Mark, Quentin D. Atkinson, and Andrew Meade. 2007. "Frequency of Word-Use Predicts Rates of Lexical Evolution throughout Indo-European History." *Nature* 449:717–20.

Polanyi, Michael. 1967. *The Tacit Dimension*. Garden City, NY: Doubleday.

Popper, Karl. 1932. "Ein Kriterium des empirischen Charakters theoretischer Systeme." *Erkenntnis* 3:426–27.

———. 1935. *Logik der Forschung: zur Erkenntnistheorie der Modernen Naturwissenschaft.* Wien: Julius Springer.

———. 1962. *Conjectures and Refutations: The Growth of Scientific Knowledge.* New York: Basic Books.

Radner, Daisie, and Michael Radner. 1982. *Science and Unreason.* Belmont, CA: Wadsworth.

Reisch, George A. 1998. "Pluralism, Logical Empiricism, and the Problem of Pseudoscience." *Philosophy of Science* 65:333–48.

Rothbart, Daniel. 1990. "Demarcating Genuine Science from Pseudoscience." In *Philosophy of Science and the Occult,* 2nd ed., edited by Patrick Grim, 111–22. Albany: State University of New York Press.

Ruse, Michael, ed. 1996. *But Is It Science? The Philosophical Question in the Creation/Evolution Controversy.* Buffalo, NY: Prometheus Books.

Shermer, Michael, and Alex Grobman. 2000. *Denying History: Who Says the Holocaust Never Happened and Why Do They Say It?* Berkeley: University of California Press.

Snow, C. P. (1959) 2008. *The Two Cultures.* With an Introduction by Stefan Collini. Cambridge: Cambridge University Press.

Stiebing, William H. 1984. *Ancient Astronauts, Cosmic Collisions, and Other Popular Theories About Man's Past.* Buffalo, NY: Prometheus Books.

Thagard, Paul R. 1978. "Why Astrology Is a Pseudoscience." *PSA* 1:223–34.

———. 1988. *Computational Philosophy of Science.* Cambridge, MA: MIT Press.

Thomas, Dave. 1997. "Hidden Messages and the Bible Code." *Skeptical Inquirer* 21:30–36.

Vollmer, Gerhard. 1993. *Wissenschaftstheorie im Einsatz, Beiträge zu einer selbstkritischen Wissenschaftsphilosophie.* Stuttgart: Hirzel Verlag.

5

Loki's Wager and Laudan's Error

On Genuine and Territorial Demarcation

MAARTEN BOUDRY

Is the demarcation problem dead, or are the rumors of its demise greatly exaggerated? The answer depends on whom you ask. Some philosophers of science have voiced the opinion that the demarcation project has been something of an embarrassment to their discipline and that terms like "pseudoscience" and "nonscience" should be erased from our philosophical vocabulary, wedded as they are to a naïve conception of science and its borderlines. Nowadays philosophy of science has recovered somewhat from this backlash against demarcation. In the wake of a growing consensus that there is no silver bullet to separate science from nonscience, philosophers have shifted their attention to more sophisticated ways of characterizing science and distinguishing it from different shades of nonscience (Nickles 2006; Hansson 2008, 2009; Pigliucci 2010).

   The major trouble with the demarcation project, as I argue in this chapter, is that it has traditionally been the banner of two distinct but often conflated intellectual projects, only one of which is pressing and worth pursuing. The genuine demarcation problem as I see it—the one with real teeth—deals with distinguishing bona fide science from pseudoscience. The second brand of demarcationism concerns the territorial boundaries separating science from such epistemic endeavors as philosophy, history, metaphysics, and even everyday reasoning.

   I argue that the territorial problem has little epistemic import, suffers from additional categorization problems, and consequently neither calls nor allows for anything more than a pragmatic and rough-and-ready solution. The normative demarcation project, by contrast, although it too has resisted a simple solution (i.e., a small set of necessary and sufficient conditions), is eminently worthy of philosophical attention, not only because it carries real epistemic import and practical urgency, but also because—fortunately—it happens to be a tractable problem. I discuss how both "demarcation" projects relate to one another and how they have often gotten mixed up. In particular, some have rashly proclaimed the death of the normative problem by performing an autopsy on the territorial problem (e.g., Laudan 1983), while others have tried to rescue the former by inadvertently resuscitating the latter (e.g., Pennock 2011).

Normative and Territorial Demarcation

To retrace the sources of confusion over the nature of the demarcation problem, it is instructive to go back to its most famous formulation. In his attempt to tackle the problem of induction, Popper initially introduced the principle of falsifiability as a neutral and territorial touchstone to separate science from other forms of knowledge. Thus, in the *Logic of Scientific Discovery* ([1959] 2002), originally published in German in 1934, he describes his falsifiability criterion as a way to distinguish "between the empirical sciences on the one hand, and mathematics and logic as well as 'metaphysical systems' on the other" (Popper [1959] 2002, 11). Obviously Popper does not dismiss logic and mathematics, but neither does he reject metaphysics outright. Indeed, he takes issue with the "derogatory evaluation" of metaphysics by the logical positivists, who famously equated it with meaningless twaddle.

In his later writings, however, the criterion of falsifiability takes on a more normative dimension. In *Conjectures and Refutations*, Popper ([1963] 2002) writes that he was troubled by the unfalsifiable character of psychoanalytic doctrines propounded by Sigmund Freud and Alfred Adler, and of certain versions of Marxist theory, comparing them unfavorably with the empirical boldness of Albert Einstein's theory of general relativity. But even so, Popper grants that psychoanalysis may contain valuable insights, even if it has not achieved scientific status (yet): "I personally do not doubt that much of what they say is of considerable importance, and may well play its part one day in a psychological science which is testable" (it isn't, and it didn't) (Popper [1963] 2002, 49).

Although falsificationism was initially framed in neutral and territorial terms, most of its proponents have followed the lead of the later Popper, wielding his yardstick to beat bad science and "pseudoscience" (an intrinsically pejorative term, Nickles 2006; see also Nickles, chapter 6, in this volume). This normative demarcation, like many other philosophical distinctions, does not yield to a simple solution (viz. one silver bullet to put an end to all nonsense), but as I argue below, that hardly means the problem is insoluble. Not only is the normative demarcation project alive and kicking (witness this volume), despite repeated assertions to the contrary, but Popper's virtue of empirical riskiness is still a key to solving the problem.

Still, what about Popper's territorial ambitions? The problem that initially puzzled him was how to distinguish science from domains of knowledge that, though valuable in their own right, belong to a different epistemic domain. What is the proper realm of science, and where exactly do we cross the border to philosophy or metaphysics, or even everyday reasoning? Territorial demarcation issues such as these, however, should be kept apart from the normative demarcation problem. Whereas territorial demarcation is concerned with a classification of knowledge, or a division of labor between different disciplines, and not with epistemic warrant per se (unless one holds that philosophy or metaphysics cannot offer knowledge at all), normative demarcation adjudicates between theories or practices we should rationally accept and those to which we should not grant any credence.

Even if the territorial borders are untraceable, as I think they are, this need not affect normative demarcation project. Before returning to the latter problem, let me gesture at some of the reasons why I think the territorial brand of demarcationism is philosophically sterile. On the one hand, there is often no way to disentangle philosophical elements from scientific theories and arguments. In philosophy, abstract reasoning and logic take the foreground, whereas in science, the emphasis is on empirical data and hypothesis testing. But scientific theories invariably rest upon certain philosophical underpinnings, and science without abstract reasoning and logical inferences is just stamp-collecting. As Daniel Dennett succinctly put it, "there is no such thing as philosophy-free science; there is only science whose philosophical baggage is taken on board without examination" (Dennett 1996, 21). On the other hand, good philosophical theories should be maximally informed by the relevant scientific knowledge and often derive support from scientific findings (philosophy of mind from neuroscience, philosophy of science from cognitive psychology and sociology, etc.).[1] In more and more contemporary philosophical discussions, logical reasoning and empirical evidence are so inextricably intertwined as to make demarcation efforts pointless and unrewarding. In light of this entangled relation, philosophers in the tradition of naturalism, which has been gaining influence over the past decades, maintain that philosophy and science are cut from the same cloth (e.g., Laudan 1990; Haack 2007). This approach does not see one discipline as prior to or wholly distinct from the other, but considers both as interdependent and continuous.[2]

In a similar vein, naturalistic epistemologists have argued that science is continuous with everyday reasoning. Modern science is a highly complex and differentiated social endeavor, but the practice of hypothesis testing and ampliative reasoning underlying science is already apparent in everyday reasoning (e.g., tracking animals, fixing a car). None of the characteristic features of modern science—the use of sophisticated technical equipment, formalization and mathematical tools, the system of peer review and public presentations, the years of formal training and practice—detach scientific reasoning from everyday knowledge acquisition. The complex institutional organization and systematic methodology of science can be seen as a highly refined and sophisticated extension of everyday reasoning, reflecting a heightened awareness of human cognitive foibles and a preoccupation with difficult, cutting-edge questions of a more theoretical nature.

In sum, though it may certainly be convenient for pragmatic purposes to distinguish science from philosophy or everyday reasoning, (i) such territorial demarcation carries nothing like the epistemic weight attached to the demarcation between science and pseudoscience, and (ii) unlike the latter project, territorial demarcation is complicated by the problems of interdependence and continuity.

The Demarcation of What?

Even if territorial demarcation is fruitless, as I think it is, this does not affect the viability of the normative demarcation between science and pseudoscience. In his (in)famous obituary of "the" demarcation project, however, Larry Laudan conflates the two kinds of demarcation, dismissing the concept of pseudoscience by dwelling on complications pertaining to territorial demarcation. Laudan (1983, 118) starts off by describing the demarcation project as an effort to "identify those *epistemic* or *methodological* features which mark off science from other sorts of belief," suggesting a neutral distinction between types of beliefs. But he then challenges the demarcation criterion of "well-testedness"—a clearly normative criterion—by listing several forms of knowledge that, although certainly well tested in his view, belong to "conventionally nonscientific fields" (Laudan 1983, 123) (e.g., singular historical claims, military strategies, literary theory, etc.). Laudan makes much of this perceived problem, and for the territorial demarcationist it may be troublesome indeed, but it is hardly what keeps the philosopher of pseudoscience awake at night. Laudan seems to assume that the two projects form a package deal, alternately shifting from the former to the latter and treating all of the associated problems as equally damaging to a single "demarcation project." This attitude is apparent in his way of dealing with past demarcation attempts. First, Laudan dismisses Popper's falsificationism as a "toothless wonder" because it accords scientific status to falsifiable but blatantly false claims, such as the claim that the earth is six thousand years old. Because it fails to perform the "critical stable-cleaning chores for which it was originally intended" (1983, 122), according to Laudan, it is a "disaster." In fact, the problem is rather trivial and can be remedied even within a falsificationist framework. To give the Popperian demarcation criterion some teeth, we need only require that, in addition to it being falsifiable, a theory must have survived repeated attempts at falsification (the Popperian notion of "corroboration"). The fact that young-earth creationism is technically "scientific" for a strict Popperian, in the sense that it is at least open to falsification, even though having been conclusively falsified, is a semantic nonissue. As we saw, however, the criterion of "well-testedness," the next shot in Laudan's historical review of demarcationism, finds no mercy with him either, this time precisely because it would count as scientific such—*dixit* Laudan—patently nonscientific claims as "Bacon did not write the plays attributed to Shakespeare." But obviously a demarcation project that would *succeed* in excluding such well-tested, historical claims from science would inevitably be another of those "toothless wonder[s]" (122) that Laudan decries in the first place (one wonders how Laudan recognizes a patently nonscientific claim, if he does not believe in any form of demarcationism). By demanding that the demarcation problem both is and is not normatively discriminating, Laudan wants to have his cake and eat it too.

Science and the Supernatural

In recent years, some philosophers and scientists have countered religious pseudoscience by wielding a demarcation criterion that is a confusing blend of territorial and normative elements. According to this principle of "methodological naturalism," science is inherently limited to providing natural explanations for the natural world, and it does not (nor can) traffic in supernatural claims. By this standard, theories like Intelligent Design (ID) creationism are immediately ruled out as science because of their covert or open reliance on supernatural causes. For example, in an official booklet of the National Academy of Sciences (1998, 124), we learn that

> because science is limited to explaining the natural world by means of natural processes, it cannot use supernatural causation in its explanations. Similarly, science is precluded from making statements about supernatural forces because these are outside its provenance.

Although it is primarily directed against pseudoscience, this natural/supernatural distinction also has territorial overtones. It exorcises questions about a broader picture of the world (e.g., Does God play any role in the universe? Is evolution blind or goal directed?) from science and relegates them to the domain of philosophy (Pennock 1999, 2011; Sober 2010).

Robert Pennock, in a paper defending methodological naturalism as a "ground rule" and "ballpark definition" of science (see also Fales, chapter 13, in this volume), starts out with rightly rebuking Laudan's view that the demarcation project is dead (see also Pigliucci, chapter 1, in this volume): "to hold that there is no difference between science and pseudo-science is to abandon any claim of insight into the analysis of knowledge or questions about distinguishing the real from the deceptive" (Pennock 2011, 195). Pennock gives a number of solid arguments for the viability of normative demarcation, but his defense of methodological naturalism quickly slips into territorial waters. This can be gleaned from Pennock's claim that science remains "scrupulously neutral" on the existence of supernatural entities (Pennock 2011, 188). God may well exist, but science has no business with him.

The appeal of methodological naturalism as a territorial demarcation is twofold. On the one hand, it gets rid of pseudosciences such as creationism and ID theory in one fell swoop. On the other hand, it makes science metaphysically innocuous, safeguarding a special domain for supernatural speculation where science is impotent, and thus establishing a modus vivendi between science and religion.[3] Alas, the solution suffers from several problems.

First, it provides a disservice to both science and philosophy. By struggling over the proper borderlines of science, this solution fuels the common misconception that only "science" possesses epistemic authority, whereas metaphysical questions, traditionally the trade of philosophers, are a matter of idle speculation only, which, interesting though it may be, can be safely ignored in scientific matters.

Second, given that the very concept of the supernatural is notoriously shaky, it is ill advised to erect any form of demarcation on its shoulders. To give substance to such a territorial demarcation claim, one needs to come up with a coherent and nontrivial definition of natural versus supernatural that does not already *presuppose* the demarcation between science and non-science. Pennock, for his part, argues that anyone who thinks that supernatural hypotheses may have testable consequences has "illegitimately [assumed] naturalized notions of the key terms or other naturalized background assumptions" (Pennock 2011, 189). But Pennock simply equates testability and naturalness and leaves us with a circular and self-serving definition of supernatural as that which is beyond scientific investigation by definition: "if we could apply natural knowledge to understand supernatural powers, then, by definition, they would not be supernatural" (Pennock 1999, 290; see also Pennock 2011; Boudry et al. 2010a; Tanona 2010). Such a definitional shortcut would not even recognize most religious miracle stories as supernatural, nor would it be helpful in dealing with typical pseudosciences. For example, if the claims of extrasensory perception (ESP) and telepathy were borne out, would we be dealing with supernatural phenomena or just elusive and poorly understood natural ones? Do aliens use advanced technology or spooky spiritual powers, as ufologists sometimes suggest? Whom do we consult to settle such matters? I argue that we need not make up our minds about these questions before deciding whether parapsychology or ufology are pseudoscientific (see below).

Third, if supernatural forces were operating in the natural world, producing tangible empirical effects, as many theists maintain, nothing would prevent scientists from empirically investigating those.[4] As I have argued elsewhere (Boudry, Blancke, and Braeckman 2010a), narrowing down the scope of science by excluding all supernatural claims from its purview is unfeasible and historically inaccurate, given that many such claims have in fact been subjected to empirical investigations (e.g., the healing power of intercessory prayer, clairvoyance, communication with angels). Upon any nontrivial definition of the term "supernatural," I see no sound reason why supernatural phenomena would be intrinsically beyond the pale of science (see Fales, chapter 13, in this volume).

Fourth, and most important for this chapter, the territorial move bypasses the real reason for the dismal epistemic status of ID creationism, which is that it exhibits more general telltale signs of pseudoscience: ID theorists refuse to flesh out their design hypothesis and use convenient immunizations that make the theory impervious to criticism; the concepts devised by ID advocates suffer from equivocations that turn their central argument into a moving target; the theory is too vague to allow for specific predictions and to achieve any form of genuine explanatory unification; ID proponents refuse to get into the details of the mechanism and method used by the designer; the bulk of ID literature consists of purely negative arguments against evolution, with the sole purpose of distorting science and sowing doubt; and so on.

The label "supernatural" is a red herring in this context because the kinds of problems listed above are neither exclusive nor intrinsic to supernatural hypotheses. In fact, all of them should sound familiar to anyone who has wandered into the strange hinterlands of science before (Fishman 2009). In the next section, I give examples of perfectly naturalistic doctrines that are guilty of precisely the same sins (particularly regarding testability and immunization), which shows that the proponents of "methodological naturalism" as a weapon against ID creationism are barking up the wrong tree.

The Revenge of Demarcationism

How does science secure epistemic warrant? No matter how we fill in the details, it should be clear that many things can go wrong in many different ways. It should come as no surprise, then, that the category of *pseudo*science (or bad science) is heterogeneous, resisting explication in terms of necessary and sufficient conditions (Nickles 2006, 194). In the skeptical literature, the term "pseudoscience" refers to nonscience posing or masquerading as genuine science. To capture this intuitive conception, Hansson (2009; see also Hansson, chapter 4, in this volume) offers the following helpful characterization of pseudoscience:

1. It pertains to an issue within the domains of science (in the wide sense).
2. It is not epistemically warranted.
3. It is part of a doctrine whose major proponents try to create the impression that it is epistemically warranted.

What is valuable about Hansson's approach is that it brackets our justifications for belief in real science and focuses on the general characterization of pseudoscience first. Despite the conceptual heterogeneity of "pseudoscience," Hansson's provision (2009, 240) that its proponents "try to create the impression that [their theory] is epistemically warranted" gives us good reason to expect some shared characteristics. In the absence of the epistemic warrant that genuine science accrues, pseudosciences are confronted with the problem of surviving the day when prophecy fails and of creating a spurious *impression* of epistemic warrant. Bona fide science is not confronted with that problem. If you have nature on your side, so to speak, you can afford to be receptive to her judgment, which is precisely what we value—among other things—in successful scientific theories. In order to survive the stern judgment of nature and the onslaught of critical arguments, however, pseudoscientists are forced to systematically evade falsification and turn apparent refutations into spurious confirmations.

This is the reason why, despite the glaring problems with his naïve falsificationism, Popper was right to champion empirical boldness as a cardinal scientific virtue. For Popper, however, particularly in his later years, the falsifiability of a theory is purely a function of its logical properties and consequence relations (Hansson 2008). But ever since the seminal work of Pierre Duhem, we know that scientific theories are tested in bundles and never in isolation. A theory is not falsifiable until it is conjoined with background assumptions, initial conditions, and auxiliary hypotheses. Depending on how we interpret Popper's logical criterion in light of these problems, it is either too restrictive, classifying some of our best theories as nonscientific, or too permissive, allowing some of the worst theories in currency (e.g., astrology) to be recognized as science (Kitcher 1982). Popper's disciple Imre Lakatos realized that every scientific "research programme" is protected against falsification by a host of auxiliary hypotheses. It is simply not true, generally speaking, that scientists abandon a theory as soon as they have witnessed an anomalous observation. Instead, they have at their disposal various ways of tweaking and adjusting auxiliary hypotheses to preserve their central hypothesis, some of which ways seem quite respectable.

Still, even after taking into account Duhem's problem of underdetermination and the complexities of science's historical development, the virtue of empirical boldness in science emerges unscathed. In particular, we still need *some* restrictions on the amount of gerrymandering that we can allow in the face of apparent refutations (Leplin 1975). One of the hallmarks of pseudoscience, as Kitcher (1982, 48) succinctly puts it, is that it has "too cozy a relation-ship with auxiliary hypotheses," applying its problem-solving strategies with "claims that can be 'tested' only in their applications," in other words, that are purely ad hoc and not independently testable.

True enough, contrary to Popper's austere falsificationist ideals, scientists do not just abandon their theory the moment they encounter a single appar-ent falsification. But no theorist can remain comfortable when running slap up against reality time and again. People may believe crazy things on flimsy grounds, but they will not buy into anything at any price (Boudry and Braeck-man 2012). A superficial ring of plausibility is a psychological sine qua non for every successful pseudoscience. Such an impression of epistemic warrant is generally created by (i) minimized risk of refutation, (ii) phony appearance of empirical boldness, or (iii) opportunities for "confirmations" without ac-tual threat of refutation. Strategies for pulling off such sleights of mind recur across the pseudoscientific domain. I present a rough typology—a tentative "*nosology* of human thought," as David Stove (1991, 187) puts it—that I have discussed in more detail elsewhere (Boudry and Braeckman 2011, 2012).

Multiple Endpoints and Moving Targets

By using conceptual equivocations and what psychologists have labeled "multi-ple endpoints" (i.e., multiple ways in which a claim may be borne out), pseudo-scientists create an asymmetry between observations capable of confirming and those that could refute a given hypothesis. In the case of conceptual equivocation, which is pervasive in astrology and doomsday predictions, one begins by endorsing a bold and strong interpretation of a claim but, when threatened with falsification, switches to a weaker and broader interpretation. In fact, typical psychic pronouncements are amenable both to a specific inter-pretation and a range of broader and more metaphorical ones (e.g. "a father figure stands behind you"). Equivocations are also found in the creationist's notion of biblical "kinds," a concept that, according to Philip Kitcher (1982, 155), is "[tailored] to suit the needs of the moment" to preserve the claim that evolution between kinds is impossible. The latter-day heirs of creationism have applied the same bait-and-switch strategy in their argument that some biological systems are "irreducibly complex," equivocating between a sound but trivial and an interesting but false version of the concept (Boudry et al. 2010b).

Shadowy Retreats

A related way to steer clear of unwelcome evidence and criticism is to remain as vague and noncommittal about one's hypothesis as possible. ID creationists steadfastly refuse to reveal anything about the mechanisms and procedures used by the alleged designer, insisting that his motives are inscrutable and that the whole affair is beyond human comprehension (this, of course, being a traditional cop-out for theists). Note that this stalemate does not derive from the supernatural character of the hypothesis, as there is nothing that prevents ID creationists from fleshing out their design hypothesis in such a way that it actually yields specific predictions (Boudry and Leuridan 2011). Pseudoscientific beliefs in general are often indeterminate and mysterious (e.g., healing crystals), which ensures that they are inaccessible to normal epistemic evaluation (Sperber 1990), and that contradictions and adverse evidence will go largely unnoticed to believers.

Conspiracy Thinking

Conspiracy thinking is a doubly convenient strategy of immunization and spurious confirmation. On the one hand, conspiracy theorists present any anomaly in the received view of some historical event as evidence of something secretive and sinister going on (Keeley 1999). On the other hand, anomalies for their own hypothesis can be explained away as being exactly what would be *predicted* on the conspiracy view. Evil conspirators, after all, can be expected to spread forged evidence and disinformation to throw us off the scent. Moreover, the very existence of critical dissenters of the conspiracy view can be construed as further evidence for the belief system. In Freudian psychoanalysis, for instance, which exhibits the same epistemic structure as a conspiracy theory (Crews 1986; Boudry and Buekens 2011), critics are suspected of being motivated by unconscious resistance and defense mechanisms, exactly as predicted by the theory.

Invisible Escape Clauses

Many pseudoscientists appear to make bold empirical statements, but when push comes to shove, they resort to special escape clauses and get-out-of-jail-free cards to forestall falsification, thus dashing expectations initially engendered by their statements. Parapsychology is notoriously abundant with such escape clauses. Examples include the idea that the presence of inquisitive minds tends to disturb psychic phenomena, known as "negative psi vibration" or "catapsi" (for a skeptical discussion, see Humphrey 1996; Wiseman 2010), or the argument that psi is "actively evasive" because its primary function is to "induce a sense of mystery and wonder" (Kennedy 2003, 67). Again, in a full-blown pseudoscience, such escape clauses are sufficiently vague and noncommittal to be conveniently ignored as long as they are not needed. By qualifying apparent falsifications with such moves while accepting confirmations at face value, again an asymmetry is created between what can confirm and refute a theory.

As should be clear by now, I think it is the resort to such ad hoc maneuvers and the refusal to flesh out one's hypothesis that makes a theory like ID creationism pseudoscientific, not the appeal to a "supernatural" cause per se (whatever that may mean). As Fishman (2009, 826) wrote, it is certainly possible for supernaturalists to resort to "*ad hoc* explanations for the absence of evidence or disconfirming evidence for the supernatural," but exactly the same strategy is open to defenders of mundane and perfectly natural claims. The more general and underlying problem is that "continued ad hoc rationalization of repeated bouts of contrary evidence betrays a commitment to preserve a desired hypothesis at all cost" (Fishman 2009, 826).

Further Problems with Falsificationism

Laudan levels one strenuous objection against demarcationism that directly concerns the normative version of the problem, and we should now be able to put it to rest. According to Laudan, the charge of unfalsifiability against creationism "egregiously confuses doctrines with the proponents of those doctrines" (1982, 17). Because it foists off unresponsiveness to falsification on the theory itself, the argument conflates ad hominem and ad argumentum. If creationists, astrologers, or Freudians are unmoved by repeated falsifications of their doctrines, this reveals something about their psychological makeup, but it does not impinge on the falsifiability of their doctrines. Laudan's complaint was echoed by Philip Quinn (1996), Adolf Grünbaum (2008), Edward Erwin (1996), and a number of other philosophers, and ironically it is firmly within the tradition of Popper's strictly logicist analysis of propositions and their observational implications.

But if Popper's logicist approach is ill equipped to deal with real-life ex-amples of genuine science, as has been shown by Duhem, Quine, Kuhn, and others, a fortiori it must fail in the swamps of pseudoscience. In many inter-esting cases, among which the ones Popper himself discussed (e.g., Freudian psychoanalysis, astrology), there is typically no procedure for separating the theory-in-itself from the cognitive and methodological behavior of its defend-ers (Cioffi 1998; see also Cioffi, chapter 17, in this volume). The upshot of this problem is that the philosopher of pseudoscience has no choice but to get involved in the sociology of the discipline at large, and the psychology of those who are engaged in it. Kitcher even goes as far as suggesting that the category of pseudoscience is "derivatively psychological," in the sense that "pseudoscience is just what [pseudoscientists] do" (1993, 196). I think the truth lies somewhere in between. For example, when a parapsychologist at-tributes a failed experiment to the disturbing vibes of skeptical observers, it is not clear whether this is just methodological misdemeanor on the part of the parapsychologist, or whether it follows from his adherence to a standard tenet of parapsychology (the "catapsi" effect). When an ID creationist juggles with an ambiguous concept like "kind" or "irreducible complexity," there is no way of telling where the proper theory ends and where the obfuscations by its defenders begin. No one has come up with a general procedure to settle such matters.[5]

In many cases, immunizing strategies have such a cozy relationship with a pseudoscientific doctrine that they are at least *provoked* by it. For some para-psychologists, the elusive and shy nature of psi is one of the central tenets of the doctrine, so that the practice of cherry-picking experiments with positive outcomes can be given a sensible theoretical rationalization. To take another example, Freudian psychoanalysis uses a host of methodological principles and concepts to inflate the inferential possibilities of psychoanalytic interpre-tation, the cumulative effect of which is that it is hard to imagine *any* form of human behavior that would be at odds with the theory (this was Popper's cor-rect intuition). But the use of such methodological licenses and conceptual wildcards is no accidental quirk of some psychoanalyst interpreters: it simply reflects Freud's division of the mind into unobservable and antagonistic enti-ties, and his rich account of the purposeful mental interactions between those systems (negation, substitution, condensation, reaction formation, inver-sion, repression, etc.) (Cioffi 1998; see also Cioffi, chapter 17, in this volume; Boudry and Buekens 2011).[6]

The problem with Laudan's and Günbaum's approach is that, although nothing in Freudian psychoanalysis or parapsychology strictly *dictates* such fallacious forms of reasoning, their pervasiveness becomes intelligible only when we consider the belief system in which they are embedded. In short, the entanglement of theory and psychology forces us to widen our scope beyond the propositional content and logical structure of pseudosciences.

Setting Out the Borders

Although I have argued that Laudan is wrong and that the normative demarcation problem is tractable, this does not mean no borderline cases exist. In particular, epistemic warrant is not constant over time, so theories may move in and out of the domain of science as new evidence accumulates and conceptual progress is made (Hansson 2009; see also Ruse, chapter 12, in this volume). A twilight zone does exist, with theories that are neither scientific nor quite pseudoscientific, but we can readily come up with clear instances of both kinds, which is all that is needed for the viability of the normative demarcation project (Pigliucci 2010). By contrast, I have argued that, in most interesting cases, demarcating science and philosophy or science and everyday reasoning is like distinguishing the flour and sugar in a piece of cake (maybe feasible, but not very rewarding). The problem of interdependence and continuity that complicates territorial demarcation, making it a largely unfruitful endeavor, is completely absent from normative demarcation. No genuine science depends on pseudoscience for its justification.[7] In fact, there are analogous normative problems in each of the "territories" neighboring science that I think deserve more attention than the territorial demarcation per se: which theories deserve to be called good philosophy, and which are merely vacuous pseudophilosophy? How to distinguish rigorous mathematics from pseudomathematical verbiage? What is the difference between insightful hermeneutics and pseudohermeneutics?

Indeed, the normative demarcation criterion arguably cuts across territorial borders, with, for example, pseudosciences and pseudohistory exhibiting shared features that make them more similar to one another than to, respectively, bona fide science and good historiography. The normative demarcation question in historical science concerns whether and how we can distinguish bona fide historiography from what David Aaronovitch (2010) has termed "voodoo history," such as unsubstantiated conspiracy theories about major historical events. The cavalier approach to empirical evidence among conspiracy theorists and their systematic use of ad hoc explanations bear uncanny resemblances to the strategies of "pseudoscientists." To the extent that one views history as part of science broadly construed, the received account of the Holocaust deserves to be called "scientific," whereas Holocaust negationism certainly does not. Although the methodological differences between experimental and historical sciences make for fascinating philosophical discussions (Cleland 2002; see also Cleland and Brindell, chapter 10, in this volume), it seems that what distinguishes nuclear physics from cold fusion theory, and Second World War history from Holocaust denial, is an epistemic issue of an altogether different order.

The same can be maintained when it comes to philosophy. For example, conceptual equivocation is as pernicious in philosophy as it is in science, and the self-protective rationale is exactly the same (Law 2011). The philosopher André Kukla has complained about the systematic vacillation in the social constructivist literature between strong and weak versions of a claim, coining the terms "switcheroos" and "reverse-switcheroos" to describe these "philosophical sins" (Kukla 2000, x). Nicholas Shackel has similarly analyzed the strategy of equivocation in postmodernist philosophy:

> Having it both ways is essential to the appeal of postmodernism, for it is precisely by apparently speaking simultaneously of two different concepts with the same word that the appearance of giving a profound but subtle analysis of a taken-for-granted concept is created. (Shackel 2005, 304)

These discussions illustrate that philosophers are facing a normative demarcation task in their own discipline. Indeed, the fuzzy borders between philosophy and science, and the commonalities of their respective pseudo-counterparts, further downplay the territorial demarcation problem. Philosophers and scientists alike should join efforts to separate the wheat from the chaff in both domains rather than staking their own territorial borders. As Massimo Pigliucci writes, one of the most fruitful interactions between science and philosophy consist of the "joint defense against the assault from pseudoscientific quarters" (2008, 11).

Conclusion

In this chapter, I have expressed little confidence in the viability of the territorial demarcation problem, and even less interest in solving it. Not only is there no clear-cut way to disentangle epistemic domains like science and philosophy, but such a distinction carries little epistemic weight. The demarcation problem that deserves our attention is the one between science and pseudoscience (and the analogous ones between philosophy and pseudophilosophy and between history and pseudohistory). Separating the wheat from the chaff in these disciplines is a problem with both epistemic import and practical urgency, particularly in the face of relentless attempts by various people—practitioners of alternative medicine, creationists of different stripes, parapsychologists—to claim scientific respectability. Naïve falsificationism has been widely (and wisely) abandoned in philosophy of science, but the value of bold theorizing, broadly construed as hospitability to critical evaluation, remains intact both in science and philosophy. Instead of hankering for a silver bullet of demarcation, desirable though such a tool would be, we have no choice but to get down in the trenches and engage ourselves with the claims and arguments of pseudoscientists, scrutinizing their doctrines carefully and pointing out specific fallacies.

ID creationism invokes supernatural entities and is guilty of a host of pseudoscientific sins, but the two issues should not be conflated. Because we have become so accustomed to supernaturalists falling for the pseudoscience trap, and because we have grown weary of creationist hypotheses that, when push comes to shove, boil down to "God did it and his ways are mysterious," we can hardly imagine any other supernatural hypothesis to be viable (the prospects, admittedly, are extremely bleak). But even if all current theories with property X happen to be pseudoscientific, this does not mean that talk of X is off limits. In this case, it may simply tell us a great deal about the (contingent) absence of evidence for supernatural phenomena and about the widespread psychological attachment to the supernatural in spite of this absence.

The appropriate way of dealing with a supernaturalist pseudoscience like ID creationism is not to relegate it to a domain where science has no authority, but to confront the conceptual and empirical problems of the theory head on. In that respect, Laudan is completely on the mark when he writes that "our focus should be squarely on the empirical and conceptual credentials for claims about the world" (1983, 125). But Laudan (as well as Popper) was wide of the mark when he reduced the demarcation job to evaluating the propositional content of the theory. In the murky hinterland of science, such a neat distinction between the theory-as-such and the way it is handled by its advocates is invariably hard to come by. Pseudoscience is too messy to be analyzed on the level of the theory-in-itself, and demarcationists need more refined instruments of analysis.

Thomas Paine once wrote that "it is error only, and not truth, that shrinks from inquiry." Because pseudoscience is propagated in the face of reason and empirical evidence (otherwise it would presumably be epistemically warranted), it engages in systematic attempts to dodge falsification and criticism, to give a spurious appearance of empirical boldness that is always belatedly disappointed, and to twist apparent falsifications into confirmations. If a theoretical endeavor pretends to be science while it exhibits these and other epistemic sins to a sufficiently egregious extent, don't we need *some* word to capture it and distinguish it from bona fide science? If Laudan thinks that "pseudoscience" is just a "hollow phrase" (1983, 125), does he have a better term in store?

In Norse mythology, the trickster god Loki once made a bet with the dwarfs, on the condition that, should he lose, the dwarfs would cut off his head. Sure enough, Loki lost his bet, and the dwarfs came to collect his precious head. But Loki protested that, while they had every right to take his head, the dwarfs should not touch any part of his neck. All the parties involved discussed the matter: some parts obviously belonged to the neck, and others were clearly part of Loki's head, but still other parts were disputable. Agreement was never reached, and Loki ended up keeping both head and neck. In argumentation theory, Loki's Wager is known as the unreasonable insistence that some term cannot be defined and therefore cannot be subject of discussion. In this chapter, I hope to have shown that denunciating the normative demarcation project is an instance of Loki's Wager, while quarreling with territorial demarcation is not.

NOTES

1. Quine famously tried to dissolve the analytic/synthetic distinction on which many would want to erect the boundaries between science and philosophy. But one need not buy into Quine's argument to question territorial demarcation.

2 .There are different ways of partitioning this broader domain of knowledge. The German word *Wissenschaft* encompasses both the natural sciences and historical disciplines, whereas "science" is usually taken to refer more narrowly to the natural sciences. The more expansive domain of "empirically informed knowledge" that encompasses both science and philosophy was coined "scientia" by the philosopher of science William Whewell.

3. Not all those who defend methodological naturalism are friendly for religion. Pigliucci

(2011) for example argues that supernaturalism is incoherent and hence "not even wrong," which is hardly a cause of comfort to the religious.

4. Of course, it is true that God is nowhere to be found in science textbooks and in the technical literature, and modern scientists clearly eschew supernatural explanations. There is good inductive ground for doing so because appeals to the supernatural have always turned out to be premature in the past, and the scientific naturalization of the world has been relentless and one-directional. The mistake of territorialists is to retrospectively translate the contingent outcome of scientific progress into self-imposed methodological strictures.

5. This often leads to the seemingly contradictory claim that Freudian psychoanalysis is both unfalsifiable and falsified: to the extent that we can isolate specific hypotheses and disentangle them from the rest of Freudian doctrine, such hypotheses may be falsified.

6. Notoriously, the conception of a deceitful and manipulative unconscious gives rise to a form of conspiracy theorizing, in which any form of contrary evidence can be interpreted as arising from unconscious resistance to psychoanalytic insights (even the "hostility" of critics).

7. Pseudoscience often feeds on real science, if only as a template to imitate. Also note that I do not deny that pseudoscientists may make serendipitous discoveries.

REFERENCES

Aaronovitch, David. 2010. *Voodoo Histories: The Role of the Conspiracy Theory in Shaping Modern History.* New York: Riverhead Books.

Boudry, Maarten, Stefaan Blancke, and Johan Braeckman. 2010a. "How Not to Attack Intelligent Design Creationism: Philosophical Misconceptions About Methodological Naturalism." *Foundations of Science* 15 (3): 227–44.

———. 2010b. "Irreducible Incoherence and Intelligent Design: A Look into the Conceptual Toolbox of a Pseudoscience." *Quarterly Review of Biology* 85 (4): 473–82.

Boudry, Maarten, and Johan Braeckman. 2011. "Immunizing Strategies & Epistemic Defense Mechanisms." *Philosophia* 39 (1): 145–61.

———. 2012. "How Convenient! The Epistemic Rationale of Self-Validating Belief Systems." *Philosophical Psychology* 25 (3): 341–64.

Boudry, Maarten, and Filip Buekens. 2011. "The Epistemic Predicament of a Pseudoscience: Social Constructivism Confronts Freudian Psychoanalysis." *Theoria* 77 (2): 159–79.

Boudry, Maarten, and Bert Leuridan. 2011. "Where the Design Argument Goes Wrong: Auxiliary Assumptions and Unification." *Philosophy of Science* 78 (4): 558–78.

Cioffi, Frank. 1998. *Freud and the Question of Pseudoscience.* Chicago: Open Court.

Cleland, Carol E. 2002. "Methodological and Epistemic Differences between Historical Science and Experimental Science." *Philosophy of Science* 69 (3): 474–96.

Crews, Frederick C. 1986. *Skeptical Engagements.* Oxford: Oxford University Press.

Dennett, Daniel C. 1996. *Darwin's Dangerous Idea: Evolution and the Meanings of Life.* New York: Simon and Schuster.

Erwin, Edward. 1996. A Final Accounting: Philosophical and Empirical Issues in Freudian Psychology. Cambridge, MA: MIT Press.

Fishman, Yonatan I. 2009. "Can Science Test Supernatural Worldviews?" *Science & Education* 18 (6–7): 813–37.

Grünbaum, Adolf. 2008. "Popper's Fundamental Misdiagnosis of the Scientific Defects of Freudian Psychoanalysis and of their Bearing on the Theory of Demarcation." *Psychoanalytic Psychology* 25 (4): 574–89.

Haack, Susan. 2007. *Defending Science-Within Reason: Between Scientism and Cynicism.* Amherst, NY: Prometheus Books.

Hansson, Sven Ove. 2008. "Science and Pseudo-science." In *Stanford Encyclopedia of Philosophy.* http://plato.stanford.edu/entries/pseudo-science/.

———. 2009. "Cutting the Gordian Knot of Demarcation." *International Studies in the Philosophy of Science* 23 (3): 237–43.

Humphrey, Nicholas. 1996. *Soul Searching: Human Nature and Supernatural Belief.* London: Vintage.

Keeley, Brian L. 1999. "Of Conspiracy Theories." *Journal of Philosophy* 96 (3): 109–26.

Kennedy, J. E. 2003. "The Capricious, Actively Evasive, Unsustainable Nature of Psi: A Summary and Hypotheses." *Journal of Parapsychology* 67 (1): 53–75.

Kitcher, Philip. 1982. *Abusing Science: The Case against Creationism.* Cambridge, MA: MIT Press.

———. 1993. *The Advancement of Science: Science without Legend, Objectivity without Illusions.* New York: Oxford University Press.

Kukla, André. 2000. *Social Constructivism and the Philosophy of Science.* New York: Routledge.

Laudan, Larry. 1982. "Commentary: Science at the Bar-Causes for Concern." *Science, Technology & Human Values* 7 (41): 16–19.

———. 1983. "The Demise of the Demarcation Problem." In *Physics, Philosophy, and Psychoanalysis: Essays in Honor of Adolf Grünbaum,* edited by R. S. Cohen and L. Laudan, 111–28. Dordrecht: D. Reidel.

———. 1990. "Normative Naturalism." *Philosophy of Science* 57 (1): 44–59.

Law, Stephen. 2011. *Believing Bullshit: How Not to Get Sucked into an Intellectual Black Hole.* Amherst, NY: Prometheus Books.

Leplin, Jarrett. 1975. "The Concept of an Ad Hoc Hypothesis." *Studies in History and Philosophy of Science* 5 (4): 309–45.

National Academy of Sciences. 1998. *Teaching About Evolution and the Nature of Science.* Washington, DC: National Academy Press.

Nickles, Thomas. 2006. "The Problem of Demarcation." In *The Philosophy of Science: An Encyclopedia,* edited by Sahotra Sarkar and Jessica Pfiefer, 1:188–97. New York: Routledge.

Pennock, Robert T. 1999. *Tower of Babel: The Evidence against the New Creationism, Bradford Books.* Cambridge, MA: MIT Press.

———. 2011. "Can't Philosophers Tell the Difference between Science and Religion? Demarcation Revisited." *Synthese* 178 (2): 177–206.

Pigliucci, Massimo. 2008. "The Borderlands between Science and Philosophy: An Introduction." *Quarterly Review of Biology* 83 (1): 7–15.

———. 2010. *Nonsense on Stilts: How to Tell Science from Bunk.* Chicago: University of Chicago Press.

Popper, Karl R. (1959) 2002. *The Logic of Scientific Discovery.* London: Routledge.

———. (1963) 2002. *Conjectures and Refutations: The Growth of Scientific Knowledge.* London: Routledge.

Quinn, Philip L. 1996. "The Philosopher of Science as Expert Witness." In *But Is It Science? The Philosophical Question in the Creation/Evolution Controversy,* edited by Michael Ruse, 367–85. Amherst, NY: Prometheus Books.

Shackel, Nicholas. 2005. "The Vacuity of Postmodern Methodology." *Metaphilosophy* 36 (3): 295–320.

Sober, Elliott. 2010. "Evolution without Naturalism." In *Oxford Studies in Philosophy of Religion,* edited by Jonathan Kvanvig, 3:187–221. Oxford: Oxford University Press.

Sperber, Dan. 1990. "The Epidemiology of Beliefs." In *The Social Psychological Study of Widespread Beliefs*, edited by George Gaskell and Colin Fraser, 25–44. Oxford: Clarendon Press.

Stove, David C. 1991. *The Plato Cult and Other Philosophical Follies*. Oxford: B. Blackwell.

Tanona, Scott. 2010. "The Pursuit of the Natural." *Philosophical Studies* 148 (1): 79–87.

Wiseman, Richard. 2010. "'Heads I Win, Tails You Lose.' How Parapsychologists Nullify Null Results." *Skeptical Inquirer* 34 (1): 36–39.

PART II

History and Sociology of Pseudoscience

6

The Problem of Demarcation
History and Future

THOMAS NICKLES

"The problem of demarcation" is Karl Popper's label for the task of discriminating science from nonscience (Popper 1959, 34; 1963, chap. 1). His own criterion remains the most frequently cited today: empirical testability or "falsifiability." Nonscience traditionally includes not only pseudoscience and metaphysics but also logic, pure mathematics, and other subjects that cannot be tested against experience, including the normative topics studied in value theory. The question is whether we can discriminate "sound science" from the impostors. Given human gullibility; given commercial, political, and legal interests; and given the diversity of the sciences and of philosophies of science, it is not surprising that no one agrees on whether there exists an adequate criterion of demarcation, that is, a reliable decision procedure for deciding whether something is a science (or, more modestly, genuinely scientific), and, if so, what that criterion is.

Pseudoscience, including some of what is today called "junk science," consists of enterprises that pretend to be scientific but fail to be testable, or that have questionable records of empirical success. Terms such as "pseudoscience" and "bad science" cover a variety of sins, including incompetent but honest work, potentially good work that is difficult to test or that has utterly failed to find empirical support, and deliberately dishonest scientific pretensions. Bad or pretended science carries many other labels: "anomalistics" (Bauer 2001), "fringe science," "pathological science" (Irving Langmuir on

J. B. Rhine's work on ESP, Park 2000, 40ff), to name just a few. "Junk science" often deliberately exploits scientific uncertainty to confuse and mislead judges, juries, and politicians, usually by substituting mere possibility for known probability (Huber 1991). It falls just short of "fraudulent science," in which scientists fudge their results or expert witnesses lie about the current state of knowledge. Physicist Robert Park (2000) lumps all these cases together as "voodoo science." He is especially concerned about claims with public currency that escape full scientific scrutiny because of official secrecy, political intervention, the legal adversary system, and the de facto adversary system employed in the media. The latter results in what Christopher Toumey (1996, 76) calls "the pseudosymmetry of scientific authority": "unbiased reporting," like expert witnessing, sometimes pretends that for every expert there is an equal and opposite expert and that there are pro and contra sides to every "controversy" that should be weighted equally.

Traditional solutions to the problem of demarcation have attempted to answer questions like these:

- What is science? (Or again, more modestly: What is it to be scientific?)
- What is special about science?
- What constitutes an empirical method and empirical knowledge?
- Which enterprise merits the greatest authority in disclosing the nature of the universe and our place in it?
- Which enterprise is most valuable in solving the problems we face as a people (health, economic, etc.) and/or in developing new technology?

And, by implication:

- Why is science important?
- What is at stake in the defense of the role of science in society?

In recent decades, the problem of demarcation has lost visibility in philosophical circles even as science and technology have gained unparalleled power and even as postmodernist groups, usually on the political left, and also economic interests and religious creationists, usually on the political right, increasingly challenge that authority. Meanwhile, science studies experts (e.g., Traweek 1988; Gieryn 1999) have been busy showing how all manner of more subtle disciplinary and cultural boundaries are constructed and maintained while blurring the old boundaries between science and technology and between internal and external factors.

Accordingly, we now recognize that demarcation issues arise within scientific research itself, usually with less public social impact than the traditional one. Here the question is usually subtler than whether a given move is minimally scientific, for example, whether it departs too far from a particular ongoing research program or the current state of play in the field, whether it is scientifically interesting, and so on (Is it good semiconductor physics? Is it good proteomics, by current lights?). Here the distinction is not between science and pseudoscience but between good science and bad science, or old-style science versus new, more promising science. Sometimes the issue is which field has the responsibility for explaining a given phenomenon (as in the history of Brownian motion) or whether one field reduces another (as in Durkheim's argument that sociological explanations of suicide do not reduce to psychological ones).

Journal editors must decide whether a given submission fits the specialty area of that journal sufficiently to send it out to referees. Government funding agencies must discriminate (demarcate) those research proposals deemed more promising today, in that specialty area, from those less so. In fact, as I argue, this question of relative fertility offers the most defensible criterion of demarcation in general social contexts such as that of creationism vs. evolutionary theory. While testability remains a useful rule of thumb, appeals to future promise are a superior indicator of what both scientists themselves and the general public (typically) want. Surprisingly, perhaps, it can be easier, and more socially acceptable, to discriminate more promising from less promising projects than to invoke a stigmatizing demarcation between science and nonscience.

The distinction between science and nonscience does not automatically favor science. For instance, in *Tractatus Logico-Philosophicus*, Ludwig Wittgenstein (1922) drew the distinction in part to protect ethics from the incursions of science. Worries about scientism—the view that science and scientists might become the ultimate authorities in most or all dimensions of society—are often expressed on secular humanistic grounds as well as on religious grounds. Even the logical positivists of the Vienna Circle used their criterion of demarcation to distinguish mathematics, logic, and the logic of science (about the only part of philosophy that Rudolf Carnap thought worth saving) from empirical science itself.

Historical Background

From the ancient Greeks to the present, Western methodologists have attempted to solve the problem of demarcation by specifying a *criterion* or intellectual filter in the form of necessary and/or sufficient conditions for *epistēmē*, *scientia*, or good science. Historically prominent criteria of demarcation draw on virtually all the main areas of philosophy. Criteria have been couched in terms of the ontological status of the objects of knowledge (e.g., Platonic Forms, Aristotelian essences), the semantic status of the products of research (science as a body of true or at least meaningful claims about the universe), the epistemological status of the products of research (science as a body of certain or necessary or reliable or warranted claims), the logical form of those claims (universal or particular, derivability of predictions from them), and value theory (the normative method that produces and/or evaluates the claims, e.g., inductive or hypothetico-deductive method, or comparison of a field with a model discipline such as physics or of a particular problem solution via the application of already exemplary solutions in Thomas Kuhn's sense: see Kuhn 1970b, 186ff).

For Aristotle, a claim is scientific if it is (a) general or universal, (b) absolutely certain, and (c) causal-explanatory. The possessor of genuine scientific knowledge has a demonstrative understanding of the first causes or essences of all things of a given kind. The logic or methodology of science and the investigative process itself are distinct from science proper. Aristotle stated his demarcation criteria primarily in terms of the qualities of the products, not the process of producing them.

Two thousand years later, Galileo Galilei, René Descartes, Isaac Newton, and other seventeenth-century natural philosophers still required virtual certainty for a claim to belong to the corpus of scientific knowledge, although metaphysical or demonstrative certainty was now giving way to practical certainty (Shapiro 1983). These early investigators also required causal-explanatory power of a sort, witness Newton's goal of finding true causes (*verae causae*) in his First Rule of Reasoning in the *Principia*; however, many of them abandoned as impossible Aristotle's demand for first causes and real essences. Within Judeo-Christian and Muslim cultures, God was now the first cause, but no one knew exactly how God accomplished the creation. The job of natural philosophy was to discover the "second" or proximate causes of phenomena. And John Locke argued that we humans are capable of knowing only "nominal essences" as opposed to "real essences." Within science itself,

Newtonian mechanists could calculate the motion of the planets in terms of the laws of motion and gravitation but failed to find either the cause or the essence of gravity. Thus, they could not provide a demonstrative chain of reasoning back to first causes (McMullin 2001).

Descartes rejected rhetoric and the other humanities subjects as a basis for a new science. Years later, in the wake of the English Civil War, the newly chartered Royal Society of London expressly excluded religion and politics from its discussions and insisted that scientific discourse be conducted in plain (nonfigurative) language. The members took Francis Bacon rather than Descartes as their secular saint, often interpreting him as a simple inductivist. On this view, to be scientific a claim must be induced from a body of previously gathered experimental or observational facts and then tested against new facts. Nature must be allowed to speak first as well as last. As Newton put it, investigation must start from phenomena known to exist, from truth, not from Cartesian hypotheses. Such a view was not antireligious, as Puritan and other investigators were quick to point out, since "the book of nature" was God's creation and supposedly more reliable and less subject to human (mis)interpretation than sacred scripture. Contrary to violating religious injunctions against prying into nature's secrets, Bacon argued that such investigation was our moral obligation, in order to improve the plight of the human race.

The thinkers of the scientific Enlightenment shaped the modern concern with demarcation. If science is to be the supreme expression of human reason and the broom that sweeps away the cobwebs of tradition, folk wisdom, and arbitrary authority, then it is crucial to distinguish the new science from pretenders, especially the Aristotelian teachings as well as old wives' tales (Amsterdamski 1975, 29). The Enlightenment legacy is that science, parliamentary democracy, and economic freedom are the sacred institutions of modern society and that their special status must be preserved and extended.

Historically, then, demarcation has typically been a conservative exercise in exclusion, an attempt to preserve the purity of modern science as the primary engine of social progress. In the modern period, it has been most often associated with strongly empiricist methodologies, which regard any claim with a suspicion proportional to its distance from experimental observation. (Mathematical subjects were legitimized in a different way.) In its early versions, demarcation was associated with foundationist epistemologies. A genuine science gives us the truth. Toward the end of the Enlightenment, Kant made it academic philosophy's job to demarcate science from nonscience— but on a priori grounds—and also to adjudicate disputes among the sciences.

The Kantian framework became especially influential because it was incorporated within the newly reformed German university system.

In the nineteenth century came widespread agreement that "Baconian" induction is an overly restrictive method, that "the method of hypothesis" (hypothetico-deductive method) is not only legitimate but also far more fruitful, more productive of scientific progress, given that certainty is an unattainable goal (Laudan 1981). The method of hypotheses cannot achieve certainty because of the fallacy of affirming the consequent, but neither could the inductive method that it largely supplanted, given well-known problems of underdetermination of universal claims.

Methodologists now realized that scientific research was a riskier enterprise than previously envisioned. It could not realistically claim to begin either from absolutely certain premises or aim to reach certain truth in its conclusions. Some nineteenth-century and virtually all twentieth-century methodologists responded to the clarified logical situation by becoming fallibilists, to some degree, and by adopting self-correcting or successive-approximation methodologies of science in place of the old foundationist ones. Since these investigators could no longer appeal to fail-safe epistemic status as the mark of substantive scientific claims, some retreated from the products to the process that produces them: a claim is scientific if and only if it is produced by a proper application of "the" scientific *method*; and a discipline is scientific if and only if it is guided by that method. The idea here is that theories may come and go, but the goals and method of science remain constant and the empirical basis solid. In its strong form, this is the idea that there exists a permanent scientific method that contains the design principles for all genuine science, past and future.[1] This conception of method enjoys currency today among some school textbook writers, administrators, and the general public but not among many scientists or science studies practitioners. Of course, process or method had been part of the Baconian, Cartesian, and Newtonian criteria all along, but the new dispensation considerably broadened what counted as a legitimate process as well as dropping the near certainty of the final product.

As the nineteenth century wore on, the basic idea of scientific method became still thinner. William Whewell, Auguste Comte, W. S. Jevons, and others minimized the importance of the process of discovery in favor of the empirical testability of the products of that process. Reversing the Bacon-Hume and Cartesian emphasis on antecedents, they asserted that it is observable consequences—predictions—that count, and that novel predictions count most (Laudan 1981, chap. 11; Nickles 1987a). Popper and the positivists would later recast this distinction in epistemic terms, as the difference between the subjective, nonrational psychological leaps of the "context of discovery" and the objective, logical inferences of the "context of justification." This move reduced scientific method to a logical minimum while retaining the idea of method as "the logic of science."

Twentieth-Century Developments

In something of a reprise of Kantian history, the problem of demarcation was a central feature of the dominant philosophies of science—logical empiricism and Popperianism—at the time when philosophy of science emerged as a professional specialty area within academic philosophy, namely the period 1925 to 1965. Most members of both schools were firmly committed to the methodological unity of science: all legitimate sciences have the same logical or methodological structure, with physics as the best example. This view, today widely rejected, provided sufficient motivation for thinking that a single criterion of demarcation for all science was adequate to do the job.

At *Tractatus* §4.11, Wittgenstein (1922) had written that "the totality of true propositions is the whole of natural science (or the whole corpus of the natural sciences)." Inspired in part by Wittgenstein, some leading positivists adopted not verified truth but empirical verifiability as their criterion of demarcation. For them demarcation was also a question of empirical meaningfulness: a claim is scientific if and only if it is empirically meaningful; and it is empirically meaningful if and only if it is empirically verifiable in principle. Moreover, the meaning-content of a claim was supposedly delimited by the specific conditions of verification. The kindred approach of operationism required that all theoretical *terms* be operationally defined in advance of theory.

Popper differed from the positivists in several different ways. First, he rejected this linguistic turn, holding that metaphysical claims can be meaningful even if empirically untestable. In fact, he said, the deepest problems of modern science often originated as metaphysical problems.[2] Second, Popper was an ardent anti-inductivist who further articulated the consequentialist position. What marks a claim as scientific is not its derivability from a large body of facts or even its high inductive probability, only that it (together with appropriate premises expressing the initial and boundary conditions) yields testable logical consequences. In his view even a general theory that has passed many severe tests (and is thus highly corroborated) still has probability zero. According to Popper (1959, app. vii), "in an infinite universe (it may be infinite with respect to the number of indistinguishable things, of or spatio-temporal regions) *the probability of any (non-tautological) universal law will be zero*" (Popper's emphasis). Moreover, a theory is not more scientific the more it appears to explain, the greater its apparent coverage. Newton's and Einstein's bold theories are scientific because they make risky empirical claims that can fail; but Marxist and Freudian theories are not scientific, according to Popper ([1934] 1959, 34; 1963, chap. 1), despite their claims to wide explanatory power, because their advocates allow nothing to count as a refutation. Far from making strong claims about reality, these theories actually exclude nothing, for the content of a law claim or theory is a function of how much it excludes. Finally, as already noted, Popper couched his demarcation criterion in terms of empirical falsifiability instead of verifiability, on the ground that general law claims can be shown false by a single counterinstance, whereas no number of confirmations can establish them as true.

Carl Hempel's influential review of the literature (1950, 1951) convincingly summarized the history of failures of the various criteria of meaning and demarcation proposed by the logical positivists, operationists, and Popper: the proposed criteria are at the same time too restrictive *and* too permissive. In agreement with W. V. Quine's attack on the analytic-synthetic distinction, Hempel concluded: "theory formation and concept formation go hand in hand; neither can be carried on successfully in isolation from the other" (113). Thereafter, these programs faded in importance, although Popper's less so. Indeed, it gradually became clear that useful method cannot be isolated from our empirical knowledge and our best guesses about the structure of the domain in question any more than concept formation can.[3]

Two later developments will complete this skeletal history. In *The Structure of Scientific Revolutions* ([1962] 1970a, 1970b), Thomas Kuhn rejected Popper's claim that scientists do or should test even their deepest assumptions. On the contrary, said Kuhn, during periods of "normal science" these assumptions (embodied in what he called the "paradigm") are constitutive of that branch of science. To subject them to criticism would destroy good science as we know it. Besides, a theory framework is never abandoned in isolation. Scientists must have something to work on in its place. It must be displaced by something better. Kuhn substituted the existence of a routine problem-solving tradition as the mark of mature science. Normal scientific problems are so highly constrained that Kuhn called them "puzzles," by analogy to crossword puzzles. The legitimate solution of such a puzzle is to apply the techniques embodied in standard puzzle solutions—Kuhn's "exemplars." Scientists solve puzzles by directly modeling them on one or more exemplars. Thus astrology is a pseudoscience not because it was unfalsifiable but because it could not sustain a normal-scientific puzzle-solving tradition. A similar response would apply to today's so-called creation science. It formulates no new research puzzles and solves no open problems. Kuhn's criterion is totally at odds with Popper's, although the two men often agreed on what counts as science (Worrall 2003).

Popper adamantly rejected Kuhn's model of scientific development as irrational. Neither dogmatic normal science nor chaotic revolutionary transitions featured his "critical approach to science and philosophy." Popper's protegé, Imre Lakatos (1970), agreed with Popper and many other critics that Kuhn had reduced science to periods of dogmatic rule by a priestly class, interrupted by occasional crisis periods of "mob rule." But Lakatos used the occasion to develop his own "methodology of scientific research programmes," and thus to emerge from Popper's shadow, on the basis of a critical examination of an entire spectrum of falsificationist positions. Lakatos agreed with Kuhn, as against Popper, that theories are not proposed and shot down in isolation. Rather, larger historical units are in play, namely, scientific research programs that plan a succession of ever-more developed theories subject to the constitutive guidelines of the program (the negative and positive heuristic), which are protected from falsification.

Lakatos was more interested in the demarcation of progressive from less successful research programs than of science from nonscience, and he made it both a matter of degree and relative to historical context. The phlogiston, caloric, and ether programs may have been the best available in their day, but anyone defending them today is surely unscientific. So the question becomes how to measure the progressiveness of a research program. Competing programs fight long battles of attrition, wrote Lakatos. A research program progresses insofar as (i) it makes novel theoretical predictions in heuristically motivated (non–ad hoc) ways, (ii) some of these predictions are confirmed, and (iii) successor theories in the program can explain why their predecessors worked as well as they did. A program degenerates insofar as it lags in these respects. Lakatos held that it is not necessarily irrational to retain allegiance to a degenerating program for an indefinite period of time, for history shows that a languishing program can achieve new vigor. Said Lakatos, there is no such thing as instant rationality.

Like Popper, Lakatos and successors such as Peter Urbach, John Worrall, and Elie Zahar attempted to purify science of ad hoc statements, that is, theory modifications that are heuristically unmotivated and that lead to no new predictions. These analysts disagreed in detail on what counts as ad hoc and why ad hoc science is bad science (Nickles 1987b). However, their emphasis on forward-looking heuristics as a basis for scientists' decision making was a major advance.

Despite the criticism, there does seems to be something right in Kuhn's view that an established science is one that supports routine problem solving. And it was Kuhn who first emphasized heuristic promise (expectation of future fertility) as the decisive factor in scientists' decisions about what to work on next. This was Kuhn's solution to the problem of new theories, or new paradigms: why would a practitioner abandon a polished older theory with an excellent empirical track record for some relatively underdeveloped, radical new ideas?

While Popper had reduced scientific method to a minimum, Kuhn virtually abandoned the notion of scientific method as useless for understanding how science works.[4] Kuhn surprised the Popperians and positivists by claiming that scientific work is far *more* routine and yet far *less* methodical and rule bound than they had imagined, given the traditional understanding of method as a set of rules. How is this possible? Kuhn contended that it is largely on the basis of their implicit, practical, expert knowledge that scientists within a specialist community agree so readily on what is good and bad science. Michael Polanyi (1958) had previously emphasized the importance of the "tacit knowledge" of experts.

Demarcation as a Social Problem

Reflecting on the steady weakening of proposed criteria of demarcation, Laudan (1983) concluded that demarcation is no longer an important philosophical problem. Popper's falsifiability criterion, he said, weakens the demarcation exercise beyond recognition. No longer does the criterion of demarcation mark out a body of belief-worthy claims about the world, let alone demonstrably true claims, let alone claims about ultimate causal essences. For on Popper's criterion every empirically *false* statement is automatically scientific. Popper completely abandoned the traditional attempt to characterize science in terms of either the epistemic or the ontological status of its products.

Laudan's view is that it is wrong to make invidious, holistic distinctions in advance about whether something is scientific. Rather, scientists proceed piecemeal, willing to consider anything and everything on its merits. Pseudoscience does not need its own separate category. It is enough to reject something as bad science.

This pragmatic move deliberately blurs the distinction between the form and content of science, that is, between the logic or method of science and empirical claims themselves. However, the move rejects the traditional demarcation problem only to raise another, at least equally difficult issue: how can philosophers of science (and other members of society) reliably discriminate good science from bad science? Laudan (like Kuhn before him) would answer that philosophers do not need to. That is a job for contemporary practicing scientists who have demonstrated their expertise. Sometimes the answers will be obvious, but sometimes they will be both piecemeal and highly technical.

There is much that I agree with in Laudan's position, but it can be difficult to apply in practical legal and political contexts. For it is hard to get agreement on what counts as a genuine scientific expert, for the same reason that it is difficult to achieve agreement on what counts as legitimate science in the first place.

Laudan (1996a, chap. 12) applied his position to the Arkansas trial of 1981–82 (*McLean v. Arkansas*) over the teaching of creationism in public school biology classes, a stance that would also apply to the more recent Dover, Pennsylvania, case of 2005 (*Kitzmiller v. Dover Area School District*).[5] Laudan agreed with the decision that creationism should not be taught as biology, but he was severely critical of every point of Judge Overton's philosophical justification of his decision. Overton appealed to Popper's falsifiability criterion to show that creationism is not science. Laudan replied that creationist doctrine itself *is* science by that criterion. It is obviously empirically testable since it has already been falsified. To be sure, its advocates have behaved in a nonscientific manner, but that is a different matter. The reason it should not be taught is simply that it is *bad* science. (Despite his criticism of Popper's criterion, Laudan apparently accepts it as one indicator of bad science, although he would agree that some false science remains important for both computational and pedagogical reasons, e.g., classical mechanics.) Michael Ruse (1982), who had invoked Popper's criterion in court testimony, responded to Laudan that, given the extant legal and social situation, Overton's reasoning was correct, for labeling it as unscientific was the only legal way to stop the teaching of "creation science" as a serious alternative to biological evolution.

It is doubtful whether Laudan's more nuanced treatment of the issue would have the same practical effect. Philosophers and scientists must make their cases to lay audiences. We cannot consider questions of demarcation apart from their social context anymore than their scientific context, and there is little opportunity for esoteric detail (often dismissed as "hair split-ting") in a court of law or in popular venues.

Should Judge Overton (or Judge John Jones in the *Dover* case) have ruled that creationism cannot be taught because it is bad science, or that it can *only* be taught as an example of bad science? (One sort of "bad science"— paradigmatic mistakes—can be valuable in teaching and learning.) Surely it would be a bad precedent for sitting judges to rule on what is good or bad science. And yet in a lesser sense they must, for the US Supreme Court's deci-sion of 1993, *Daubert v. Merrill-Dow Pharmaceuticals*, makes judges the gate-keepers for keeping unsound science out of the courtroom. *Daubert* requires judges, scientific laypersons though they be, to consider whether the alleged scientific claims have been tested, whether the claims have been subjected to peer review, the error rate of the claims, and whether the relevant scientific community accepts the claims, although full consensus is not required. Thus the US legal system itself incorporates something like Popper's criterion.

A related complication is that legal reasoning differs in important ways from scientific reasoning as well as philosophical reasoning, so one should not expect full convergence between scientific and legal modes of thought and action. (Ditto for reasoning in the political and public policy spheres.) For example, scientific conclusions are typically guarded and open to future revision in a way that legal decisions are not. Legal judgments are final (ex-cept for appeal) and must be made within a short time span on the basis of the evidence and arguments adduced within that time, whether or not suf-ficient scientific knowledge is available (Foster et al. 1993; Lynch and Jasanoff 1988). The value of a scientific claim or technique often resides in its heuristic potential, not in its known truth or correctness, whereas the judicial system wants and needs the truth now. Scientists seek general understanding of phe-nomena, whereas judges and attorneys must achieve rapid closure of particu-lar disputes. Scientific conclusions are often statistical (with margins of error given) and not explicitly causal, whereas legal decisions are typically causal and normative (assigning blame), individual, and nonstatistical, although cases involving smoking, cancer, and such have begun to broaden the law's conception of scientific reasoning. In the United States and elsewhere, many legal proceedings, both criminal and civil, are explicitly adversarial, whereas scientific competition is adversarial only in a de facto way. Scientists rely most heavily on evidential reasons, whereas law courts require all evidence to be introduced via testimony and to be accepted (or not) on that authority. The rules of evidence also differ. Judges must decide, in binary fashion, whether a given piece of evidence is admissible at all and whether a given witness is admissible as a scientific expert. When there is a jury, the judge instructs the jury what it may and may not take into consideration, and many juries are not even allowed to take notes during a trial. In some respects, legal reasoning is more conservative than "pure" scientific reasoning since lives may be immediately at stake; whereas in science, as Popper (1977, 152) says, our theories "die in our stead."

Third, the current situation is further complicated by the shifting use of the terms "junk science" and "sound science." In the highly litigious context of the United States, "junk science" originally meant dubious claims defended by hired expert witnesses in liability lawsuits, especially against wealthy corporations. While the increasing number of scientifically frivolous lawsuits does indeed threaten the financial stability and the innovative risk taking of corporations, in recent years corporate executives and powerful politicians have corrupted the terminology by labeling as "junk science" any scientific claim or methodology that threatens their interests and as "sound science" any claim that favors them (Rampton and Stauber 2001).

A Summary of Philosophical Difficulties with Demarcation

(1) The ancient "problem of the criterion" was the following argument for global skepticism. To know that we possess knowledge, we must have a criterion of truth (or knowledge). But we can never know that we have the correct criterion. For either the criterion pretends to be self-justifying (hence we have vicious logical circularity) or else it depends on a deeper criterion, whence the problem arises all over again in an even deeper and less accessible form (whence we begin a vicious logical regress). Fallibilists today reject the argument. It assumes an untenable foundationism that equates knowledge with absolute certainty. The implication for demarcation is that "is a science" cannot be equated with "is a body of absolute truths about the universe."

(2) Nor does retreating to "science is a body of well-supported beliefs" find clear sailing. While we do want to hold that science provides the best and most reliable take on the world that we possess to date, we do not want to fall into the trap of treating science as just another belief system, even an

empirically supported one based partly on faith, and hope that today's conclusions will stand up. For it is the very mark of progress that scientific beliefs keep changing—by subtraction and transformation as well as by simple addition. As is well known, its opponents are inclined to regard evolutionary biology (for instance) as merely an alternative belief system, comparable to their own, with Charles Darwin as its prophet and author of its sacred scripture.[6] Thus we need to say something stronger about methods and about future promise. Evolutionary theory, climate science, and so on, are not (or not just) "beliefs."

Another approach is to begin with clear and agreed-on examples of science and nonscience and proceed to determine which essential features are present in all sciences but lacking elsewhere. This approach faces several difficulties. (3) It assumes basic agreement on paradigm cases. In today's social context, that begs the very question at issue. There are other epistemological difficulties as well (Laudan 1986). A reply is that it is not necessary to convince opponents, only neutral parties, as was successfully done with both Judge Overton in Arkansas and Judge Jones in Pennsylvania. (4) Most analysts now reject the essentialism of traditional approaches to demarcation and their underlying assumption of the unity of science. As much work in science studies has revealed, the various progressive sciences are characterized by different goals and methods or practices. To the reply that testability is so weak that the implied essentialism is innocuous, a rejoinder is that it is so weak that it fails to fulfill one traditional function of demarcation, namely, to provide a deep understanding of the "nature" of science. Yet, given the diversity of the sciences, Laudan's talk of the demise of the old problem seems fair if that means finding a universal criterion that is also profound. (5) The approach has failed to yield workable specific proposals. Testability, as we have seen, is both too wide and too narrow. Popper's own "critical approach" wavers between treating science as very special and making everything in society evidence based and otherwise subject to criticism, a form of scientism. (6) In particular, insofar as a criterion based on past science is given teeth, it is almost sure to obstruct future scientific investigation. One does not have to swallow Kuhn's characterization of scientific revolutions to appreciate that the sciences continue to evolve, to ramify, to diversify—to redefine themselves. Enforcing any such criterion would likely harm future science more than help it. To adopt a criterion that legislates for all future science falls into the trap of thinking that we are today in the privileged position of having finally escaped from history, that we are capable of surveying all possible forms of progressive science, past and future, of knowing that there are no unconceived alternatives, whether substantive, methodological, or axiological (Stanford 2006). Lord Kelvin would have been shocked by both the substance, standards, and goals of Werner Heisenberg's quantum mechanics; and had the moderns retained Aristotle's demarcation criterion, they would not have become modern. Modern science would never have gotten off the ground. A reply is that such a demarcation exercise itself can be updated as the sciences change, yet still have force at any given time.

The traditional retreat from substance to method faces further difficulties. (7) There is no such thing as "the" scientific method, let alone agreement on what exactly it is. (8) To make matters worse, in contexts of public controversy, today's attempts to provide a more nuanced conception of research practices do not seem to help. Studies of the public engagement with science tend to show that the more members of the general public learn about how science is really done, the less confidence they have in it. They seem stuck in an all-or-nothing, foundationist conception of science and knowledge. (9) The general idea behind the retreat to method in politics and law is that it is easier to reach agreement about a fair procedure than about substantive distribution of social goods, guilt, or blame. And in today's social context, "fairness" is a rhetorical appeal often used against science rather than in its favor. At this point, scientific expertise and popular democracy do not mix well. Thus, it is intellectually completely irrelevant that a majority of Americans have more confidence in their religious beliefs than in evolution. (10) The appeal to method has become so weak that Laudan is surely correct in saying we still need to take seriously the idea of science as a body of substantive claims. After all, a major problem with the creationist approaches is that they cannot get any model or theory off the ground that is even remotely compatible with the basics of modern physics and chemistry, let alone geology or biology.

Some Reasons for Optimism

So we are back to square one. Or are we? Surely we ourselves can avoid the "all or nothing" trap and appreciate that the above considerations argue for a pluralistic approach. Demarcation should proceed on several fronts, no one of which is intellectually decisive but which, together, provide sufficient purchase for practical purposes on neutral playing fields. What began as a logical or metaphysical issue ends up being a concern modulated by pragmatic reasons (Resnik 2000). While there is some truth to the reported demise of the traditional demarcation problem, that is partly because it has given way to multiple demarcation issues, intellectual and practical. This volume, and the work cited herein, is evidence that philosophers are now giving more attention to practical policy implications than in the past.

There is room for optimism, for several useful if fallible indicators of nonscience and bad science, when taken together, yield reliable results. Past empirical success certainly counts for something, and empirical testability remains valuable as a rule of thumb for projects extending into the future. But, beyond that, we can usually add important indications of future promise, based, in large part, on the testimony of established experts in that and related domains, people who have established records of good research design, for example. These considerations can usually be made intelligible to a lay audience in a neutral setting without having to rely heavily on appeals to tacit knowledge.

In discussing theory choice within science, a number of authors have mentioned fertility as a desideratum. Lakatos's criteria for the progress of a research program and Laudan's rate of recent problem-solving progress are important examples. But these, like many references to fertility, remain largely retrospective assessments. They are important, to be sure, but rather conservative. Since we don't want to unduly curb the scientific imagination, I follow Kuhn here to suggest that estimation of future fertility, or what I have termed "heuristic appraisal" (Nickles 2006), can be even more important, both within technical scientific contexts (the problem of new theories, new research initiatives, and so-called bandit problems more generally) and in social contexts. Philip Kitcher (1982, 48ff) is among those who both include fecundity explicitly among the criteria of scientific success and extend it to estimation of the likelihood of future breakthroughs. Projections into the future are risky, of course, but unavoidable at the frontiers of research. After all, all decision making, including demarcation, concerns future action in an uncertain world.

Despite these helpful gestures toward future fertility, we poorly understand how such judgments are made in various scientific, funding, and policy contexts, and how the judgment processes might be improved. The fact remains that the vast bulk of philosophical writing about scientific decision making remains cast in the mold of "theory of justification" or "confirmation theory" from the positivist era with its retrospective emphasis on empirical track record. It does poorly with assessment of undeveloped ideas that may harbor breakthroughs. No creative discipline can be rigorously evidence based every step of the way.

Nor should we defensively fear transforming currently accepted theories. Suppose that new results, say from Evo-Devo research, showed that current evolutionary theory needs significant revision, in other words, that the current theory is false in detail. Creationists would be delighted, thereby entirely missing the point. For if this were to be shown, it would be the result of exciting scientific advances that manifest the fertility of current lines of investigation and the expertise of those conducting the research. It would hardly be a triumph to the credit of creationists. Even when scientific research shows current understandings to be wrong, this reinforces the value of the methods of scientific investigation (broadly understood). It does not undermine them. For science itself can be considered a complex adaptive system. Just as, in earlier times, appeal to "the scientific method" trumped substantive criteria, so, today, a far more nuanced appreciation of scientific techniques, practices, and expertise should trump the "belief system" conception of science.

The approaches to demarcation defended here do not guarantee smooth sailing, of course, in all contexts. There is rightly much concern in philosophy and in social studies of science more generally about how to determine who is a relevant expert and how nonexperts can gain a sufficient voice in policymaking in a democratic society (Collins and Evans 2007; Pigliucci 2010, chap. 12). However, these issues do not overlap completely. It is often rather easy, on the grounds defended here, to pick out legitimate sciences from the frauds.

For one thing, although it is not absolute, the belief-practice distinction helps to remove a persistent confusion over who is an expert. At the far frontiers of research, where no one knows exactly what lies beyond, there are no experts in the sense of those who know or even reliably believe *that* such-and-such is true; but there clearly are experts in the sense of those who know *how* to proceed with frontier research, who are able to furnish comparative heuristic appraisals of the competing proposals, and who possess the flexibility to revise these appraisals appropriately as research proceeds. At this writing, no one really knows how life originated from nonlife on Earth, yet scientific specialists in several fields are clearly better equipped than anyone else to investigate this question. Their propositional knowledge in this area is currently slim, but their investigative know-how is extensive and holds promise of future results. It is noteworthy that creationists, including Intelligent Designers, have made zero progress on this issue. A quick response to the "It's only a theory" jibe at evolution is that "Intelligent Design Theory," as it currently exists, is not even a sketch of a crude model or mechanism, let alone a developed, successful theory in the scientific sense.

"Creation science," as so far developed, cannot hold a candle to research in the various branches of evolutionary biology. Compare the research progress in both areas since Darwin's time. More to the present point, think of future promise. For the foreseeable future, we can justifiably expect more progress *every month* in evolutionary science, including evolutionary-developmental biology or "Evo-Devo" and "the extended synthesis" (Pigliucci and Müller 2010), than that so far produced by *the entire history* of modern "creation science"! We should not waste precious science teaching time on projects that remain so unpromising.

Acknowledgments

This chapter is a heavily revised reprint of "Demarcation, Problem of" from *The Philosophy of Science: An Encyclopedia*, vol. 1, edited by Sahotra Sarkar and Jessica Pfeifer, 188–97, New York: Routledge, 2006. The editors and I thank the Taylor & Francis Group for permission. And I thank the editors and reviewers for helpful suggestions.

NOTES

1. Nickles (2009) argues that this is in fact a secularized creationist idea in supposing that the scientific method can serve as a designing agent that implicitly and omnisciently contains all future discoveries.

2. See Agassi (1964). Popper's point can be extended to future science, since untestable, imaginative speculations may make us aware of possibilities that were unconceived, even unconceivable before, and these are sometimes necessary for major scientific breakthroughs (Stanford 2006).

3. Compare the development of artificial intelligence from Newell and Simon's General Problem Solver through knowledge-based systems to genetic algorithms. See Nickles (2003a), which also addresses the "No Free Lunch" theorems of Wolpert and Macready.

4. Feyerabend (1975) also later rejected method in order, unlike Kuhn, to challenge the authority of modern science and its traditional constraints on inquiry.

5. More recently, Laudan has written an entire book on legal epistemology critiquing the judicial system (Laudan 2006).

6. Rouse (2003, 119) notes the irony that, despite Kuhn's and science studies' challenge to "textbook science," the leading philosophical models of science remain representational and hence lend encouragement to the creationists' fideistic conception of science as just another belief system—and hence to their conception of science education.

REFERENCES

Agassi, Joseph. 1964. "The Nature of Scientific Problems and Their Roots in Metaphysics." In *The Critical Approach to Science and Philosophy*, edited by Mario Bunge, 189–211. New York: Free Press.

Amsterdamski, Stefan. 1975. *Between Experience and Metaphysics*. Dordrecht: Reidel.

Bauer, Henry. 2001. *Science or Pseudoscience?* Urbana: University of Illinois Press.

Collins, Harry, and Robert Evans. 2007. *Rethinking Expertise*. Chicago: University of Chicago Press.

Feyerabend, Paul. 1975. *Against Method*. London: New Left Books.

Foster, Kenneth, David Bernstein, and Peter Huber, eds. 1993. *Phantom Risk: Scientific Inference and the Law*. Cambridge, MA: MIT Press.

Gieryn, Thomas. 1999. *Cultural Boundaries of Science: Credibility on the Line*. Chicago: University of Chicago Press.

Hempel, Carl. 1950. "Problems and Changes in the Empiricist Criterion of Meaning." *Revue Internationale de Philosophie* 11:41–63. Reprinted in Hempel (1965), 101–19, as merged with Hempel (1951).

———. 1951. "The Concept of Cognitive Significance: A Reconsideration." *Proceedings of the American Academy of Arts and Sciences* 80:61–77. Reprinted in Hempel (1965), 101–19, as merged with Hempel (1950).

———. 1965. *Aspects of Scientific Explanation*. New York: Free Press.

Huber, Peter. 1991. *Galileo's Revenge: Junk Science in the Courtroom*. New York: Basic Books.

Kitcher, Philip. 1982. *Abusing Science: The Case Against Creationism*. Cambridge, MA: MIT Press.

Kuhn, Thomas. 1962. *The Structure of Scientific Revolutions*. Chicago: University of Chicago Press.

———. 1970a. "Logic of Discovery or Psychology of Research?" In Lakatos and Musgrave 1970, 1–23. Reprinted in Kuhn's *The Essential Tension*, 266–92. Chicago: University of Chicago Press.

———. 1970b. "Postscript-1969." In *The Structure of Scientific Revolutions*, 2nd ed. Chicago: University of Chicago Press.

Lakatos, Imre. 1970. "Falsification and the Methodology of Scientific Research Programmes." In Lakatos and Musgrave 1970, 91–195.

Lakatos, Imre, and Alan Musgrave, eds. 1970. *Criticism and the Growth of Knowledge*. Cambridge: Cambridge University Press.

Laudan, Larry. 1981. *Science and Hypothesis*. Dordrecht: Reidel.

———. 1986. "Some Problems Facing Intuitionistic Meta-Methodologies." *Synthese* 67:115–29.

———. 1996a. *Beyond Positivism and Relativism*. Boulder, CO: Westview Press.

———. 1996b. "The Demise of the Demarcation Problem." In Laudan 1996a, 210–22. (Originally published 1983.)

———. 2006. *Truth, Error, and Criminal Law: An Essay in Legal Epistemology*. New York: Cambridge University Press.

Lynch, Michael, and Sheila Jasanoff, eds. 1988. *Social Studies of Science* 28 (5–6). Special issue on "Contested Identities: Science, Law and Forensic Practice."

McMullin, Ernan. 2001. "The Impact of Newton's *Principia* on the Philosophy of Science." *Philosophy of Science* 68:279–310, followed by three commentaries and McMullin's reply.

Nickles, Thomas. 1987a. "From Natural Philosophy to Metaphilosophy of Science." In *Kelvin's*

Baltimore Lectures and Modern Theoretical Physics: Historical and Philosophical Perspectives, edited by Robert Kargon and Peter Achinstein, 507–41. Cambridge, MA: MIT Press.

———. 1987b. "Lakatosian Heuristics and Epistemic Support." *British Journal for the Philosophy of Science* 38:181–205.

———. 2003a. "Evolutionary Models of Innovation and the Meno Problem." In *International Handbook on Innovation*, edited by Larisa Shavinina, 54–78. Amsterdam: Elsevier Scientific.

———, ed. 2003b. *Thomas Kuhn*. Cambridge: Cambridge University Press.

———. 2006. "Heuristic Appraisal: Context of Discovery or Justification?" In *Revisiting Discovery and Justification: Historical and Philosophical Perspectives on the Context Distinction*, edited by Jutta Schickore and Friedrich Steinle, 159–82. Dordrecht: Springer.

———. 2009. "The Strange Story of Scientific Method." In *Models of Discovery and Creativity*, edited by Joke Meheus and Thomas Nickles, 167–207. Dordrecht: Springer.

Park, Robert. 2000. *Voodoo Science: The Road from Foolishness to Fraud*. Oxford: Oxford University Press.

Pigliucci, Massimo. 2010. *Nonsense on Stilts: How to Tell Science from Bunk*. Chicago: University of Chicago Press.

Pigliucci, Massimo, and Gerd Müller, eds. 2010. *Evolution: The Extended Synthesis*. Cambridge, MA: MIT Press.

Polanyi, Michael. 1958. *Personal Knowledge: Toward a Post-Critical Philosophy*. Chicago: University of Chicago Press.

Popper, Karl. 1959. *The Logic of Scientific Discovery*. London: Hutchinson. Popper's own translation and expansion of his *Logik der Forschung* (Vienna: Julius Springer, 1934).

———. 1963. *Conjectures and Refutations*. New York: Basic Books.

———. 1977. "Natural Selection and the Emergence of Mind." Darwin Lecture, reprinted in *Evolutionary Epistemology, Rationality, and the Sociology of Knowledge*, edited by Gerard Radnitzky and W. W. Bartley III, 139–56. LaSalle, IL: Open Court, 1987.

Rampton, Sheldon, and John Stauber. 2001. *Trust Us, We're Experts!* New York: Tarcher/ Putnam.

Resnik, David. 2000. "A Pragmatic Approach to the Demarcation Problem." *Studies in History and Philosophy of Science* 31:249–67.

Rouse, Joseph. 2003. "Kuhn's Philosophy of Scientific Practice." In Nickles 2003b, 101–21.

Ruse, Michael. 1982. "Pro Judice." *Science, Technology, & Human Values* 7:19–23.

Shapiro, Barbara. 1983. *Probability and Certainty in Seventeenth-Century England: A Study of the Relations between Natural Science, Religion, History, Law and Literature*. Princeton, NJ: Princeton University Press.

Stanford, Kyle. 2006. *Exceeding Our Grasp: Science, History, and the Problem of Unconceived Alternatives*. New York: Oxford University Press.

Toumey, Christopher. 1996. *Conjuring Science: Scientific Symbols and Cultural Meanings in American Life*. New Brunswick, NJ: Rutgers University Press.

Traweek, Sharon. 1988. *Beamtimes and Lifetimes: The World of High Energy Physicists*. Cambridge, MA: Harvard University Press.

Wittgenstein, Ludwig. 1922. *Tractatus Logico-Philosophicus*. London: Routledge and Kegan Paul. (Original German edition 1921.)

Worrall, John. 2003. "Normal Science and Dogmatism, Paradigms and Progress: Kuhn 'versus' Popper and Lakatos." In Nickles 2003b, 65–100.

7

Science, Pseudoscience, and
Science Falsely So-Called

DANIEL P. THURS AND RONALD L. NUMBERS

On July 1, 1859, Oliver Wendell Holmes, a fifty-year-old Harvard Medical School professor and littérateur, submitted to being "phrenologized" by a visiting head reader, who claimed the ability to identify the strength of character traits—amativeness, acquisitiveness, conscientiousness, and so forth—by examining the corresponding bumps on the head. Holmes found himself among a group of women "looking so credulous, that, if any Second Advent Miller or Joe Smith should come along, he could string the whole lot of them on his cheapest lie." The astute operator, perhaps suspecting the identity of his guest, offered a flattering assessment of Holmes's proclivities, concluding with the observation that his subject "would succeed best in some literary pursuit; in teaching some branch or branches of natural science, or as a navigator or explorer." Holmes found the event so revealing of contemporary gullibility that shortly thereafter he drew on his experience in writing his *Atlantic Monthly* feature "The Professor at the Breakfast-Table" (Holmes 1859, 232–43; Lokensgard 1940). He began by offering a "definition of a *Pseudo-science*," the earliest explication of the term that we have found:

> A Pseudo-science consists of a nomenclature, with a self-adjusting arrangement, by which all positive evidence, or such as favors its doctrines, is admitted, and all negative evidence, or such as tells against it, is excluded. It is invariably connected with some lucrative practical application. Its professors and

practitioners are usually shrewd people; they are very serious with the public, but wink and laugh a good deal among themselves. . . . A Pseudo-science does not necessarily consist wholly of lies. It may contain many truths, and even valuable ones.

Holmes repeatedly punctuated his account by denying that he wanted to label phrenology a pseudoscience; he desired only to point out that phrenology was *"very similar"* to the pseudosciences. On other occasions Holmes categorically included phrenology—as well as astrology, alchemy, and homeopathy—among the pseudosciences (Holmes 1859, 241–42; Lokensgard 1940, 713; Holmes, 1842, 1).

Before pseudoscience became available as a term of reproach, critics of theories purporting to be scientific could draw on a number of older deprecatory words: humbuggery, quackery, and charlatanism. But no phrase enjoyed greater use than "science falsely so-called," taken from a letter the apostle Paul wrote to his young associate Timothy, advising him to "avoid profane and vain babblings, and oppositions of science falsely so called" (1 Tim. 6:20). Although "science" had appeared in the original Greek as *gnōsis* (meaning knowledge generally), English translators in the sixteenth and seventeenth centuries chose "science" (a synonym for knowledge). By the mid-eighteenth century (at the latest), the phrase was being applied to disagreeable natural philosophy. In 1749, for example, the philosophically inclined Connecticut minister Samuel Johnson complained that "it is a fashionable sort of philosophy (a science falsely so-called) to conceive that God governs the world only by a general providence according to certain fixed laws of nature which he hath established without ever interposing himself with regard to particular cases and persons" (Hornberger 1935, 391; Numbers 2001, 630–31).

Even after the appearance of the term "pseudoscience," "science falsely so-called" remained in wide circulation, especially among the religious. In the wake of the appearance of Charles Darwin's *On the Origin of Species* (1859) and other controversial works, a group of concerned Christians started the Victoria Institute, or Philosophical Society of Great Britain; they dedicated it to defending "the great truths revealed in Holy Scripture . . . against the opposition of Science, falsely so called." The religiously devout Scottish anti-Darwinist George Campbell, the eighth Duke of Argyll, dismissed Thomas H. Huxley's views as "science falsely so called." After Huxley's good friend John Tyndall used his platform as president of the British Association for the Advancement of Science in 1874 to declare all-out war on theology, one Presbyterian wit dubbed Britain's leading scientific society "The British Association for the Advancement of 'Science, Falsely So-Called.'" Ellen G. White, the founding prophet of Seventh-Day Adventism, condemned "science falsely so-called"—meaning "mere theories and speculations as scientific facts" opposed to the Bible—over a dozen times (Numbers 2006, 162; Argyll 1887b; Livingstone 1992, 411; Comprehensive Index 1963, 3:2436; E. White 1888, 522; Numbers 1975).

Few epithets, however, have drawn as much scientific blood as pseudoscience. Defenders of scientific integrity have long used it to damn what is not science but pretends to be, "shutting itself out of the light because it is afraid of the light." The physicist Edward Condon captured the transgressive, even indecent, implications of the term when he likened it to "Scientific Pornography" (Brinton 1895, 4; Condon 1969). Such sentiments might make it appear that tracing the history of pseudoscience would be an easy enough task, simply requiring the assembly of a rogue's gallery of obvious misconceptions, pretensions, and errors down through the ages. But, in fact, writing the history of pseudoscience is a much more subtle matter, especially if we eschew essentialist thinking.

If we want to tell the history of pseudoscience, we have to come to grips with the term's fundamentally rhetorical nature. We also need to take a historically sensitive track and focus on those ideas that have been rejected by the scientific orthodoxy of their own day. But here, too, problems arise. For most of the history of humankind up to the nineteenth century, there has been no clearly defined orthodoxy regarding scientific ideas to run afoul of, no established and organized group of scientists to pronounce on disputed matters, no set of standard scientific practices or methods to appeal to. However, even in the presence of such orthodoxy, maintaining scientific boundaries has required struggle. Rather than relying on a timeless set of essential attributes, its precise meanings have been able to vary with the identity of the enemy, the interests of those who have invoked it, and the stakes involved, whether material, social, or intellectual. The essence of pseudoscience, in short, is how it has been used.

The Invention of Pseudoscience

English speakers could have paired "pseudo," which had Greek roots, and "science," which entered English from Latin by way of French, at any time since the medieval period. However, pseudo-science (almost universally written with a hyphen before the twentieth century) did not become a detectable addition to English-language vocabularies until the early 1800s. Its greater circulation did not result from a sudden realization that false knowledge was possible. Instead, it involved larger shifts in the ways that people talked, including a greater tendency to append "pseudo" to nouns as a recognized means of indicating something false or counterfeit. This habit was apparent as early as the seventeenth century, but became particularly common during the nineteenth (*Oxford English Dictionary*, s.v. "pseudo").

Even more significant, increased usage paralleled important changes in the concept of science. Pseudoscience appeared at precisely the same time during the early part of the 1800s that science was assuming its modern meaning in English-speaking cultures to designate knowledge of the natural world. The more the category of science eclipsed and usurped significant parts of those activities formerly called natural philosophy and natural history, the more rhetorical punch "pseudoscience" packed as a weapon against one's enemies. By the same token, even taking account of its many possible meanings, pseudoscience gave people the ability to mark off scientific pretense and error as especially worthy of notice and condemnation, making science all the more clear by sharpening the outlines of its shadow and opening the door to attestations of its value in contrast with other kinds of knowledge. In this sense, pseudoscience did not simply run afoul of scientific orthodoxy—it helped to create such orthodoxy.

Pseudoscience began its career in the English-speaking world rather modestly. While certainly in general use during the early and mid-1800s, it was still somewhat rare, particularly in contrast to the latter decades of the century and the 1900s. This is visible, for instance, in full-text searches for the term in American magazine articles from 1820 to 1920 (American Periodical Series Online 2012). In general, the results of such a survey show no usage at first, then a steadily increasing appeal, with a slight surge in the 1850s and a dramatic increase in the 1880s, leading to a comparatively high and somewhat constant level of use around the turn of the century. By this time, pseudoscience was becoming an international term of opprobrium. The French used the same word as the English did; however, other nationalities coined cognates: *Pseudowissenschaften* in German, *pseudoscienza* in Italian, *seudociencia* in Spanish, *pseudovetenskap* in Swedish, *pseudowetenschap* in Dutch, and *псевдонаука* in Russian (Larousse 1866–79; Littré 1873–74). Americans, however, seemed to have been fondest of the term, which makes the American context particularly interesting for examining the rise and evolution of pseudoscience.

During the 1830s and 1840s, a wide variety of novel ideas appeared on the American intellectual landscape that some people thought strange. These novelties included religious groups such as Mormons and Millerites and social reform movements associated with women's rights and abolitionism. A host of new scientific and medical approaches, ranging from the do-it-yourself botanical cures of Thomsonianism to the minute doses of homeopathy, also circulated. Probably the most emblematic of the crop of *-isms, -ologies,* and *-athies* that flourished in the antebellum soil was phrenology. In its most popular form, which linked the shape of the skull (and the cerebral organs underneath) to the details of an individual's personality, phrenological doctrine was spread across the nation by a cadre of devoted lecturers and head readers and by a thick stack of cheap literature (Thurs 2007).

For many of its skeptics, phrenology provided one of the primary examples of pseudoscience and, sometimes with special emphasis, "*pseudo science.*" As we indicate above, the very first reference we have found to "pseudo-science" appeared in 1824 and was directed obliquely at phrenology ("Sir William Hamilton on Phrenology" 1860, 249; *Medical Repository of Original Essays and Intelligence* 1824, 444). Still, even amidst antebellum debate over phrenological ideas, invocations of pseudoscience remained fairly few. Phrenology was identified as pseudoscience more frequently during the latter portions of the nineteenth century than it was in its heyday. This was partly because in the early years a strong scientific orthodoxy remained more hope than actuality. Historian Alison Winter has argued that individual scientific claims during this period had to be established without the help of an organized community of practitioners with shared training, beliefs, and behaviors (1998, 306–43). The same was true of attempts to exile ideas from science. Even among the most notable members of the American scientific scene, there was less-than-universal agreement about the status of novel ideas. One of the most prominent men of science of the period, Yale's Benjamin Silliman, was publicly friendly to phrenology, albeit in one of its more scholarly forms.

The most important explanation for the relative rarity of charges of pseudoscience in the early parts of the nineteenth century was that science remained a somewhat amorphous term. By the second quarter of the 1800s, it had largely taken the place of earlier names, such as natural philosophy and natural history, for the study of the natural world. But enough of its former connection to reliable and demonstrable knowledge in general remained that science went well beyond the natural and included a huge swath of areas, from theology to shorthand. Such enormous extent was supported by contemporary methodological standards that made it much easier to include fields within porous scientific boundaries than to exclude them. This fuzziness actually made science difficult to use in many cases, and Americans appealed to "the sciences" collectively or to individual sciences, such as chemistry or geology, more frequently than during later eras (Thurs 2007, 24). It was also difficult to know exactly how to describe scientific transgression. Popular rhetoric had not quite settled on "pseudoscience" as the means to do that. Many Americans also denounced "pseudo-chemists," "pseudo-induction," "pseudo-observation," and, in the case of a new religious phenomenon with scientific pretensions, "pseudo-spiritualism." In Britain, the mathematician Augustus de Morgan contributed the term "paradoxer" to describe those whose ideas "deviated from general opinion, either in subject-matter, method or conclusion" ("Scientific Agriculture" 1856, 93; Raymond 1852, 839; Garwood 2007, 70).

The fuzziness of science meant that many ways of categorizing knowledge with false pretensions to truth did not have any direct link with the scientific at all. Defenders of scientific integrity routinely denounced "mountebanks" and "pretenders." Even some of the American practitioners who agitated most for a more controlled and organized scientific community, including the secretary of the Smithsonian Institution, Joseph Henry, and the director of the United States Coast Survey, Alexander Dallas Bache, talked about expelling "charlatans" and "quacks" from the scientific world (Slotten 1994, 28). Indeed, by far the most common term for error in the guise of science was "quackery." "Quackery" was particularly common in discussions of medicine, but the scope of the category was not tightly linked to the medical or scientific alone. The definition offered by English physician Samuel Parr, as paraphrased by the American antiquackery crusader David Meredith Reese, identified "every practitioner, whether educated or not, who attempts to practise imposture of any kind" as a quack. And though Reese noted in his *Humbugs of New-York* that the "epithet is often restricted within narrow limits, and is attached ordinarily only to those ignorant and impudent mountebanks, who, for purposes of gain, make pretensions to the healing art," he included abolitionism within the sphere of humbuggery (Reese 1838, 110–11).

Likewise, many Americans offered broad definitions of the pseudoscientific that did little more than identify it with pretensions to an unwarranted scientific status. Astrology and alchemy were two widely used historical examples of pseudoscience because they resembled actual sciences in name, but were not truly scientific (or so it was universally assumed). When it came to explaining pretension, some people focused on the "servile sophistry of pseudo-science" and on the perversion of legitimate scientific terminology (Ure 1853, 368). An 1852 article in *Harper's New Monthly Magazine* decried the "dextrous use of some one long new-coined term" (Raymond 1852, 841). Such verbiage was another widely recognized sign of pseudoscience, as was the tendency to theorize too quickly or leap to conclusions. In other cases, observers attributed false claims cloaked in science to ethical lapses. To such leading American men of science as Henry and Bache, moral fiber was an essential possession of the true scientific practitioner (Slotten 1994, 29).

This lack of specificity paralleled the fuzziness of science itself. At the same time, there were some emerging ideas about the nature of pseudoscience, quackery, charlatanism, or mountebankery applied specifically to scientific matters that were beginning to draw lines between science and not-science in new and ultimately modern-sounding ways. For instance, one antebellum characteristic of pretense in the scientific or medical worlds was the improper influence of the hope for material gain. Both David Meredith Reese and Oliver Wendell Holmes pointed to the corrupting influence of money in the creation and diffusion of quackery and pseudoscience (Reese 1838, 111; Holmes 1859). Such claims helped to create a zone of pure scientific knowledge or medical practice.

By far the most important distinction emerging between science and not-science was the one between scientific and popular knowledge. This distinction played a profoundly important role in the control of scientific knowledge and practice by aspiring leaders of science and medicine. Since such so-called professionalization was more advanced in Britain than in the United States, one of the earliest public links between pseudoscience and "popular delusions" or the "slowness in the capacity of the popular mind" occurred in an evaluation of homeopathy in the Edinburgh-based *Northern Journal of Medicine.* American pens inscribed less strident connections between pseudoscience and untutored interest in scientific subjects, thanks in part to their incompatibility with the majority of antebellum public rhetoric; but it was evident beneath the surface. In a letter to Joseph Henry in 1843, John K. Kane, the senior secretary of the American Philosophical Society, worried that "all sorts of pseudo-scientifics" were on their way to try to win the recently vacated position as head of the Coast Survey by way of enlisting Henry's help to encourage Bache to make a bid for the position. Henry himself lamented in a letter to Bache the recent "avalanch of pseudo-science" and wondered how the "host of Pseudo-Savants" could "be controlled and directed into a proper course" (Kane 1843, 5:451; Henry 1844, 6:76–77).

Science and Pseudoscience in the Late Nineteenth Century

The last quarter of the nineteenth century witnessed a marked growth in references to pseudoscience, both in American periodicals and in English-speaking culture generally. Method remained a powerful means of identifying pseudoscience, particularly the tendency to leap to conclusions "in advance of the experimental evidence which alone could justify them." Likewise, the presence of potential profit continued to signal the presence of pseudo-science. One correspondent in an 1897 issue of *Science* noted that the recent discovery of x-rays had "already been made a source of revenue by more than one pseudo-scientist" (Sternberg 1897; Stallo 1882). At times, a moral note appeared in broadsides against pseudoscience. Daniel G. Brinton's presiden-tial address at the annual meeting of the American Association for the Ad-vancement of Science (AAAS) in 1894 strongly linked pseudoscience to the fundamental dishonesty and snobbery of "mystery, concealment, [and] oc-cultism." Mary Baker Eddy's Christian Science found itself both denounced as a pseudoscience (and a science falsely so-called) and divided by "pseudo Scientists," who questioned the prophet (Brinton 1895; Nichols 1892; Brown 1911; "Separation of the Tares" 1889; Schoepflin 2003).

Emerging distinctions between science and religion became one of the primary fault lines in the growing gap between science and not-science. But it was not the only one. Other potent divisions in American public culture included those between pure and applied science and between the generation of scientific knowledge and its popularization. All these new rhetorical habits enhanced the value of ejecting ideas and people from the scientific fold, and therefore the invocation of pseudoscience. They also paralleled a number of important shifts in widespread ways of talking about pseudoscience. An 1896 letter to the editor of *Science* from one reader criticized "'practical' or pseudo-science" men, by which he meant those uneducated in theory (Fernow 1896, 706). Equations between pseudoscience and popular science also became far more common than earlier in the century. As men of science (or "scientists" as they were increasingly being called) began to privilege research over the diffusion of science—which in their opinion often simplified and degraded pure scientific knowledge—popularization often found itself associated with pseudoscience. An 1884 article in the *New York Times* leveled as one of its primary complaints against the "pseudo-science" of phrenology that books and pamphlets about it were "within reach of everybody" ("Character in Finger Nails" 1884).

The growing sense of "popular science" as something distinct from science itself that informed such characterizations of science and not-science went hand in hand with evolving ideas about the nature of the public that was consuming popularized science. One of its chief characteristics in late nineteenth-century depictions was its credulity. A variety of commentators noted, particularly in the years before 1900, that the scientific discoveries of the previous hundred years had been so dramatic and extensive that ordinary people had become "ready to accept without question announcements of inventions and discoveries of the most improbable and absurd character" or that the "general public has become somewhat overcredulous, and untrained minds fall an easy prey to the tricks of the magazine romancer or to the schemes of the perpetual motion promoter" (Woodward 1900, 14). According to some observers, "an army of pseudo-scientific quacks who trade upon the imperfect knowledge of the masses" had grown up alongside legitimate purveyors of orthodoxy. In this sense, the enormous and bewildering power of science was itself to blame for the spread of pseudoscientific ideas. But the "evil influence of a sensational press" also played a deleterious role ("Time Wasted" 1897, 969).

Such concerns reflected genuine worries, but they also paralleled the growth of the commercialized mass media as a new cultural force, as well as the creation of a new kind of mass public such publications made possible. By the early 1900s, many Americans had adopted the habits and methods of the advertising industry to promote everything from public health campaigns to religious revivals. An article in an 1873 issue of the *Ladies' Repository* decried the supposed tendency of many magazine readers to "swallow any bolus that speculative doctors of chances may drop into their gullets," particularly when an article began "'Dr. Dumkopf says'" (Carr 1873, 125). The geologist and science popularizer Joseph LeConte similarly complained about mistakes "attested to by newspaper scientists, and therefore not doubted by newspaper readers" (LeConte 1885, 636). For champions of science, such conditions could appear to be prime breeding grounds for error. One observer worried about "plausibly written advertisements," particularly for medicines, that falsely invoked science and thus threatened to dupe and unwary public with pseudoscientific claims (Sternberg 1897, 202). The battle between science and pseudoscience in the press occasionally produced a call to arms among professional practitioners. In 1900, the president of the AAAS recognized in his annual address that while scientists' "principle business is the direct advancement of science, an important, though less agreeable duty, betimes, is the elimination of error and the exposure of fraud" (Woodward 1900, 14).

As the invocation of pseudoscience began to proliferate in the emerging gaps between the scientific and popular, it also started to appear with greater frequency along another emerging distinction, namely the one between the physical and social sciences. Encouraging this boundary was an increasing tendency to link science with physical nature, a move that often left the study of humans in a kind of limbo. One character in a serialized story originally published in the British *Contemporary Review* and later reprinted for American audiences in *Appleton's Monthly* lamented that, though potentially important, social science was "at present not a science at all. It is a pseudo-science" (Mallock 1881a, 660; Mallock 1881b, 531). Sometimes, the harshest critics were social science practitioners themselves, a fact that might indicate some small measure of anxiety about the status of their field and the need to root out unacceptable practices. In 1896, the American economist-cum-sociologist Edward A. Ross claimed that ethics was a pseudoscience "like theology or astrology" because it sought to combine the mutually exclusive perspectives of the individual and the group. Albion W. Small, holder of the first American chair in sociology (at the University of Chicago), asserted, with a nod to the growing boundary between the scientific and popular, that his own field was "likely to suffer long from the assumptions of pseudo-science" because "sociologists are no more immune than other laymen against popular scientific error." Others charged that social pseudoscience, particularly in economics, was misusing the tools of physical science, concealing "its emptiness behind a breastwork of mathematical formulas" (E. Ross 1896; Small 1900; *Science* 1886, 309).

Still, it was discussion of science and religion that seemed to generate the most invocations of pseudoscience during the late 1800s. A large amount of pseudoscientific rhetoric spilled out of the contemporary debate over evolution and its implications. In 1887, Thomas Henry Huxley himself published an article entitled "Scientific and Pseudo-Scientific Realism," the first salvo in an extended exchange between Huxley and George Campbell, the eighth Duke of Argyll, an outspoken opponent of Darwinism. After a reply by Campbell, Huxley penned a second article simply called "Science and Pseudo-Science," to which the duke responded with an essay entitled "Science Falsely So Called." All these appeared in the British magazine the *Nineteenth Century*.

Huxley's contributions were reprinted across the Atlantic in the *Popular Science Monthly*; his second article was additionally reprinted in the *Eclectic Magazine* (Huxley 1887a, 1887b, 1887c, 1887d, 1887e; Argyll 1887a, 1887b). The primary issue in this exchange was the status of natural law and its relationship with the divine. The Duke of Argyll claimed that natural laws were directly ordained and enforced by God. To Huxley, by contrast, any suggestion that the uniformity of natural law implied the existence of divine providence was illegitimate and pseudoscientific.

Huxley, however, was in the minority. Many practicing men of science were wary of supernatural interference in the natural world by the late 1800s, but few of them used the term "pseudoscience" to tar such belief. Instead, the vast majority of cries of pseudoscience came from the opponents of evolution. One author in the conservative *Catholic World* depicted the Huxleys, Tyndalls, and Darwins of the world as the "modern Cyclops, who in forging their pseudo-sciences examine nature, but only with one eye" ("Socialism and Communism" 1879, 812–13). Other defenders of orthodox science and traditional religion, Catholic and Protestant, decried the "materialistic or pseudo-scientific skepticism of the day" or denounced the "pseudo-scientific sect" of Darwinian evolutionists (*Presbyterian Quarterly and Princeton Review* 1873; "Darwin on Expressions" 1873, 561). Sometimes, anti-evolutionists refined their attacks by equating the pseudoscientific nature of evolution with "foreign tendencies which are alien from science or philosophy," including materialism and atheism (Hewitt 1887, 660–61). Other critics sometimes claimed, without convincing evidence, that evolution did not meet the approval of most men of science, or at least the most prominent ones, thereby invoking a sense of orthodoxy for their cause.

Pseudoscience in the New Century

Between the late nineteenth and early twentieth centuries, the mechanisms both for enforcing and communicating orthodoxy in scientific matters grew to new heights in the United States and in Britain. Graduate training, specialized journals, and membership in exclusive organizations all helped to establish a substratum of agreed-on practices, facts, and concepts that most scientists learned to share. The increased professionalization of science paralleled more stringent boundaries around the scientific, including distinctions between scientists and laypeople and between legitimate scientific knowledge and scientific error and misunderstanding. Less permeable divisions were also emerging within science among the various disciplines, shaped by the need to master the expanding volume of specialized knowledge in any one area of work. Such an environment provided considerable encouragement for invocations of pseudoscience. It remained a prominent feature of talk about such areas as the social sciences and popular ideas. Over the course of the early 1900s, the term also began to take important new features that would eventually dominate the rhetoric of pseudoscience, particularly during the last third of the century.

One area of continuity in early twentieth-century discussions of pseudo-science involved method (see Nickles, chapter 6, in this volume). To show that textual criticism was a pseudoscience, a 1910 article in the journal of the Modern Language Association of America made what was by that time a venerable equation between the pseudoscientific and the violation of inductive reasoning (Tupper 1910, 176). Over the next several decades, ideas about scientific methodology changed in some significant ways, particularly in much more positive assessments of the role of theory. But though the methods of science changed, pseudoscience remained in violation of them. By far the most important shift in ideas about the methodology of science, however, was the emerging concept of the "scientific method." The proliferation of "scientific method" in public discussion implied a growing sense that science operated in special ways distinct enough to require its own name. It was, in short, a product of stronger boundaries around science, just like pseudoscience. Appropriately, a stricter view of "the established methods of science" was even more intimately linked with the pseudoscientific than during the previous century. A 1926 article in *California and Western Medicine* depicted nonorthodox medical ideas as an "attack upon the scientific method not alone in medicine but in all fields of knowledge" (Macallum 1916, 444; Frandsen 1926, 336). In a more neutral mood, it was also possible to see a uniquely scientific method as what actually joined science and pseudoscience. The anthropologist Bronislaw Malinowski called magic "a pseudo-science" not because he thought it was illegitimate but because it had practical aims and was guided by theories, just like science (Malinowski 1954, 87).

Talk about scientific method grew especially intense in the 1920s in discussions of the scientific status of the social sciences, which continued to find themselves depicted as scientific outsiders. A 1904 article in the *American Journal of Sociology* complained about charges leveled by "workers in other sciences" that sociology was pseudoscientific. In *The Public and Its Problems*, the influential philosopher and psychologist John Dewey pointed out that no methodology could eliminate the distinction between "facts which are what they are independent of human desire" and "facts which are what they are because of human interest and purpose.... In the degree in which we ignore this difference, social science becomes pseudo-science." Some critics described psychology as "the pseudo-science of thought." Others found in Sigmund Freud an example of the archetypical "pseudo-scientist" (Small 1904, 281; Dewey 1927, 7; Hearnshaw 1942, 165; Macaulay 1928, 213). As social scientists reacted to the ferment in their own rapidly changing disciplines, they sometimes painted the ideas of opponents in pseudoscientific tones. Advocates of Franz Boas's new ideas about cultural anthropology sometimes claimed that "the old classical anthropology . . . is not a science but a pseudo-science like medieval alchemy." By midcentury one critic, exasperated by "the deference paid to the pseudo-sciences, especially economics and psychology," declared that "if all economists and psychiatrists were rounded up and transported to some convenient St. Helena, we might yet save something of civilization" (L. White 1947, 407; "Pseudo-science" 1952).

In discussions of race in particular, the concept of pseudoscience proved a useful tool for redrawing scientific maps for a new century. In a 1925 letter, W. E. B. Du Bois claimed that talk about racial disparity was "not scientific because science is more and more denying the concept of race and the assumption of ingrained racial difference." However, he lamented that "a cheap and pseudo-science is being sent broadcast through books, magazines, papers and lectures" asserting "that yellow people and brown people and black people are not human in the same sense that white people are human and cannot be allowed to develop or to rule themselves" (Aptheker 1973, 1:303). More than twenty years later, another critic of racial stereotypes asked "what greater evidence of the use of pseudo-science can we ask than that afforded by Nazi doctrines of the 'superiority' of *Das Herrenvolk*?" (Krogman 1947, 14).

The increasingly sharp boundaries of science made the scientific appear more distinct and separate from what it was not, but they also cast a sharper shadow. Playing off descriptions of science as something unified and set apart were portrayals of pseudoscience that made it resemble a shadow science. On occasion such pseudoscience became a kind of semicoherent, though still deeply flawed, collection of pseudodisciplines with their own practitioners, sources of support, and methods of working. In 1932, the journalist and acidic wit H. L. Mencken suggested that every branch of science had an evil twin, "a grotesque Doppelgänger," which transmuted legitimate scientific doctrines into bizarre reflections of the truth. Though Mencken did not gather these doppelgängers together into a single pseudoscientific horde, others did address the "assumptions and pretensions of the hydra-headed pseudo-science." The sense of a wide variety of pseudoscientific ideas originating from an essential core culminated in a tendency to see pseudoscience not simply as the scattered errors of true science but as an example of "anti-science," a concept that would become widespread among those worried about popular misconception during the rest of the century (Mencken 1932, 509–10; Tait 1911, 293; Frandsen 1926, 336–38).

Pseudoscience and Its Critics in the Late Twentieth Century

The rhetorical trends of the previous century and a half laid the foundations for an explosion of talk about pseudoscience in America and elsewhere during the last third of the 1900s. After a small increase during the 1920s, usage of the term rose dramatically in the pages of English-language print media from the late 1960s onward, dwarfing the previous level of public invocation. Such talk was especially aimed at a series of unorthodoxies that appeared to erupt into popular culture after World War II, including the astronomical theories of Immanuel Velikovsky, sightings of UFOs and their occupants, and reports of extrasensory perception (ESP). Such notions were not necessarily more transgressive than, say, phrenology, but they did occur against the backdrop of the greater establishment of science, including the massive infusion of material support for research from the federal government, particularly the military. That establishment helped to create an environment that encouraged protecting the boundaries of science against invasion. Scientists now had much more to lose. The considerable increase in support for scientific work also helped to establish a heightened sense of scientific orthodoxy. A highly developed system of graduate education provided the required credentials to would-be scientists and socialized them into certain shared practices, beliefs, and pieces of knowledge. Professional journals, policed by the peer-review process, ensured a similar synchronization, at least on basic matters of fact, theory, and method.

An enhanced sense that there was a scientific orthodoxy, as well as mechanisms for ensuring one, resulted in a much more strongly bounded concept of science. Against this backdrop Edward Condon, who had himself constructed a career within the military-industrial-academic establishment, denounced pseudoscience as "scientific pornography." Where consensus broke down, charges of pseudoscience could mobilize a communal distaste for transgression against one's rivals. Social scientists, ethnoscientists, sexologists, sociobiologists, and just about anyone working in psychology and psychiatry were susceptible to be labeled pseudoscientists. Emerging new fields, clamoring for respect and support, were often met with pseudoscientific accusations. While skeptics portrayed the so-called Search for Extraterrestrial Intelligence, or SETI, as beyond the pale of scientific respectability, proponents such as the Cornell astronomer Carl Sagan asserted that SETI had moved from "a largely disreputable pseudoscience to an interesting although extremely speculative endeavor within the boundaries of science" (1975, 143).

Alongside the construction of a much stronger and more organized scientific establishment emerged a more structured means of communicating scientific consensus to the general public, whether they were reading textbooks in the expanding public school system, scanning the daily newspaper, or watching the evening news on television. But some observers inside and outside the scientific community continued to worry that popular treatments of science were as much a source of pseudoscientific ideas as they were a means to combat them. Complaints about the media's handling of scientific topics provided a constant drumbeat. Even more insidious were the publications unfettered by the sorts of respectability that often kept mainstream media in check. Condon's 1969 denunciation of scientific pornography laid responsibility squarely on "pseudo-science magazine articles and paper back books," which sold by the tens of thousands and even millions (6–8).

Since the early 1950s concerned citizens had been agitating for the formation of "one organization that could represent American science in combating pseudo-science." After decades of delay, champions of science finally banded together in 1976 to police the public sphere. Aroused by the popularity of Immanuel Velikovsky's *Worlds in Collision* (1950), Erich von Däniken's *Chariots of the Gods* (1968), and Charles Berlitz's *The Bermuda Triangle* (1974)—to say nothing of Uri Geller's spoon bending and Jeane Dixon's prophesying—a group of skeptics under the leadership of philosopher Paul Kurtz formed the Committee for the Scientific Investigation of Claims of the Paranormal (CSICOP), renamed the Committee for Skeptical Inquiry (CSI) in 2006, and began publication of a pseudoscience-busting journal that they soon called the *Skeptical Enquirer*. One supporter, Carl Sagan, who became a crusader against pseudoscience during the last third of the twentieth century, asserted in the journal that "poor popularizations of science establish an ecological niche for pseudoscience" and worried that there was a "kind of Gresham's Law by which in popular culture the bad science drives out the good" (Miles 1951, 554; Kurtz 2001; Sagan 1987, 46; Abelson 1974, 1233).

Inspired by CSICOP and the slightly older Association française pour l'information scientifique, publisher of *Science et pseudo-sciences,* similar groups sprang up around the world. By 1984, organizations of skeptics had formed in Australia, Belgium, Canada, Ecuador, Great Britain, Mexico, the Netherlands, New Zealand, Norway, and Sweden. During the next quarter century, the anti-pseudoscience movement spread to Argentina, Brazil, China, Costa Rica, the Czech Republic, Denmark, Finland, Germany, Hungary, India, Israel, Italy, Japan, Kazakhstan, Korea, Peru, Poland, Portugal, Russia, South Africa, Spain, Sri Lanka, Taiwan, Venezuela, and elsewhere. Many of the societies in these lands published their own magazines or newsletters. Concern often paralleled the eruption of some activity regarded as pseudoscientific, such as the Falun Gong in China and the Indian government's plan to introduce "Vedic astrology" as a legitimate course of study in Indian universities ("State of Belief 1984; International Committees 1984, 97; "Show-and-Tell" 2000; Ramachandran 2001; Jayaraman 2001; Committee for Skeptical Inquiry 2012).

Beyond equating the pseudoscientific and popular, late twentieth-century invocations of pseudoscience continued along the lines set down during the 1800s and early 1900s. The transgression of proper methodology remained a primary means of identification. Sagan claimed in 1972 that the reason some people turned toward UFOs or astrology was "precisely that they are often beyond the pale of established science, that they often outrage conservative scientists, and that they seem to deny the scientific method" (1972, xiii).

New methodological authorities also appeared. In the 1930s the Viennese philosopher Karl Popper begin writing about "falsifiability" as a criterion "*to distinguish between science and pseudo-science.*" He knew "very well," he later said, "that science often errs, and that pseudo-science may happen to stumble on the truth," but he nevertheless thought it crucial to separate the two (Popper 1963, 33; Popper 1959; Collingwood 1940; Lakatos 1999). Michael Ruse's invocation of falsifiability to distinguish between science and religion in a high-profile creation-evolution trial in Little Rock, Arkansas, in the early 1980s prompted Larry Laudan, another well-known philosopher of science, to charge his colleague with "unconscionable behavior" for failing to disclose the vehement disagreements among experts regarding scientific boundaries in general and Popper's lines in particular. By emphasizing the nonfalsifiability of creationism to deny its scientific credentials, argued Laudan, Ruse and the judge had neglected the "strongest argument against Creationism," namely, that its claims had already been falsified. Laudan dismissed the demarcation question itself as a "pseudo-problem" and a "red herring." Ruse, in rebuttal, rejected Laudan's strategy as "simply not strong enough for legal purposes." Merely showing creation science to be "bad science" would have been insufficient in this case because the U.S. Constitution does not ban the teaching of bad science in public schools (Numbers 2006, 277–68; L. Laudan 1983; Ruse 1988, 357).

Just as they had during the late 1800s, differences between science and religion loomed large in characterizations of pseudoscience, although in precisely the opposite way than they had before. Rather than signaling the overaggressive separation of scientific and religious concerns, Americans during the second half of the twentieth century more often linked pseudoscience to the illegitimate mixture of science and religion. Charges of pseudoscience aimed at a wide variety of targets, from creationism to UFOs to federal standards for organic food, all of which were denounced as involving religious motivations rather than scientific ones. By the turn of the twenty-first century, particularly in the context of a number of politically charged debates involving science, there was also a pronounced tendency to see pseudoscience as arising from the intrusion of political concerns onto scientific ground. In controversies over global warming, stem-cell research, Intelligent Design (ID), and even the demotion of Pluto from planetary status (because the final decision was made by vote), partisans depicted their opponents as proceeding from political motivations and thus distorting pure science. In reaction to the comments of President George W. Bush that appeared to open the possibility of including information on ID in public schools, critics charged him with raising a "pseudoscience issue" and "politicizing science," which "perverted and redefined" the true nature of scientific knowledge. Assertions of a "Republican war on science" from the left of the political spectrum echoed this sentiment (C. Wallis 2005, 28; Alter 2005, 27; Sprackland 2006, 33; Mooney 2005).

The most dramatic development in portrayals of pseudoscience after midcentury was the emergence of what we might call "Pseudoscience" with a capital "P" and without a hyphen. This reflected a growing sense during the 1920s and 1930s that pseudoscientific beliefs were not simply scattered errors to be exorcized from the boundaries of science, but rather a complex system of notions with their own set of boundaries, rather like an "alternative" version of science. The loss of the hyphen was a subtle indication of this transition, insofar as it weakened a seeming dependence on the scientific and suggested something more than simply false science. From the late 1960s on, many skeptical scientists and popularizers explicitly depicted links among a large collection of unusual topics, including "everything from PK (psychokinesis, moving things by will power) and astral projection (mental journeys to remote celestial bodies) to extraterrestrial space vehicles manned by web-footed crews, pyramid power, dowsing, astrology, the Bermuda triangle, psychic plants, exorcism and so on and so on." In the midst of public debate over Intelligent Design, journalist John Derbyshire blasted the "teaching of pseudoscience in science classes" and asked "why not teach the little ones astrology? Lysenkoism? . . . Forteanism? Velikovskianism? . . . Secrets of the Great Pyramid? ESP and psychokinesis? Atlantis and Lemuria? The hollow-earth theory?" Though lists differed, they often revolved around a similar core of unorthodoxies, including what one author characterized in 1998 as the "archetypical fringe theory," namely, belief in UFOs (Pfeiffer 1977, 38; Derbyshire 2005; Dutch 2012).

Supporters of science also wrote encyclopedic condemnations, most notably Martin Gardner's (1952) pioneering *In the Name of Science*, subsequently published as *Fads & Fallacies in the Name of Science*, which gathered a variety of subjects under the general banner of pseudoscience. Indeed, universalized depictions of pseudoscience became a convenient and clearly articulated target for those dedicated to crusading against antiscience in all its forms. From the 1970s on, CSICOP's *Skeptical Enquirer* proved to be one of the most important locations in which pseudoscience was forged, elaborated on, and stridently denounced. In 1992, a like-minded organization in southern California, the Skeptics Society, began publishing a second major magazine devoted to "promoting science and critical thinking," *Skeptic*, published and edited by the historian of science Michael Shermer. To focus on medical matters, CSICOP in 1997 helped to launch *The Scientific Review of Alternative Medicine and Aberrant Medical Practices*, followed five years later by a sister journal, *The Scientific Review of Mental Health Practice: Objective Investigations of Controversial and Unorthodox Claims in Clinical Psychology, Psychiatry, and Social Work.*

But pseudoscience did not just provide a nicely packaged enemy; it also provided an object for more neutral study. As early as 1953, the History of Science Society proposed adding a section on "Pseudo-Sciences and Paradoxes (including natural magic, witchcraft, divination, alchemy and astrology)" to the annual *Isis* critical bibliography. Later, history and philosophy of science

(HPS) departments featured the study of pseudoscience. "Another important function of HPS is to differentiate between science and pseudo-science," announced the University of Melbourne. "If HPS is critical of the sciences, it is even more so when dealing with pseudo-sciences and the claims they put forth to defend themselves." In the late 1970s, scholarly studies of pseudoscience—by scientists as well as historians, philosophers, and sociologists of science—began appearing in increasing numbers ("Proposed System of Classification" 1953, 229–31; R. Wallis 1979; Hanen, Osler, and Weyant 1980; Radner and Radner 1982; Collins and Pinch 1982; Leahey and Leahey 1983; R. Laudan 1983; Ben-Yehuda 1985; Hines 1988; Aaseng 1994; Zimmerman 1995; Friedlander 1995; Gross, Levitt, and Lewis 1966; Bauer 2001; Park 2000; Mauskopf 1990). Sociologists of science associated with the "strong programme" in the sociology of knowledge at the University of Edinburgh were especially influential in encouraging this development. Among the cardinal tenets of this initiative was impartiality "with respect to truth and falsity" of scientific claims; in other words, the same type of explanations would be applied to "true and false beliefs" alike. Thus encouraged, reputable historians of science devoted entire books to such topics as phrenology, mesmerism, parapsychology, and creationism. By the early twenty-first century, Michael Shermer was able to bring out a two-volume encyclopedia of pseudoscience (Bloor 1991; Mauskopf and McVaugh 1980; Cooter 1984; Numbers 2006; Winter 1998; Gordin 2012; Shermer 2002; L. Laudan 1981).

Despite attempts to situate it in a less negative context, pseudoscience almost always remained a term of denunciation. Still, it did capture something real. People interested in unusual topics had begun to link them together. An examination of one extensive bibliography suggests a growing tendency, particularly during the 1960s and 1970s, to combine multiple unorthodoxies into a single volume. This practice had its roots in the 1920s, particularly in the work of the former journalist and failed novelist Charles Fort. In his *Book of the Damned* and in several subsequent volumes, Fort catalogued stories about a wide range of unusual phenomena, including strange aerial objects; rains of frogs, fish, and other unusual things; psychic events; accounts of spontaneous human combustion; and other phenomena he claimed had been ignored or "damned" by orthodox science (Leith 1986; Fort 1919). His efforts ultimately inspired the formation of the Fortean Society in 1932, as well as the publication of a number of self-described Fortean magazines that continued the compilation of strange phenomena, and a range of aspiring ufologists and parapsychologists. Fort's ideas also bled into popular culture. Many of the topics covered by Fort appeared, sometimes in nearly identical terms, in episodes of the *X-Files* during its nine-year run on television (1993–2002). Devotees of the unusual have typically avoided the term "pseudoscience" in favor of "alternative," "forbidden," or "weird" science. They have also emphasized what one observer has labeled a "kinder, gentler science," more accessible than mainstream science (A. Ross 1991, 15–74). In recent decades, critics of alternative science have created their own synonyms for pseudoscience, including "antiscience," "cargo-cult science," and "junk science" (Feynman 1997, 338–46; Holton 1993; Huber 1991). But all such rhetoric, along with the grandparent of them all—"pseudoscience"—remains closely connected to the preservation of scientific boundaries and the protection of scientific orthodoxy.

Acknowledgments

This is a somewhat abridged version of Daniel Patrick Thurs and Ronald L. Numbers, "Science, Pseudo-science, and Science Falsely So-called," in *Wrestling with Nature: From Omens to Science*, ed. Peter Harrison, Ronald L. Numbers, and Michael H. Shank (Chicago: University of Chicago Press, 2011), 281–305. We thank Katie M. Robinson for her assistance in preparing this version for publication.

REFERENCES

Aaseng, Nathan. 1994. *Science versus Pseudoscience*. New York: Franklin Watts.

Abelson, Philip H. 1974. "Pseudoscience." *Science* 184:1233.

Alter, Jonathan. 2005. "Monkey See, Monkey Do: Offering ID as an Alternative to Evolution Is a Cruel Joke." *Newsweek*, August 15.

American Periodical Series Online. 2012. "Key Facts." Accessed October 6, 2012. http://www.proquest.com/products_pq/descriptions/aps.shtml.

Aptheker, Herbert, ed. 1973. *The Correspondence of W. E. B. DuBois*. Vol. 1. Amherst: University of Massachusetts Press.

Argyll, Duke of. 1887a. "Professor Huxley on Cannon Liddon." *Nineteenth Century* 21:321–39.

———. 1887b. "Science Falsely So Called." *Nineteenth Century* 21:771–74.

Bauer, Henry H. 2001. *Science or Pseudoscience: Magnetic Healing, Psychic Phenomena, and Other Heterodoxies*. Urbana: University of Illinois Press.

Ben-Yehuda, Nachman. 1985. *Deviance and Moral Boundaries: Witchcraft, the Occult, Science Fiction, Deviant Sciences, and Scientists*. Chicago: University of Chicago Press.

Bloor, David. 1991. *Knowledge and Social Imagery*. 2nd ed. Chicago: University of Chicago Press. (Originally published 1976.)

Brinton, Daniel G. 1895. "The Character and Aims of Scientific Investigation." *Science* 1:3–4.

Brown, William Leon. 1911. *Christian Science Falsely So Called*. New York: Fleming H. Revell.

Carr, E. F. 1873. "A Theory, an Extravaganza." *Ladies' Repository* 12:125.

"Character in Finger Nails." 1884. *New York Times*, December 5.

Collingwood, R. G. 1940. *An Essay on Metaphysics*. Oxford: Clarendon Press.

Collins, Harry, and Trevor Pinch. 1982. *Frames of Meaning: The Social Construction of Extraordinary Science*. London: Routledge.

Committee for Skeptical Inquiry. 2012. "International Skeptical Organizations." Accessed October 6, 2012. http://www.csicop.org/resources/international_organizations.

Comprehensive Index to the Writings of Ellen G. White. 1963. Mountain View, CA: Pacific Press.

Condon, Edward. 1969. "UFOs I Have Loved and Lost." *Bulletin of the Atomic Scientists* 25 (December): 6–8.

Cooter, Roger. 1984. *The Cultural Meanings of Popular Science: Phrenology and the Organization of Consent in Nineteenth-Century Britain*. Cambridge: Cambridge University Press.

"Darwin on Expressions." 1873. *Littell's Living Age*, n.s., 2:561.

Derbyshire, John. 2005. "Teaching Science." *National Review*, August 30. http://old.national review.com/derbyshire/derbyshire-archive.asp.

Dewey, John. 1927. *The Public and Its Problems*. New York: Henry Holt.

Dutch, Steven. 2012. "The Great Season: 1965–1981." Accessed October 6, 2012. http://www.uwgb.edu/dutchs.

Fernow, B. E. 1896. "Pseudo-science in Meteorology." *Science* 3:706–8.

Feynman, Richard P. 1997. *"Surely You're Joking Mr. Feynman!" Adventures of a Curious Character*. New York: W. W. Norton. (Originally published 1985.)

Fort, Charles. 1919. *The Book of the Damned*. New York: Boni & Liveright.

Frandsen, Peter. 1926. "Anti-scientific Propaganda." *California and Western Medicine* 25:336–38.

Friedlander, Michael W. 1995. *At the Fringes of Science*. Boulder, CO: Westview Press.

Gardner, Martin. 1952. *In the Name of Science*. New York: G. P. Putnam's Sons.

Garwood, Christine. 2007. *Flat Earth: The History of an Infamous Idea*. London: Macmillan.

Gordin, Michael D. 2012. *The Pseudoscience Wars: Immanuel Velikovsky and the Birth of the Modern Fringe*. Chicago: University of Chicago Press.

Gross, Paul R., Normal Levitt, and Martine W. Lewis, eds. 1996. *The Flight from Science and Reason*. New York: New York Academy of Sciences.

Hanen, Marsha P., Margaret J. Osler, and Robert G. Weyant, eds. 1980. *Science, Pseudo-science and Society*. Waterloo, ON: Wilfrid Laurier University Press.

Hearnshaw, L. S. 1942. "A Reply to Professor Collingwood's Attack on Psychology." *Mind* 51:160–69.

Henry, Joseph. 1844. Joseph Henry to Alexander Dallas Bache, April 16. In *The Papers of Joseph Henry,* edited by Marc Rothenberg et al., 6:76–77. Washington, DC: Smithsonian Institution Press, 1992.

Hewitt, Augustine F. 1887. "Scriptural Questions." *Catholic World* 44:660–61.

Hines, Terence. 1988. *Pseudoscience and the Paranormal*. Amherst, NY: Prometheus Books.

Holmes, Oliver Wendell. 1842. *Homoeopathy, and Its Kindred Delusions*. Boston: William D. Ticknor.

———. 1859. "The Professor at the Breakfast-Table." *Atlantic Monthly* 3:232–42.

Holton, Gerald. 1993. *Science and Anti-science*. Cambridge, MA: Harvard University Press.

Hornberger, Theodore. 1935. "Samuel Johnson of Yale and King's College: A Note on the Relation of Science and Religion in Provincial America." *New England Quarterly* 8:378–97.

Huber, P. W. 1991. *Galileo's Revenge: Junk Science in the Courtroom*. New York: Basic Books.

Huxley, Thomas Henry. 1887a. "Science and Pseudo-science." *Nineteenth Century* 21:481–98.

———. 1887b. "Science and Pseudo-science." Reprinted. *Popular Science Monthly* 31:207–24.

———. 1887c. "Science and Pseudo-science." Reprinted. *Eclectic Magazine,* n.s., 45:721–31.

———. 1887d. "Scientific and Pseudo-scientific Realism." *Nineteenth Century* 21:191–204.

———. 1887e. "Scientific and Pseudo-scientific Realism." Reprinted. *Popular Science Monthly* 30:789–803.

International Committees. 1984. *Skeptical Inquirer* 9:97.

Jayaraman, J. 2001. "A Judicial Blow." *Frontline* 18 (12). http://www.frontlineonnet.com/fl1812/18120970.htm.

Kane, John K. 1843. John K. Kane to Joseph Henry, November 20. In *The Papers of Joseph Henry*, edited by Nathan Reingold et al., 5:451. Washington, DC: Smithsonian Institution Press, 1992.

Krogman, Wilton Marion. 1947. "Race Prejudice, Pseudo-science and the Negro." *Phylon* 8:14–16.

Kurtz, Paul. 2001. "A Quarter Century of Skeptical Inquiry: My Personal Involvement." *Skeptical Inquirer* 25:42–47.

Lakatos, Imre. 1999. "Lecture One: The Demarcation Problem." In *For and Against Method*, edited by Matteo Motterlini, 20–31. Chicago: University of Chicago Press.

Larousse, Pierre. 1866–79. *Grand dictionnaire universel de XIXe siècle*. Paris: Administration du Grand dictionnaire universel.

Laudan, Larry. 1983. "The Demise of the Demarcation Problem." In *Physics, Philosophy, and Psychoanalysis: Essays in Honor of Adolf Grünbaum*, edited by R. S. Cohen and Larry Laudan, 111–27. Dordrecht: D. Reidel.

———. 1981. "The Pseudo-science of Science?" *Philosophy of the Social Sciences* 11:173–98.

Laudan, Rachel, ed. 1983. *The Demarcation between Science and Pseudo-science*. Blacksburg, VA: Center for the Study of Science in Society, Virginia Polytechnic Institute and State University.

Leahey, Thomas, and Grace Leahey. 1983. *Psychology's Occult Doubles: Psychology and the Problem of Pseudoscience*. Chicago: Nelson-Hall.

LeConte, Joseph. 1885. "Rough Notes of a Yosemite Camping Trip—III." *Overland Monthly*, 2nd ser., 6:636.

Leith, T. Harry, ed. 1986. *The Contrasts and Similarities among Science, Pseudoscience, the Occult, and Religion*. 4th ed. Toronto: York University, Department of Natural Sciences, Atkinson College.

Littré, Émile, ed. 1873–74. *Dictionnaire de la langue française*. Paris: Libraire Hachette.

Livingstone, David N. 1992. "Darwinism and Calvinism: The Belfast-Princeton Connection." *Isis* 83:408–28.

Lokensgard, Hjalmar O. 1940. "Oliver Wendell Holmes's 'Phrenological Character.'" *New England Quarterly* 13:711–18.

Macallum, A. B. 1916. "Scientific Truth and the Scientific Spirit." *Science* 43:439–47.

Macaulay, Rose. 1928. *Keeping Up Appearances*. London: W. Collins Sons.

Malinowski, Bronislaw. 1954. *Magic, Science and Religion*. Garden City, NY: Doubleday.

Mallock, W. H. 1881a. "Civilization and Equality." *Contemporary Review* 40:657–72.

————. 1881b. "Civilization and Equality." Reprinted. *Appletons' Journal*, n.s., 11:526–38.

Mauskopf, Seymour H. 1990. "Marginal Science." In *Companion to the History of Modern Science*, edited by R. C. Olby et al., 869–85. London: Routledge.

Mauskopf, Seymour H., and Michael R. McVaugh. 1980. *The Elusive Science: Origins of Experimental Psychical Research*. Baltimore: Johns Hopkins University Press.

Medical Repository of Original Essays and Intelligence. 1824. Untitled review. n.s., 8:444.

Mencken, H. L. 1932. "Nonsense as Science." *American Mercury* 27:509–10.

Miles, Samuel A. 1951. Letter to the editor. *Science* 114:554.

Mooney, Chris. 2005. *The Republican War on Science*. New York: Basic Books.

Nichols, C. F. 1892. "Divine Healing." *Science* 19:43–44.

Numbers, Ronald L. 1975. "Science Falsely So-Called: Evolution and Adventists in the Nineteenth Century." *Journal of the American Scientific Affiliation* 27 (March): 18–23.

————. 2001. "Pseudoscience and Quackery." In *The Oxford Companion to United States History*, edited by Paul S. Boyer, 630–31. New York: Oxford University Press.

————. 2006. *The Creationists: From Scientific Creationism to Intelligent Design*. Expanded ed. Cambridge, MA: Harvard University Press.

Park, Robert L. 2000. *Voodoo Science: The Road from Foolishness to Fraud*. New York: Oxford University Press.

Pfeiffer, John. 1977. "Scientists Combine to Combat Pseudoscience." *Psychology Today* 11:38.

Popper, Karl. 1959. *The Logic of Scientific Discovery*. New York: Basic Books. (Originally Published in 1934 as *Logik der forschung*.)

————. 1963. *Conjectures and Refutations: The Growth of Scientific Knowledge*. London: Routledge and Kegan Paul.

Presbyterian Quarterly and Princeton Review. 1873. Review of *Inductive Inquiries in Philosophy, Ethics, and Ethnology* by A. H. Dana. 2:561.

"Proposed System of Classification for *Isis* Critical Bibliography." 1953. *Isis* 44:229–31.

"Pseudo-science." 1952. *National and English Review* 139.

Radner, Daisie, and Michael Radner. 1982. *Science and Unreason*. Belmont, CA: Wadsworth.

Ramachandran, R. 2001. "Degrees of Pseudo-science." *Frontline* 18 (7). http://www.frontline onnet.com/fl1807/18070990.htm.

Raymond, Henry J. 1852. "Editor's Table." *Harper's New Monthly Magazine* 4:839–43.

Reese, David Meredith. 1838. *Humbugs of New-York*. New York: Weeks, Jordan.

Ross, Andrew. 1991. *Strange Weather*. New York: Verso.

Ross, Edward A. 1896. "Social Control." *American Journal of Sociology* 1:513–35.

Ruse, Michael. 1988. "Pro Judice." In *But Is It Science? The Philosophical Question in the Creation/Evolution Controversy*, edited by Michael Ruse, 357. Amherst, NY: Prometheus Books.

Sagan, Carl. 1972. Introduction. In *UFO's—A Scientific Debate*, edited by Carl Sagan and Thornton Page, xi–xvi. Ithaca, NY: Cornell University Press.

————. 1975. "The Recognition of Extraterrestrial Intelligence." *Proceedings of the Royal Society of London*, ser. B, 189:143–53.

————. 1987. "The Burden of Skepticism." *Skeptical Inquirer* 12:46.

Schoepflin, Rennie B. 2003. "Separating 'True' Scientists from 'Pseudo' Scientists." In *Christian Science on Trial: Religious Healing in America*, 82–109. Baltimore: Johns Hopkins University Press.

Science. 1886. Review of *Mathematical Economics* by Wilhelm Launhardt. 8:309–11.

"Scientific Agriculture." 1856. *Country Gentleman* 7:93.

"Separation of the Tares and the Wheat." 1889. *Christian Science Journal* 6:546.

Shermer, Michael, ed. 2002. *The* Skeptic *Encyclopedia of Pseudoscience*. Santa Barbara, CA: ABC Clio.

"Show-and-Tell Time Exposes Pseudo-science." 2000. *People's Daily Online*, March 26. http://english.peopledaily.com.cn/english/200003/26/eng20000326N120.html.

"Sir William Hamilton on Phrenology." 1860. *American Journal of Insanity* 16:249.

Slotten, Hugh Richard. 1994. *Patronage, Practice, and the Culture of American Science: Alexander Dallas Bache and the U.S. Coast Survey*. Cambridge: Cambridge University Press.

Small, Albion W. 1900. "The Scope of Sociology." *American Journal of Sociology* 6:42–66.

———. 1904. "The Subject-Matter of Sociology." *American Journal of Sociology* 10:281–98.

"Socialism and Communism in 'The Independent.'" 1879. *Catholic World* 28:808–17.

Sprackland, Robert George. 2006. "A Scientist Tells Why 'Intelligent Design' Is NOT Science." *Educational Digest* 71:29–34.

Stallo, J. B. 1882. "Speculative Science." *Popular Science Monthly* 21:145–64.

"The State of Belief in the Paranormal Worldwide." 1984. *Skeptical Inquirer* 8:224–38.

Sternberg, George M. 1897. "Science and Pseudo-science in Medicine." *Science* 5:199–206.

Tait, Peter Guthrie. 1911. "Religion and Science." In *Life and Scientific Work of Peter Guthrie Tait*, edited by Cargill Gilston Knott, 293. Cambridge: Cambridge University Press.

Thurs, Daniel Patrick. 2004. "Phrenology: A Science for Everyone." In *Science Talk: Changing Notions of Science in American Culture*, 22–52. New Brunswick, NJ: Rutgers University Press.

"Time Wasted." 1897. *Science* 6:969–73.

Tupper, Frederick, Jr. 1910. "Textual Criticisms as Pseudo-science." *PMLA* 25:164–81.

Ure, Andrew. 1853. *A Dictionary of Arts, Manufactures, and Mines*. New York: Appleton.

Wallis, Claudia. 2005. "The Evolution Wars." *Time*, August 15, 28.

Wallis, Roy, ed. 1979. *On the Margins of Science: The Social Construction of Rejected Knowledge*. Keele: University of Keele.

White, Ellen G. 1888. *The Great Controversy between Christ and Satan*. Mountain View, CA: Pacific Press.

White, Leslie A. 1947. "Evolutionism in Cultural Anthropology: A Rejoinder." *American Anthropologist* 49:400–13.

Winter, Alison. 1998. *Mesmerized: Powers of Mind in Victorian Britain*. Chicago: University of Chicago Press.

Woodward, R. S. 1900. "Address of the President." *Science* 12:12–15.

Zimmerman, Michael. 1995. *Science, Nonscience, and Nonsense: Approaching Environmental Liberacy*. Baltimore: Johns Hopkins University Press.

8

Paranormalism and Pseudoscience as Deviance

ERICH GOODE

Many nineteenth-century intellectuals, philosophers, and social scientists, such as Herbert Spencer, August Comte, and Karl Marx, adopted a rationalistic view of human behavior: they argued that an increase in society's level of education and the dissemination of the scientific method and scientific knowledge would result in the disappearance of what they regarded as mysticism, occult beliefs, pseudoscience, and other superstitious nonsense—and in their eyes, that included religious dogma. They would have been baffled to witness the persistence and vigor—even more so the resurgence—of late twentieth- and early twenty-first-century belief in supernatural claims.

The polls conducted by public opinion organizations indicate an increase in extrascientific beliefs among the American public during the past two or three decades. The most recent Pew Forum on Religious and Public Life survey results on whether respondents had ever been "in touch with the dead" increased from 17 percent in 1990 to 29 percent in 2009, and those saying that they had ever seen or been in touch with a ghost doubled from 1990 (9 percent) to 2009 (18 percent). Two-thirds (65 percent) of the respondents in the 2009 poll said that they had had at least one of eight supernatural experiences, such as holding a belief in reincarnation, in astrology, or in an actual ghostly encounter. Other polling agencies, such as Harris, Gallup, and Zogby, turn up similar findings. In its 2010 report, *Science and Engineering Indicators*, the National Science Board, a subdivision of the National Science Foundation, summarized such polls, concluding that three-quarters of the American public hold at least one pseudoscientific belief, in such phenomena as ESP, haunted houses, ghosts, telepathy, clairvoyance, astrology, and communication with the dead. Clearly, the belief that nonscientific, supernatural, and paranormal powers and forces are genuine has grown substantially since the late twentieth century.

"Pseudoscience" is a derogatory term skeptics use to refer to a cluster or system of beliefs whose adherents, scientists argue, mistakenly claim is based on natural laws and scientific principles; adherents of such belief systems cloak their views in the mantle of science. Paranormalism invokes supernatural powers—those that scientists believe are contrary to or contradict the laws of nature. The difference between the two is that, according to scientists and philosophers of science, the proponents of pseudoscience masquerade their beliefs and practices as if they were science, while adherents of paranormalism may or may not concern themselves with the integration of their claims into conventional science; a major proportion of them "just know" that extra-scientific or occult forces act on the material and psychological realm, regardless of what scientists say is impossible. Pseudoscientists are strongly oriented to the scientific establishment and attempt to debunk, overturn, or incorporate their claims into traditional science; in contrast, many paranormalists do not much care what traditional science argues. The two heavily overlap, however: most pseudosciences invoke paranormal forces, and many paranormal thought systems dovetail into schemes of pseudoscientific thought. Sociologically, in delineating science from nonscience, it is important to make two fundamental qualifications. One is that, to the contemporary uninformed lay observer, much cutting-edge science resembles paranormal thought; it seems exotic, far-out, distantly removed from the immediately verifiable world of the empirical here and now. To the layperson, phenomena such as worm holes, black holes, superstrings, alternate dimensions, time and space relativity, quarks, and mesons represent inaccessible, inconceivable, almost fanciful phenomena, although scientists say they have empirical methods by which to determine their existence and influence. To use a historical example, when Alfred Wegener devised his theory of continental drift, he was unaware of *how* continents "drifted," and many of his geological peers ridiculed the theory and the theorist. Only later did scientists accumulate evidence on plate tectonics that explained the mechanism of continental drift (Ben-Yehuda 1985, 124). In 1859, when Charles Darwin published *On the Origin of Species*, he was completely unaware of the science of genetics—genetic variation being the engine of natural selection and hence evolution—which emerged only decades afterward, further substantiating Darwin's theory. In any case, some speculation is necessary for scientific advances, and the community of scientists eventually hit on ways to verify or falsify novel ideas and either incorporate their implications into a recognized body of knowledge or dismiss them as fatally flawed. The same cannot be said for paranormal or pseudoscientific notions.

Another qualification in drawing distinctions between science and pseudoscience, and science and paranormalism, is that the sciences are based on *established, contemporary* knowledge about how nature works. Some systems of thought currently regarded as quack theories and pseudoscience were once accepted by some, many, even most, respected scientists in past centuries—for instance, the theory of criminogenic "atavisms" promulgated by Cesare Lombroso; the phrenology of Franz Joseph Gall, which argued that bumps on the head, indicating the uneven development of brain matter, determined character and human behavior; not to mention alchemy (turning lead into gold) and astrology, which early on attracted then-respectable investigators' support. Indeed, the very notion of *being a scientist* is determined by contemporary criteria since scientific methodology and programs of scientific study were not established until two or three centuries ago. Hence, much of our notion of what is pseudoscience and paranormalism is based on hindsight. Today, we consider such investigations pseudoscience, but since what we refer to as science did not exist, the premises of many of them were accepted as valid by the most educated sectors of society at the relevant time.

Five Types of Pseudoscientific Belief Systems

Sociologically, we look at extrascientific beliefs by focusing on how they are generated and sustained. The routes through which these processes take place are many and varied. Perhaps five are most likely to be interesting to the skeptical observer; for each, we should ask the basic question: "Who is the believer's social constituency?" And for each, the answer is significantly different.

First are beliefs that depend on a *client-practitioner relationship*: they are validated by paid professionals who possess presumed expertise that is sought by laypersons in need of personal assistance, guidance, an occult interpretation of reality of their lives, or a demonstration of paranormal proficiency.

Astrologers and other psychics exemplify this type of paranormalism. The social constituency of the astrologer and the psychic is primarily the client and, secondarily, other astrologers and psychics.

Second are paranormal belief systems that begin within a religious tradition, and are sustained by a religious institution, that existed long before there was such a thing as the contemporary version of a scientist. Such beliefs sustain, and continue to be sustained by, one or more identifiable religious organizations. Creationism is a prime example here. In 1947, the US Congress passed the Fourteenth Amendment, which protected individual liberties from state action, invoking the First Amendment barring the establishment of religion. The legislation led to a series of decisions barring classroom religious instruction, school-sponsored prayer, mandatory Bible reading, and anti-evolution laws (Larson 2007, 22–23). In 1961, in response to these challenges, Virginia Tech engineering professor Henry Morris published *The Genesis Flood*, which "gave believers scientific-sounding arguments supporting the biblical account of a six-day creation within the past ten thousand years" (Larson 2007, 23). The book launched the creation science or scientific creationism movement, which seeks a "balanced treatment" between evolution and creationism in the public school science curriculum. In 1987, in *Edwards v. Aguilard*, the Supreme Court ruled that creation science "was nothing but religion dressed up as science" (Larson 2007, 24), thereby mobilizing supporters of creationism to generate its offspring—Intelligent Design. The vast majority of scientists believe that creation science is a pseudoscience; a 2009 Pew Research Center poll of accredited scientists found that 97 percent believed in evolution, of which 87 percent said that they believe in *nontheistic* evolution, that is, that evolution took place and God had nothing to do with it. Scientific creationists do not so much propose a scientific, empirically founded explanation of the origin of species as try to poke holes in evolutionary science. Creation science and its progeny, Intelligent Design, are classic examples of pseudoscience.

Third is a form of pseudoscience that is kept alive by a core of researchers who practice what seems to be the *form* but not the *content* of science. Many of its adherents are trained as scientists, conduct experiments, publish their findings in professional, science-like journals, and maintain something of a scientific community of believers, but most conventional scientists reject their conclusions. As we will see momentarily, parapsychology offers the best example here. Unlike astrologers and psychics, parapsychologists do not have clients and are not hired for a fee; they are independent researchers and theorists. While a substantial number of laypersons may share the beliefs parapsychologists claim to validate in the laboratory, these protoscientists or pseudoscientists form the sociological core of this system of thinking. For the parapsychological researcher, the social constituency is that tiny band of other professional parapsychologists and, ultimately, they hope, the mainstream scientific community. We see the professionalism of parapsychology in the fact that in 1957, Dr. J. B. Rhine, a psychologist and researcher who first systematically investigated "psi" (paranormal or psychic powers) formed the Parapsychology Association, the field's first professional organization. To be admitted, a candidate must have conducted professionally recognized scientific research on parapsychology and hold a doctorate in a credentialed field at an accredited university. Today, the association enrolls more than one hundred full members and about seventy-five associate members—tiny by the standards of most academic disciplines, but productive with respect to output. These are several indications of a professionalized field, as well as the fact that parapsychologists comprise a protoscientific (or pseudoscientific) *community*. What mitigates *against* their scientific status, however, is that mainstream scientists do not accept the validity of their research or conclusions, their practitioners do not publish in recognized scientific journals, and their work is not supported by reputable funding agencies.

Fourth are paranormal belief systems that can be characterized as grassroots in nature. They are sustained not so much by individual theorists, a religious tradition or organization, a client-practitioner relationship, or a core of researchers, as by the grass roots—the broad-based public. In spite of the fact that it is strongly influenced by media reports and the fact that there are numerous UFO organizations and journals, the belief that unidentified flying objects (UFOs) are "something real" has owed its existence largely to a more-or-less spontaneous feeling among the population at large. The ufologist's constituency is primarily other ufologists and, secondarily, the society as a whole.

And last are paranormal beliefs that originate from the mind of a social isolate, a single person with an unusual, implausible, scientifically unworkable vision of how nature works. The isolate's message is presumably directed mainly at scientists, although any connection with the scientific community is tenuous or nonexistent. Scientists refer to these people as "cranks." Here, the social constituency of the crank usually does not extend beyond himself (nearly all cranks are men). Cranks do not really address their message to the scientific community since they do not engage in science-like activities or associate with other scientists; their goal is to *overturn* or *annihilate* conventional science, not contribute to it. The crank usually advances theories that are completely implausible to most scientists, or irrelevant or contrary to the way the world operates, or impervious to empirical test. Cranks tend to work in almost total isolation from orthodox scientists. They have few, if any, fruitful contacts with genuine researchers and are unaware of, or choose to ignore, the traditional canons of science such as falsifiability and reproducibility. They tend not to send their work to the recognized journals; if they do, it is rejected for what scientists regard as obvious, fundamental flaws. They tend not to be members of scientific academies, organizations, or societies. And they tend not to receive grants or fellowships or awards from scientific organizations. In short, they are not members of the scientific *community*. Cranks also have a tendency toward *paranoia*, usually accompanied by *delusions of grandeur* (Gardner 1957, 12–14): they believe that they are unjustly persecuted and discriminated against because they are geniuses, because their ideas are so important and revolutionary they would threaten the scientific establishment. Cranks argue that practitioners of entire fields are ignorant blockheads who are blinded by pig-headed stubbornness and stupidity. Only they themselves, the true visionaries, are able to see the light. Consequently, they must continue to fight to expose the truth.

Were proponents of novel theories that were eventually validated and accepted regarded as cranks? Did any physicist in 1905 regard Albert Einstein as a crank for proposing his theory of relativity? One physicist says no, that Einstein's contemporaries did not and would not have branded him a crank (Bernstein 1978). To begin with, Einstein published his ideas in a recognized journal of physics. Second, his theory of relativity passed the test of the "correspondence" principle, that is, it proposed exactly *how* it corresponded with, fit into, or extended existing and established theory. In other words, Einstein's theory was very clear on just where Newton's principles ended and where his own theory began. In contrast, crank theories "usually start and end in midair. They do not connect in any way with things that are known" (12). Third, most crank theories "aren't even wrong." Says physicist Jeremy Bernstein, "I have never yet seen a crank physics theory that offered a novel quantitative prediction that could be either verified or falsified." Instead, they are "awash in a garble of verbiage . . . , all festooned like Christmas decorations." Einstein's paper was very clear about its predictions; it virtually cried out to be empirically tested (13). To support his case, a crank may claim, "they laughed at Galileo, too!" But the fact is, Galileo was a member of the intelligentsia or scientific community, such as it was, in the 1600s, tested falsifiable hypotheses, and no one except the Catholic Church "laughed" at—or, more accurately, punished—him.

In a way, then, cranks want it both ways; cranks have a love-hate relationship with established science. On the one hand, they do not play by the rules of conventional science. But on the other, they are sufficiently removed from social contact with those who set those rules that they are either unaware of what those rules are or are deluded into thinking that such rules are mere technicalities that can be swept away by the tidal wave of truth; they alone possess that truth. They want to *annihilate* the prevailing theories of established science. Scientists are wrong, I am right, they are ignorant, I am well-informed, seems to be the prevailing position of the crank. The hubris or *arrogance* of the person possessed of superior wisdom and knowledge seems to suffuse the crank's self-presentation. But on the other hand, the crank also lusts to be *accepted* by the scientific fraternity. Otherwise, why do they send established scientists their writings? Cranks deeply and sincerely believe that, through the presentation of their evidence and the sheer power of their argument, they will convince the scientific powers that be that they are right.

Case Study 1: Astrology

Astrology is a system of thought that argues that the position, movement, and size of heavenly bodies—principally the sun, moon, and solar system planets—influence the personality, behavior, and destiny of humans on Earth. The origins of astrology can be traced back to the religions of Babylonia and other ancient civilizations some five thousand years ago. These civilizations sought a means of predicting weather and seasonal variations and what they meant for the lives of their populaces and their rulers' reins. Astrology remained a respectable academic and intellectual discipline, with ties to astronomy, meteorology, and medicine, until the Scientific Revolution, which began in 1543 with the publication of two revolutionary empirical treatises, one on astronomy, by Nicolaus Copernicus, and the other on anatomy, by Andreas Vesalius. Subsequently, the curtain was drawn on astrology's scientific status. Still, Isaac Newton, one of the greatest mathematicians and physicists of all time, was a firm believer in the validity of astrology, though he managed to keep his belief in the validity of the scientific and astrological systems of thought separate.

In its report, *Science and Engineering Indicators*, the National Science Foundation (2006) stated that astrology is a pseudoscience. In the United

States, polls indicate that roughly a third of the public believes that astrology is "very" or "sort of" scientific, while two-thirds express the view that it is "not at all" scientific. This belief is significantly correlated with education, however: in a 2008 GSS survey conducted by the National Opinion Research Corporation (NORC), eight in ten (81 percent) respondents with a graduate education say that astrology is not at all scientific as compared with half (51 percent) with only a high school education. Respondents who scored *highest* on the factual knowledge measures were *most* likely to believe that astrology is "not at all" scientific (78 percent), while those who scored at the *bottom* of the knowledge scale were the *least* likely to believe this (45 percent). The National Science Board, under the auspices of the National Science Foundation (NSF), sponsored thirteen surveys between 1979 and 2008 that demonstrated an increase in the percentage of Americans who believe that astrology is at least "not at all" scientific—from 50 to 63 percent. (These findings and a discussion of them are in the National Science Board/NSF, *Science and Engineering Indicators: 2010.*) Astrology represents one of the few pseudoscientific belief systems that in recent years has manifested a decline among the general public. NSF concludes that, given its lack of an evidentiary basis and its invocation of powers unknown by scientists and outside the realm of the natural laws, astrology is a pseudoscience.

Pseudoscientists attempt to validate their scientific credentials; in fact, the interested observer can study astrology in much the same way that she can study economics, English literature, or chemistry: more than one program offer advanced degrees—though at fringe and nonaccredited rather than mainstream universities. The back cover of the book *Astrology for Enlightenment* announces that the author, Michelle Karén (2008), received a D. Astrol. S.—presumably, a doctorate in astrological sciences, or the equivalent of the PhD—from the Faculty of Astrological Studies in London. Moreover, the book copy states, Karén studied at the Manhattan Institute of Astrology. This indicates that the craft entails esoteric knowledge or wisdom not possessed by the ordinary layperson, and that the craft can be taught and learned in at least two institutions of higher learning; most scientists would label such a system of thought as a pseudoscience.

Like psychics, astrologers face a serious problem of credibility. To be believed, they must formulate their prognostications in such a way that they correspond with what their clients want to hear, find remarkable, and observe to take place in the material world. In other words, there is a "recipe" for a successful reading, and astrologers are successful to the extent that they follow that recipe. For clients, the technically scientific status of astrology is secondary to the fact that, to them, readings should "feel right," and feeling right is based on personal meaning, not merely empirical validity (Hyman 2007, 32). To put the matter another way, predictions and assessments must *seem* true more than be scientifically testable. Still, the client must feel that a reading appears to be verified in the material world, though if sufficiently vague, a wide range of assessments will be borne out.

There are two sets of factors that make an astrological reading or interpretation "feel right." The first set is related to what the practitioner *does* and the second is related to what the client or audience *expects* or *believes*. In reality, these two dovetail with one another. The practitioner gives the client what he expects.

Psychologist Ray Hyman (2007) examined the techniques of "cold" readings conducted by astrologers and psychics. A cold reading is one in which the practitioner conducts a psychic assessment of someone she has never met before. The astrologer tends to open with the "stock spiel," a statement of the standard character assessment that fits nearly everyone, but which clients feel applies to them uniquely. Many clients are astounded at the accuracy with which their special and unique characteristics have been described. "How could she know that?" they ask. "She must have psychic powers to know so much about me." In a cold reading, astrologers usually move on to assess clients under the assumption that people tend to be motivated by the same forces and factors that moves them to seek consultation. Birth, health and illness, death, love, marriage, loved ones, and money usually generate most of the problems for which we want an answer. In other words, there is a "common denominator" in all clients (Hyman 2007, 34); astrologists and psychics generally work on "variations on one or more themes." Astrologers also work with certain clues that are found in the client's appearance or remarks, providing evidence on which to base an assessment. Some of these clues include age, inferred social class, weight, posture, manner of speech, eye contact, and so on. In other words, the psychic sizes up the client with respect to the statistical or actuarial likelihood that certain problems or issues will be of concern to him. And last, through skillfully designed probing questions, astrologers formulate and test "tentative hypotheses" based on the client's answers. Reactions (eye movement, body language, tone of voice) will supply information to the practitioner that the client is unaware he is supplying. Clients walk away from a consultation without being aware that everything the reader told them was what they themselves said, in one way or another, to the reader (34–35).

Sociologically, testing the validity of paranormal claims is secondary. Social scientists are interested in the *enterprise* of paranormalism, and that enterprise makes extensive use of techniques and reasoning processes that resonate or correspond with what clients believe to be true. As such, they are sociological or cultural products. Their origin, validation, and dynamics can be understood through social forces such as socialization, interaction, stratification, hierarchy, deviance, conformity, and persuasion. Nonetheless, the resort of practitioners to devices that conjure up their supposed scientific credentials despite their threadbare empirical and ontological status is also revealing.

Case Study 2: Parapsychology

You receive a phone call from a friend about whom you were just thinking; simultaneously, you and your closest friend blurt out exactly the same sentence; someone predicts that an event will take place, and it does. "That must have been ESP!" you declare. Does such a power exist? Can two people communicate with one another without the use of words? Can we clearly picture objects miles away, in our mind, without devices of any kind? Is it possible to predict the future? Or "see" events that took place in the past that we did not witness and no one has told us about? Can we bend spoons with the power of our minds?

Although observers define parapsychology variously, Tart (2009, 89–97) discusses the "big five" of parapsychology as "telepathy," or mind-to-mind communication; "clairvoyance," the ability to "see" or perceive things beyond sensory reach, without the aid of technology or information; "precognition," seeing the future (along with "postcognition"—the ability to see the past); "psychokinesis" (PK), or the ability to move physical objects solely with one's mind; and "psychic healing." The terms other than "psychokinesis" are also frequently referred to as "extrasensory perception" (ESP) or, less commonly (and a bit confusingly), "clairvoyance." Psychokinesis is sometimes referred to as "telekinesis"; clairvoyance (again, confusingly) sometimes refers to seeing the future and the past. These parapsychological powers that supposedly make such things happen are widely known as "psi" (Tart, 2009, 89). The essence of psi is mind-to-matter and mind-to-mind influence or communication or perception (Hines 2003, 113–50; Tart 2009, chs. 7, 8, and 9).

A very high proportion of the public believes that ESP and other parapsychological powers exist. In 2005, the Gallup organization conducted a survey that asked respondents whether they believe in "extra-sensory perception, or ESP." Four in ten (41 percent) of the respondents said the power of ESP is real; in the United States alone, this adds up to more than 100 million adults. About a third of persons surveyed (31 percent) said that they believe in "telepathy" or the ability of some people to "communication between minds without using the traditional senses." For "clairvoyance, or the power of the mind to know the past and predict the future," the figure was slightly above a quarter of the sample (26 percent). A fifth (21 percent) said that they believe that "people can communicate with someone who has died," and in an earlier *Newsweek* poll, for "telekinesis, or the ability of the mind to move or bend objects using just mental energy," the figure was 17 percent. Interestingly, correlations between social characteristics and these beliefs seem to be weak or practically nonexistent. Worldwide, the number believing in parapsychological powers almost certainly adds up to approximately 3 or 4 billion people. A belief this widespread demands attention.

With respect to parapsychology, as we saw, the beliefs of the rank-and-file or grass roots is less relevant than the small social grouping whose members are engaged in conducting systematic research on this subject. There are a few hundred parapsychologists (professionals with PhDs) around the world who use the techniques of conventional science—that is to say, controlled experiments—to conduct research designed to test or verify the existence of psi. As we saw, comparatively few scientific creationists are professional scientists with PhDs in relevant fields, and very few of them conduct what scientists would refer to as research and publish their findings verifying the truth of the biblical account of creation in mainstream science journals. And nearly all ufologists are self-taught in their chosen field; virtually none has an advanced degree in any field related to ufology.

In contrast, parapsychology researchers conduct systematic, scientific investigation of the reality of psi, a particular type of paranormal power. While astrologers and psychics claim to possess psi themselves, parapsychologists study or examine psi in others. The research methods of parapsychologists are far more science-like than is true of the practitioners of any other area of paranormalism. These facts make this belief system interesting for a variety of reasons.

As we have seen, the man and woman on the street rely much more on anecdotes and personal experience in drawing their conclusions than on the results of systematic research. In fact, they are likely to regard the controlled experiments that parapsychologists conduct as overly technical and restrictive. In contrast, the professional parapsychological researcher argues that the controlled experiment is one of the most essential tools in establishing the validity of psi. Note that the parapsychologist does not *necessarily* believe that psi exists (although the vast majority do), but practically all believe that systematic research is the only means of testing its reality.

Some observers (Irwin 2004; Radin 1997; Tart 2009) argue that the research *methods* of parapsychologists are no less rigorous and "scientific" than those of conventional, mainstream psychologists. And, they say, if the research were to deal with a conventional subject, the *findings* of these studies would be convincing to most scientists, at least to most social scientists. But scientists find two problems with parapsychological research.

One is that parapsychologists offer no convincing conventional explanation for *why* their findings turn out the way they do. Moving objects with the mind? "How?" the conventional scientist asks. What's the *mechanism* by which someone can bend a spoon without touching it? *How* do subjects view faraway objects without the aid of instruments? *In what way* do minds "communicate"? What *causes* psi or parapsychological powers? The problem is, parapsychologists give no answers that satisfy conventional scientists—some of them include particles with "imaginary mass and energy" called *psitrons*, the force of quantum mechanics, "wave packets," synchronicity, as-yet unexplored energy fields, and so on—bound as scientists are to a materialistic cause-and-effect perspective. In principle, supporters of parapsychology claim, the same objection can be raised for some features of conventional or cutting-edge science. For instance, why do "superstrings" exist? No physicist has any idea, but most believe in them. Scientists remain unconvinced by such invocations, regarding them as little more than mumbo-jumbo—but they undergird the convictions of the committed parapsychologist.

Physicists make use of material forces such as velocity, mass, friction, gravity, and heat; biologists invoke molecules, cells, genetics, biochemistry, and anatomy; social psychologists and sociologists speak of socialization, peer influence, prestige, power, and social sanctions. These concepts, forces, or factors can be readily understood in a straightforward, naturalistic, cause-and-effect fashion. But poke too far into the structure of any natural and especially social science, and forces that are "first movers," and hence cannot be explained, begin to appear. Hence, paranormalists say, why should science be privileged over the explanations they advocate?

What are the parapsychologists' cause-and-effect explanations for psi? Even if their studies of empirical regularities demonstrating that *something* is going on were accepted, what *material* explanations for such effects do para-psychologists offer? As I stated, some resort to theories of electromagnetic forces (Irwin 2004), "energy field" explanations, the action of "elementary particles" (170), or quantum mechanics (Radin 1997, 277–78, 282–86). In a like manner, Rhonda Byrne, a spiritualist author, argues in *The Secret* and *The Power* that a godlike agent will provide anything we want if we truly believe it will. The causal instrument? Again, Byrne invokes physical mechanisms such as magnetism, quantum mechanics, and other forces of theoretical physics. Think of it and it's yours; it's the "law of attraction." Byrne's *The Secret* (2006) appeared on the hardback, advice bestseller list for more than three years, and *The Power* (2010) once stood at number one on that list—so clearly hers is a welcome, popular, and comforting argument. Again, scientists regard such entreaties as so much hocus-pocus. Some parapsychologists "treat psi as a negative 'wastebasket' category . . . , atheoretic anomalies in need of an explanation" (Truzzi, 1987, 5:6). But none has an explanation of how these forces generate or cause the effects their findings point to that is plausible to most scientific observers. In the words of Dean Radin (who holds a doctor-ate in educational psychology), "the only thing we can do is to demonstrate correlations. . . . *Something* is going on in the head that is affecting *something* in the world" (Radin 1997, 278; Brown 1996). To most scientists, this asser-tion is not sufficient until a convincing explanation is supplied, and the causal mechanisms, when they are spelled out, simply do not articulate with the ob-servations paranormalists supply; hence, they must be attributable to fraud or measurement error.

Traditional scientists have a second problem with granting a scientific sta-tus to parapsychology: its inability to *replicate* findings, or what Truzzi calls "psi on demand" (1982, 180). As we saw earlier, scientists take replication seri-ously. When a scientist produces a finding in an experiment or a study, if the principle on which that finding rests is valid, another scientist should be able to conduct the same research and come up with the same finding. Findings should be repeatable, experiment after experiment, study after study. (Repli-cation is taken more seriously in the natural sciences than the social, however, and more seriously in psychology than sociology.) If entirely different results are obtained in repeat experiments, something is wrong with either the experi-ment or the finding. Radin (1997, 33–50) argues that parapsychology does not display replication any less than traditional science. Yet, in addition, he claims that psi is elusive, subtle, and complex and that our understanding of it is in-complete. Hence, experiments demonstrating psi are difficult to replicate.

Conventional scientists are not likely to find Radin's argument convincing because parapsychology is an experimental field, and parapsychologists have often been unable to replicate the findings of their experiments. In some experiments, psi "effects" appear, while in other almost identical experiments, they do not. Psi seems fragile and elusive. The assumption that forces are consistent throughout the universe is the bedrock of science. To most scientists, the lack of a plausible explanation and the inability to replicate research findings are serious deficiencies in parapsychology that "will probably prevent full acceptance" of the field by the general scientific community (Truzzi 1982, 180). Many observers refer to the field as a "pseudoscience" (e.g., Hines 2003). When mainstream scientists say that the field of parapsychology is *not* scientific, they mean that no satisfying naturalistic cause-and-effect explanation for these supposed effects has yet been proposed and that the field's experiments cannot be consistently replicated. "Is there such a thing as mind over matter? Can energy or information be transferred across space and time by some mysterious process that on the face of it seems to confound the principles of biology and physics?" asks journalist Chip Brown, echoing the sentiment of Susan Blackmore, a PhD in psychology and a scientific critic of parapsychology. "Most scientists believe the answer is no—no, no, no, a thousand times no" (Brown 1996, 41).

Extrascientific Belief Systems as Deviance

When prestigious and influential social actors label people and beliefs as wrong, this has important consequences, most notably in the sphere of deviance. One major consequence of promulgating extrascientific beliefs is that, in spite of their popularity, they are not valorized or validated by society's major institutions; in fact, their adherents are likely to be chastised, censured, condemned. That heaven and earth were created by God ex nihilo ("out of nothing") in six days, six thousand years ago; that the position of the sun, moon, and stars at one's birth determine one's fate in life; that psi (psychic or spiritual power) exists and that its possessors can communicate with others, mind-to-mind, and/or can move inanimate objects; and that the proverbial "little green men" have landed on Earth are not dominant beliefs. Influential parties in influential institutions tend to label believers as "kooks," "weirdoes," or eccentrics, and are treated or reacted to accordingly. Evolution is the very foundation of modern biology; if creationist theory is correct, nearly every page of nearly every current biology textbook would have to be scrapped and rewritten. Belief in creationism is a statement about how nature operates; it is contrary to some of the first principles of science—that is, that stars, planets, and living beings can be created, fully formed, out of nothing. The same radical revision of contemporary science would be necessary if parapsychology, astrology, and the extraterrestrial origin of UFOs were valid. When psychiatrist John Mack published *Abduction*, a book asserting that his patients really were abducted by aliens (1994), some of his Harvard colleagues in psychiatry pressured the university administration to fire him. (The attempt failed.) Creationists are virtually absent among the university faculties of the country's most prestigious departments of geology and biology; the same can be said of parapsychologists and advocates of astrology among university faculty generally. Teachers who endorse creationism in their courses are told to keep their religious views out of the classroom. None of the most prestigious, influential museums in the country valorize the truth value of paranormal belief systems, which, their directors and advisory boards would say is close to zero; this is especially true of creationism. The nation's two most prestigious and influential newspapers—the *New York Times* and the *Washington Post*—marginalize and "deviantize" paranormalism, as do nearly all the top magazines. In the country as a whole, the majority of the electorate would question the fitness for office of candidates running for president or a seat in the senate who endorse one or another version of the beliefs discussed in this chapter. An economist basing her quarterly or long-range predictions on seers, psychics, or astrologers is likely to be mocked and joked about in the media and by most of the public. And, in spite of the fact that, almost by its very nature, theism contains a strain of paranormalism, most of the representatives and members of society's mainstream, ecumenical religious bodies do not accept literal biblical creationism as valid.

These regularities tell us that pseudoscientific and paranormal beliefs are deviant: espousing assertions that scientists regard as contrary to the laws of nature tends to provoke a negative reaction *among* certain parties or social actors, *in* certain social circles or institutions. These assertions are also almost certainly empirically wrong, but that is not the only or the main point; what the sociologist knows for certain is that they are unconventional and lacking in legitimacy—again, *to* those parties, *in* those circles.

Perhaps the most interesting feature of parapsychology is that it is an excellent example of a deviant science. No doubt from a traditional scientist's perspective, this makes it a pseudoscience, but to a sociologist, this means that the publications of its practitioners tend to be condemned or ignored by mainstream scientists. Dean Radin's question, "Why has mainstream science been so reluctant merely to admit the existence of psi?" (1997, 202) says it all for the supporter of parapsychology. Conventional scientists reject "extraordinary anomalies," and, given the laws of physics, chemistry, and biology, the findings of parapsychology represent anomalies; they cannot be incorporated into the existing theoretical framework. Hence, they must be debunked or neglected—not because they are anomalous, but because they contradict mechanisms scientists know to be verified and true.

McClenon (1984, 128–63) conducted a survey among the council members and selected section committee representatives of the American Association for the Advancement of Science (AAAS). His sample was made up of elite scientists in that they are in positions of leadership and hence can influence whether parapsychology is granted full scientific legitimacy. His final sample (N=339) includes social as well as natural scientists. Overall, a minority—29 percent—consider ESP "an established fact or a likely possibility" (138), actually a surprisingly high proportion. Moreover, disbelief in ESP is strongly correlated with denying legitimacy to the very subject of its investigation. Thus, "parapsychologists are labeled as deviant because scientists do not believe in the anomaly that they investigate" (145). Being a skeptic versus a believer is also related to reporting one or more personal paranormal experiences (150). Still, half as many of these scientists report having had an ESP experience (26 percent), a figure that is over half for the American population as a whole (58 percent).

McClenon (1984, 164–96) also conducted interviews with parapsychologists, attended their meetings, and read their journals. Before conducting his study, he hypothesized that the field of parapsychology was a kind of science-like cult whose members righteously defend their belief in psi and actively proselytize outsiders to their position. Contrary to his expectations, parapsychologists do not believe that proselytizing is necessary and feel that, eventually, because of the rigor of their research methods and the robustness of their findings, the "truth will be revealed" (165). In this respect, parapsychologists are similar to mainstream scientists. Still, he found, the vast majority of parapsychological research is excluded from the mainstream natural science and psychology journals. In fact, says one of McClenon's interviewees, the *best* work in the field "can't get published there. The editors reject it because it was conducted by people within the field of parapsychology. . . . The editors of most [mainstream] journals aren't that knowledgeable about parapsychology. They don't know what to look for in a piece of research" (167). According to the field's proponents, the best work is published in the specialty or parapsychology journals, thereby contributing "to the oblivion to which this body of information has been committed" (167–68). Mainstream scientists would argue that parapsychology's "best" is not good enough, by conventional science's perspective, to get published in these journals.

Parapsychological research almost never appears in mainstream science journals, and in spite of the scientific rigor the experimenters claim for its research methods, parapsychology "has no professor/graduate-student training like that which exists for the rest of science" (McClenon 1984, 171). Parapsychologists "are often discriminated against in academic circles and find it difficult to gain legitimate teaching positions, promotion, and tenure." As a result, few are in academic positions that are necessary to train graduate students (171). The prospective parapsychologist "is advised to become something else" (172): "Conceal your interest in parapsychology," they are told. "Get a doctorate in whatever subject interests you. Then you can be of value to the field" (173). Again, traditional scientists argue that in the absence of strong evidence that the tenets of parapsychology are valid, this freeze-out is entirely positive.

Since the 1980s, systematic, empirical research in the United States on parapsychological phenomena has waned considerably. Parapsychologists cite bias by academics and traditional researchers against the notion that paranormal powers exist; in contrast, scientists argue that the decline is a simple case of falsification—no such powers exist, and recent research has documented that fact. The "aura" or "energy field" around living subjects detected by Kirlian photography has been demonstrated to have been caused by natural forces, such as moisture and electrical currents; the ESP and PK effects turned up by the major labs investigating the efficacy of paranormalism have been shown to be tiny and well within the margin of researcher error, minor statistical variation, or cheating; performers such as Uri Geller who supposedly demonstrated ESP and PK have been unmasked as charlatans (Melton 1996, 987–88). Labs from Princeton to Stanford have closed their doors, found their funding dried up, and had their sponsors become increasingly absent. "The status of paranormal research in the United States is now at an all-time low" (Odling-Smee 2007, 11). The axis of research on parapsychology has shifted to Europe, particularly the United Kingdom (Irwin 2004, 248–49). And though it is virtually impossible for parapsychologists to publish in traditional science journals, a substantial number of science-like journals cater specifically to parapsychological research and paranormalism.

To repeat, parapsychology is unquestionably a deviant discipline; that can be operationalized by virtue of the fact that academics in conventional fields refuse to accord to it the respectability and legitimacy that theirs is granted. But its status as a pseudoscience is upheld more, its supporters claim, by the fact that it affirms explanatory power for forces most naturalistic scientists do not believe exist than its lack of methodological rigor. As such, parapsychology remains an anomalous pseudoscience, much in need of more conceptual clarification and empirical attention than it has received. In sum, pseudoscientific belief systems are deviant to the extent that their claims are not valorized by mainstream institutions: the curriculum of the public educational system, the most influential media outlets, major museums, the theology of the highest-ranking religious bodies. And the representatives of the major political and economic institutional spheres who make use of or promulgate predictions or prognoses readily identified as pseudoscientific or paranormal would be ridiculed, delegitimated, or relieved of their positions of power and influence. Hence, in spite of their popularity among the public at large, the sociologist regards pseudoscientific belief systems as deviant.

REFERENCES

Ben-Yehuda, Nachman. 1985. *Deviance and Moral Boundaries: Witchcraft, the Occult, Science Fiction, Deviant Sciences and Scientists*. Chicago: University of Chicago Press.

Bernstein, Jeremy. 1978. "Scientific Cranks: How to Recognize One and What to Do Until the Doctor Arrives." *American Scholar* 47 (March): 8–14.

Brown, Chip. 1996. "They Laughed at Galileo, Too." *New York Times Magazine*, August 11, 41–45.

Byrne, Rhonda. 2006. *The Secret*. New York: Atria/Beyond Words.

———. 2008. *The Power*. New York: Atria.

Gardner, Martin. 1957. *Fads and Fallacies in the Name of Science*. 2nd rev. ed. New York: Dover.

Goode, Erich. 2011. *The Paranormal: Who Believes, Why They Believe, and Why It Matters*. Rev. ed. Amherst, NY: Prometheus Books.

Hines, Terence. 2003. *Pseudoscience and the Paranormal*. 2nd ed. Amherst, NY: Prometheus Books.

Hyman, Ray. 2007. "Cold Reading: How to Convince Strangers that You Know All About Them." In *Paranormal Claims: A Critical Analysis*, edited by Bryan Farha, 31–42. Lanham, MD: University Press of America.

Irwin, Harvey J. 2004. *An Introduction to Parapsychology*. 4th ed. Jefferson, NC: McFarland.

Karén, Michelle. 2008. *Astrology for Enlightenment*. New York: Atria Books/Simon & Schuster.

Larson, Edward J. 2007. *The Creation-Evolution Debate: Historical Perspectives*. Athens: University of Georgia Press.

Mack, John. 1994. *Abduction: Human Encounters with Aliens*. New York: Scribner.

McClenon, James. 1984. *Deviant Science: The Case of Parapsychology*. Philadelphia: University of Pennsylvania Press.

Melton, J. Gordon, 1996. "Parapsychology." In *Encyclopedia of Occultism & Parapsychology*, 4th ed., edited by J. Gordon Melton, 986–88. Detroit: Gale Research.

Odling-Smee, Lucy. 2007. "The Lab that Asked the Wrong Questions." *Nature*, March 1, 10–11.

Radin, Dean. 1997. *The Conscious Universe: The Scientific Truth of Psychic Phenomena*. New York: HarperEdge.

Tart, Charles T. 2009. *The End of Materialism: How Evidence of the Paranormal Is Bringing Science and Spirit Together*. Oakland, CA: Noetic Books.

Truzzi, Marcello. 1982. "J.B. Rhine and Pseudoscience: Some Zetetic Reflections on Parapsychology." In *J.B. Rhine: On the Frontiers of Science*, edited by K. Ramakrishna Rao, 171–91. Jefferson, NC: McFarland.

———. 1987. Introduction to *Advances in Parapsychological Research*, edited by Stanley Krippner, 5:4–8. Jefferson, NC: McFarland.

9

Belief Buddies versus Critical Communities

The Social Organization of Pseudoscience

NORETTA KOERTGE

Pseudosciences purport to conform to the methodological norms of scientific inquiry; yet upon analysis by nonbelievers, the claims are deemed to clearly violate science and often common sense. Pseudoscientific belief systems are puzzling, but not just because they appear to be false or implausible. After all, the history of both science and common sense is full of erroneous beliefs. The propositions in pseudoscience are often described as bizarre or weird because we have trouble understanding why their adherents, who often appear to be quite sensible in most respects, nevertheless defend these systems so staunchly.

Philosophers of science have attempted to describe what's wrong with pseudoscience, first, by contrasting the structure of its claims with that of legitimate scientific hypotheses, and second, by comparing the typical reasoning patterns of pseudoscientists with the norms of scientific reasoning.

This chapter proposes an additional difference. Since the time of the Scientific Revolution, most scientific research has taken place within the context of scientific institutions that organize conferences and support peer-reviewed dissemination of new research. These well-organized critical communities supplement the efforts of individual scientists by promoting both positive and negative feedback. Pseudoscientists, on the other hand, feeling the stigma attached to their beliefs, often seek out only supportive allies.

Existential Angst: A Personal Introduction

My first encounter with pseudoscience happened at age eight. After the funeral of my Aunt Velma, my mother was trembling with anger. "Perhaps the doctors could have saved her," she said. "The preacher convinced Velma that his laying on of hands would work and consulting with doctors would be a sign that she didn't really have faith in God." My mother then sighed. "But he means well. He's a good man, and the church is being very kind now to Velma's little kids. It's just ignorance. People don't know any better."

My early years in a farm community were marked by that contradiction. Good people, the kindest people I knew, believed awful things. But there was hope. We just needed better education, especially in science. Dupont's slogan at the time was "Better things for better living—through chemistry!" Learning about scientific methods could improve our ways of reasoning, as well as our medical procedures.

Studying philosophy of science in London during the age of Karl Popper certainly refined my ideas about the so-called scientific method, but my overall optimism about our ability to improve society through critical thinking and piecemeal engineering persisted. True, some of my fellow British students were a bit soft on the topic of ghosts. And back in America, the widespread interest in crystal power, auras, and chakras among my friends working for gay liberation and feminist reforms was disconcerting. Yet some of their scholarly critiques of mistaken scientific views about sex and gender were valuable. As we rid science of prejudice, was it not reasonable to hope that previous victims of that prejudice would eventually come to see the value of scientific reasoning?

Then in the middle of the first decade of the twenty-first century, I attended a three-part symposium here in Bloomington called "The Truth about 9/11." At that time I had not encountered these alternative perspectives, some of which suggested that controlled demolitions had destroyed buildings at the World Trade Center in New York. I was taken totally by surprise. What shocked me to the core was not the dubious quality of the arguments and suspicions raised. Rather, it was the intellectual backgrounds and credentials of the leaders of the Truther movement and the supporters present here in my hometown. These included not only scientists and engineers, an editor of a journal in applied mathematics, an author of a book critical of traditional theological language, but even former graduate students and colleagues in *philosophy of science*! Once again I discovered a deep chasm between my way of viewing the world and that of respected friends. And whatever was going on could not simply be explained as a lack of exposure to classes in critical thinking and the epistemic norms of science.

A Multidimensional Approach to Understanding Pseudoscience

The present volume nicely illustrates the recent history of efforts to characterize pseudoscience. Philosophers first attempted to give a clear, simple demarcation. Pseudoscientific theories were unfalsifiable while scientific theories were testable. Or perhaps we should talk instead about progressive versus degenerating research programs. But when one looked in detail at the actual development of science, too many examples of successful science seemed to spend a good part of their history on the wrong side of the demarcation line (evolutionary biology was a favorite example; see Ruse, chapter 12, in this volume). Some commentators have responded by refining the classification system or introducing a spectrum, so that now we have good science, bad science, and awful science, or mature science, revolutionary science, crank science, the borderlands of science, and nonsense on stilts (see Pigliucci 2010). Then there is the distinction between nomological and historical sciences (see Cleland and Brindell, chapter 10, in this volume), which reminds us of the wide variety of pseudohistorical systems that often make dubious use of scientific claims in support of conspiracy theories.

Textbooks in scientific reasoning may take a cluster concept approach somewhat similar to the characterization of psychological disorders in the *Diagnostic Service Manual*. Thus the 2011 edition of Carey's *A Beginner's Guide to Scientific Method* describes eight scientific fallacies, such as "ad hoc rescues" and "empty jargon." Although scientists sometimes commit these fallacies, they are especially prevalent in pseudoscience. Carey lists other telltale signs of pseudoscientific theories, which tend not to be internally consistent or self-correcting over time, and the tendency of pseudoscientists to view skepticism as narrow-mindedness.

Other diagnoses of defects in pseudoscience rely on Kahneman, Slovic, and Tversky's classic work in cognitive science and behavioral economics (1982), which documents the multiple difficulties even well-educated people have with probability concepts and the propensity of humans to commit mistakes in reasoning, such as confirmation bias and the *post hoc ergo propter hoc* fallacy, in which people jump to conclusions about causal connections. Political scientists talk about the paranoid style of the people who subscribe to conspiracy theories and their feelings of powerlessness. Studies show that supporters of one kind of pseudoscience are statistically more likely to have unorthodox views on seemingly unrelated subjects. In his essay on Alfred Russel Wallace's fascination with spiritualism, Michael Shermer (2001) defines a heretic personality type, someone who is attracted to views contrary to received opinion no matter what the topic.

Popular works designed to help the general public identify pseudoscience offer checklists that increasingly supplement accounts of epistemological errors with comments about political and social factors. Whereas Carl Sagan's (1995, 210) original Baloney Detection Kit warned against appeals to authority ("in science there are no authorities"), in Michael Shermer's (2001) version of the kit the first two questions are "How reliable is the source?" and "Does this source often make similar claims?" Brian Dunning's (2008) web video "Here Be Dragons" includes *cui bono* as an important tool for the skeptic. Sherry Seethaler, a journalist and instructor in science communication at the University of California San Diego, describes how science news is often distorted by the popular media. In her 2009 book *Lies, Damned Lies, and Science*, she shows how not only headlines but even the lead paragraphs can be seriously misleading. One of her tools for the discerning reader is to compare the last paragraph of a science news story, where qualifications often appear, with the attention-grabbing first sentences!

Even this brief overview of current perspectives on pseudoscience teaches us two things: first, an understanding of the prevalence and persistence of these belief systems will include insights from social psychology and the study of institutions in addition to the familiar epistemological and cognitive factors. Second, if we want our educational system to help reduce the personal and social harms that can accrue from pseudoscience, we will need to modify not only science pedagogy (see Pigliucci 2002) but also our approach to introductory philosophy of science and critical reasoning classes. But rather than opine on such big questions, let me turn instead to yet another differentiating factor between science and pseudoscience. Recent integrated studies of history and philosophy of science have drawn attention to the unique institutions of science.

Belief Buddies—Man, the Chattering Gossipmonger

Much attention is currently being paid to the new social media. But people have always been avid sharers of stories about the latest mastodon hunt or where best to find morel mushrooms. Back in my days on the farm, neighbors talked with great concern about how the damn scientists testing their damn atomic weapons were messing up the weather. Folklorists study the spread of legends and rumors in all societies. Stephanie Singleton (2008) has documented the prevalence of genocidal suspicions about the origins of AIDS amongst African Americans in Harlem. Shocking personal anecdotes gain attention, whether they be in a pub, a laundromat, or a faculty common room. Every "liars' bench" or "Stammtisch" has its star performers—some are opinion leaders, others elicit yawns. But every participant feels some pressure to come up with a good story—topics popular in my small Indiana university town range from cougar sightings in the next county, to the premature babies born near where Westinghouse used to dump PCBs in the creek, to rumors that the new dean is going to freeze hiring in the humanities.

People with similar interests may then spontaneously coalesce into groups that communicate more or less regularly. The topics of conversation may center on plans to form a watercolor society or how nice it would be to have more bicycle paths. But what is relevant to us here are informal collectives that I will call belief buddies. These groups collect and disseminate information on issues where scientific information and approaches are more or less relevant. They often feel that their views are neglected or stigmatized in society at large. As a result, these belief buddies consciously attempt to affirm contributions that further their agenda; dissent is discouraged lest it lead to a splintering of the group.

Sometimes the emphasis on being constructive is perfectly appropriate. When the Green Drinks group meets to discuss the local problem of invasive species, this may not be the occasion for an extended philosophical discussion about whether all nonnative plants are invasive and vice versa. The purpose is to help members identify and remove kudzu and purple loosestrife from their yards and woodlands, not to clarify key ecological concepts. However, if a 9/11 Truther group invites a visitor in to report on interviews with first responders who claim that the blasts and changes in pressure they experienced in the WTC towers could only have been caused by planned explosions, it is surely appropriate to entertain questions about whether these witnesses had actually ever been inside buildings brought down by controlled demolitions as well as how much experience they had had with big fires in skyscrapers. But belief buddies may not welcome criticism, no matter how cogent to the topic at hand. Their job is to convey information that supports their core project and to reassure beleaguered constituents.

From Belief Buddies to Critical Communities

Traditional accounts of the rise of modern science emphasize experimentation, the mechanical philosophy, and applied mathematics. But here I want to focus on the advent of the new scientific societies: the Royal Society in 1663 (formerly known as the Invisible College for the promoting of Physico-Mathematical Experimental Learning) and the Académie des Sciences in 1666. Earlier organizations, such as the Accademia dei Lincei founded in 1603, had disbanded, in part because of disputes over the Copernican theory. Today we often take for granted the crucial role played by peer review and international dissemination of findings through publication. And although we may sometimes complain about the conservatism of funding agencies, the process of evaluating grant proposals surely does help sharpen research questions and deter investigators from blind alleys.

Much of the critical ethos of these early scientific societies was a heritage of practices in the medieval university, which formalized Aristotle's habit of listing and then criticizing views of his predecessors before arguing for his own position (see Grant 2001, 2005). The seventeenth-century societies added several key elements: an emphasis on novel results, be they experimental or theoretical, and on taking personal responsibility for their reliability. Since scientific knowledge is typically a common good, we sometimes find priority disputes a bit embarrassing. The credit system in science does encourage scientists to publish promptly, but not before supplying evidence for their claims, because they realize that an audience is ready to provide either criticism or corroboration. Popper (1994) talks about the "friendly-hostile cooperation" of scientists, and David Hull (1990) lists three C's of the scientific process: curiosity, checking, and credit.

Arguments from counterfactual histories are not probative, but here are two familiar cases that can illustrate the positive function of scientific institutions. First, consider Galileo Galilei, a complete scientist—an innovative experimentalist who understood how to test idealizations, an expert in mixed mathematics and an inventor of practical measuring devices, a good philosopher and polemicist, and most of all a pioneer in the development of a new physics. We still speak of the principle of Galilean relativity. Yet in his *Dialogue Concerning the Two Chief World Systems*, Galileo committed a monumental blunder, one that still puzzles historians. His proposed theory of the tides, which was so central to his defense of the Copernican system that he once planned to make it the title of the book, violated that very principle of relative motion! My point is not that Galileo made a mistake. Rather I ask the following: what if over the two decades that Galileo worked on the tidal theory, he could have tried it out in a conference of scientists? What if his book manuscript had been refereed by peers instead of church inquisitors? Galileo was ahead of his contemporaries, but I believe his contemporaries could have understood his new ideas well enough to challenge his account of the tides *if* the appropriate critical community had existed.

My second example is the case of Ignaz Semmelweis, the physician who discovered a proximate cause of childbed fever (CBF) in 1847, but was largely unsuccessful in disseminating his results, despite the fact that he worked in the Allgemeine Krankenhaus in Vienna, one of the most prestigious teaching hospitals at the time. Ever since Carl Hempel introduced this case as a textbook example of the hypothetico-deductive method, students have learned about Semmelweis's various attempts to explain the dramatically higher mortality rate in the medical students' ward as compared with the neighboring ward for midwives. The two wards served as a natural controlled experiment; to prevent crowding, patients were admitted to the wards on alternate days, and the wards were large enough that changes in mortality rates were obvious. (They easily pass today's tests for statistical significance.)

Semmelweis concluded that CBF was caused by the introduction into the blood stream of cadaveric particles from autopsies or ichorous particles from suppurating wounds. It could be prevented by careful hand washing in chlorinated lime solutions before examining the mothers. Despite their enormous practical importance, Semmelweis's discoveries had little impact. Historians have offered a variety of explanations, ranging from his personality and minority status as a Hungarian to the prevalence at the time of so-called anticontagionism. But all agree that Semmelweis should have given talks and published earlier. He finally brought out a book in 1861 with the crucial statistics as well as the results of preliminary experiments with rabbits. However, the scientific findings were interspersed with personal anecdotes and complaints, and the last section even accused his critics of murder. At the time, there were some venues for doctors to circulate research findings, but as is still the case, medical practitioners did not have a primary identification as scientists. Without the expectation and strong institutional support for publication and peer review, Semmelweis's contributions fell through the cracks.

Schizophrenia or Incompatible Social Roles?
Alchemy and the Scientific Revolution

Social psychologists define the "fundamental attribution error" as the tendency to overvalue dispositional or personality-based explanations for the observed behaviors of others while undervaluing situational explanations for those behaviors. Something like this is surely operating when we are too quickly satisfied by explanations of pseudoscience in terms of the epistemic ineptitude or credulous personality traits of its adherents. If we wish to understand Robert Boyle's or Isaac Newton's interest in the transmutation of base metals into gold and other alchemical pursuits, it is neither intellectually satisfying nor historically accurate simply to say, "Well, they kept their dedication to the new experimental philosophy in one pocket and their commitment to the chrysopoeia project in another." But as I read the rapidly expanding literature on this topic, in each case their basic views on the methods and principles of natural philosophy were homogeneous. For Boyle, the crucial difference between his role as a contributing chemist and aspiring adept had to do with conflicting norms about secrecy and the sharing of information. Newton, especially in his role as master of the mint, was also adamant about the need for secrecy lest the availability of newly synthesized gold undermine the British currency!

At Boyle's time, there was broad agreement that natural products were to be understood in terms of some combination of the four Aristotelian elements and active principles such as Mercury, Sulfur, and Salt. Whether one's research was directed primarily toward the discovery of new medicines, dyes, metallurgical processes, or the Philosopher's Stone, it was a period of rapidly expanding knowledge of new compounds, new reactions, and new techniques, both qualitative and quantitative. Traditionally, historians of chemistry drew a sharp line between the pioneers of modern chemistry, such as Boyle, and the leftover alchemists on the fringe. However, new research that looks carefully at unpublished manuscripts shows that Boyle was not only deeply involved in alchemical projects, but also had a theoretical framework that integrated his so-called mechanical philosophy with the possibility of finding an elixir that would transmute lesser metals into gold. To emphasize the continuity and broad overlap of both people and ideas in the late seventeenth century, historians such as Newman (2006), Newman and Principe (2002), and Principe (1998) introduce the term "chymistry."

Yet one bright line of demarcation remains. The "community" of alchemists at the time was loose knit and closeted. Alchemical manuscripts had long promoted secrecy by omitting key steps in recipes and through the use of mythic metaphors. There were various reasons for furtiveness. Sometimes there were laws or political pressures to protect the local currency, and so the discoverer wanted to prevent the influx of new gold. Others believed that the secret of transmutation was so metaphysically and spiritually profound that it should only be shared with the adept. And then there were those whom Boyle called "cheats" and "pretenders" who wanted to manipulate patrons for their own gain. (Erich Goode [2000] describes a similar split in the motivation amongst occultists today. Some are sincere idealists; others are snake oil salesmen.)

Boyle followed the custom of disguising the processes he had used in publications and made his private notes and letters more secure by encrypting them in varying codes (Principe 1998). But he also imagined an organization that would support more efficient communication and debate. In an unpublished, incomplete manuscript called *Dialogue on the Transmutation and Melioration of Metals*, he describes a "Noble Society" where a debate exists between Lapidists, anti-Lapidists, and a third group that is agnostic about the existence of the Philosophers' Stone. The Lapidists presented a variety of arguments in favor of their chrysopoeia project. They admitted that the projective operation of the Stone was mysterious to them and so they could not explain how it would work, but argued that this was no grounds for skepticism. Brewers, for instance, do not cease making beer just because people did not fully understand fermentation. As to the concerns about how a tiny piece of the Stone could transform large quantities of metals, it was pointed out that a small amount of rennet can induce the curdling of a big container of milk. Much of the discussion revolved around the trustworthiness of so-called transmutation histories, with competing anecdotes about personal experiences with both frauds and illustrious gentlemen. It was eyewitness accounts that seemed to carry the burden of proof (see Principe 1998, app. 1).

Since only fragments of the *Dialogue* exist, the details of Boyle's own views are not always clear, although he was certainly in the Lapidist camp. Here and in other places he saw the need for secrecy, but also included the passionate argument of Eugenius, a member of the neutral group, who asked why there should be secrecy in regard to "Medical *Arcana*, . . . which would be highly beneficial to Mankind" (Principe 1998, 67). Boyle described the discussion that follows as surprisingly brisk, but unfortunately the next paragraphs are missing.

Let us now return to our inquiry about the inadequacies of pseudoscience. Would it have been possible for someone like Boyle, a leading member of the Royal Society, also to found a Noble Society for the Scientific Study of Alchemy? And would it have led to quicker progress in our understanding of chemistry? A full exploration of these questions would lead quickly to counterfactual history, but I will venture two remarks. First, given Boyle's elitist views that only *adepts* should be privy to knowledge about the Philosophers' Stone, his Noble Society would have had to operate more like the Manhattan Project or CIA than like the Royal Society with its public demonstrations and transactions. Second, I think one can easily make the argument that chymistry at the time of Boyle and Newton would have benefited from a wider dissemination and critical discussion of alchemical research. Here's just one example. White phosphorus was first prepared by a German alchemist, Henning Brand, while heating urine in the absence of air. Brand sold his secret but some details leaked out, and Boyle managed to replicate the procedure and then showed how phosphorus could be used to ignite sulfur-coated wood splints. In this case, through a lucky happenstance, secrecy did not impede progress for very long.

More typical were the frustrating hours people spent puzzling over references to the so-called Greene Lyon or the triumphal Chariot of Antimony, alchemical names for reactions and active compounds. A fundamental research strategy of alchemists was to resolve natural materials into simpler, purer components and then recombine them to synthesize new materials with striking new properties. The lack of success of the chrysopeia project should not distract us from the plethora of new discoveries, ranging from laboratory equipment and quantitative methods to new reagents and purer chemicals. Would not all of these initiatives have benefited from an institution like Boyle's Noble Society? Yet it did not happen. Boyle and other chymistry colleagues in the Royal Society brought their personal views on natural philosophy and empirical inquiry to their individual chrysopeic pursuits, but had no critical community to support those efforts.

Pseudoscience, Peer Review, and Publication

Studies in the sociology of science show how science education inculcates practitioners with a respect for norms such as objectivity, logical consistency, and organized skepticism. And we have seen above examples from the history of science that illustrate the importance of institutionalized peer review and publication. In today's media-rich environment, it is easy for individuals working in areas deemed to be pseudoscience to circulate their ideas, which may then elicit lots of positive or negative comments on blogs, but finding a forum for a thoughtful, systematic evaluation of one's views is more difficult. So the question arises, what would happen if there were critical communities supporting pseudoscientific inquiry? Or is there something inherent in such enterprises that precludes or makes unlikely such organizational structures? (An example of a countervailing factor would be the high value alchemists placed on secrecy.)

To investigate this issue, let us look briefly at two serious contemporary attempts to institutionalize fringe science, or what Henry Bauer calls "anomalistics." In his 2001 book *Science or Pseudoscience: Magnetic Healing, Psychic Phenomena, and Other Heterodoxies*, Bauer provides useful charts comparing natural science, social science, and anomalistics with respect to both intellectual resources and the availability of organized social networks. Bauer distinguishes between "scientific heresies," which are beyond the pale, and "borderland subjects," which deserve a hearing. He has written books on examples of each, first a thorough critique of Immanuel Velikovsky's bizarre ideas about supposed collisions of the earth with Venus and Mars, which contrasts sharply with his own more sympathetic accounts of the "Enigma of Loch Ness" and what critics call AIDS denialism. After retiring as professor of chemistry and science studies, Bauer served as editor of the *Journal of Scientific Exploration*, which publishes refereed articles in areas such as ufology, cryptozoology, and psychokinesis. Although the journal includes varying points of view, its mission is to provide a forum for articles that would probably not be published in regular science journals. Let us now look at how this attempt to institutionalize pseudoscience is working.

The year 2011 marked the thirtieth annual meeting of the journal's parent organization, the *Society for Scientific Exploration*.[1] The program illustrates a variety of attempts to advance research in a wide range of areas that would usually be labeled pseudoscience. Several talks deal with the difficulties experimenters face when trying to reproduce paranormal phenomena. "Questioning the Place of Subjective States and Conscious Awareness in Anomalous Healing" describes preliminary results suggesting that volunteers with no previous experience or belief in healing were just as effective in curing cancer in mice as subjects who experienced a spiritual connection. "The Decline Effect: Exploring Why Effects Sizes Often Decline Following Repeated Replications" calls for a public repository of experimental designs and all findings, regardless of outcome. Based solely on the abstracts posted on the web, both of these papers struck me as possibly useful contributions to people who find the reports of psychic phenomena plausible enough that they have decided to pursue research in this area.

It is more difficult to find anything positive to say about "Dark Matter, Torsion and ESP Reception," which first reports that psychic reception is more accurate when the constellation Virgo is overhead and then proposes an explanation in terms of so-called torsion fields once proposed by Russian physicists, as well as a cluster of talks on nonlocality, consciousness, and the double-slit experiment. Even so, there is no denying the serious tone and style of the conference papers, although the subject matter is obviously controversial, to say the least. Almost every paper makes reference to terms common in various pseudosciences, such as "remote viewing," "reincarnation," "precognition," and "subtle energy." Predictably, on the third day there are a couple of papers complaining about dogmatism in traditional science.

We now turn to a second example of institutionalized fringe science. In contrast to the anomalistics discussed above, the articles in the *Journal of Condensed Matter Nuclear Science* at first seem completely orthodox in all respects. True, they contain frequent references to LENR, which turns out to be an acronym for Low-Energy Nuclear Reactions, which the reader eventually learns is popularly called "cold fusion." The borderline legitimacy of this research program is made quite clear in the preface of the February 2011 issue of the journal. This collection of two dozen papers originally delivered at a March 2010 meeting of the American Chemical Society was scheduled to be brought out as a book by AIP Publishing, a branch of the American Institute of Physics. In the end the publisher declined, giving no reason, but the articles are now available on the journal's website as JCMNS Volume 3.[2]

As a philosopher I was impressed by the candid admissions of the experimental and theoretical difficulties that cold fusion studies face. For example, in the lead article "What Is Real about Cold Fusion and What Explanations Are Plausible?" the authors begin by accepting the effect of the generation of anomalous amounts of heat in palladium/deuterium electrolytic cells as real, but the bulk of the paper is devoted to a discussion of confounding factors, such as impurities in the electrodes, the possible absorption of fusion products in the cell, and inaccuracies in the use of CR-39 plastic flats for neutron detection. They also analyze various possible nuclear reactions that might produce such heat. The usual proposals involving small deuterium clusters do not work, but the authors hypothesize that a mechanism involving superclusters might be viable.

The editor, Jan Marwen, sums up the situation this way in the preface: "LENR does not appear to fit into current scientific understanding, and it raises uncomfortable questions about current and historical understandings of nuclear physics. The path forward will require new openness, receptivity, and tolerance. It may require flexibility on the part of orthodox physics to learn from LENR researchers. It may also require LENR researchers to learn from orthodox physics." The implied epistemic symmetry of this appraisal has rhetorical appeal, but it provides no guidance to questions about research priorities or policy decisions.

Having looked briefly at two organizations dedicated to the study of anomalous phenomena, what conclusions can we draw about the beneficial effects of peer review? The two cases have significant differences. Cold fusion began as part of orthodox science with the Fleischmann/Pons experiments in 1989 and has gradually moved more and more toward the fringe, although the researchers involved have traditional scientific credentials. The miscellaneous topics discussed by the *Scientific Exploration* folks, on the other hand, have always been viewed with suspicion and their proponents in some cases have little relevant scholarly training.

The two cases also have striking similarities. Although there is some complaining about the stigma attached to their enterprise, both groups openly recognize that the phenomena they claim to be investigating are difficult to instantiate. Sometimes heat is produced in Pd/D cells and sometimes it isn't. Sometimes ESP works and sometimes it doesn't. And although some members of each group go on ahead to theorize about possible mechanisms to explain the anomalous phenomena, others work hard in trying to design experimental setups that will make the phenomenon reproducible.

Conclusion

I have argued that one characteristic differentiating typical science from typical pseudoscience is the presence of critical communities, institutions that foster communication and criticism through conferences, journals, and peer review. But can the formation of such an institution really make a pseudoscience more intellectually viable? (And do we want people getting tenure for publications in anomalistics journals, even though they are peer reviewed?)

One might hope that by coordinating efforts to go down blind alleys quickly and efficiently, the adherents of lost causes would collect refutations that would change their minds. However, might it not work in the opposite way? Might the existence of a seemingly respectable support group actually rally the troops and impede the process of correcting false claims and discarding untestable ones?

These questions cannot be given a general answer. Although I think that the operation of critical communities is an important attribute of science, I am not proposing it as a litmus test with which to demarcate science from pseudoscience. Both the Discovery Institute, which promotes Intelligent Design, and the International Center for Reiki Training mimic the institutional aspects of science (and thereby seek legitimacy), yet neither seem more promising than Boyle's alchemical society. But even in those cases, I think, there is a benefit to institutions that encourage critical discussion, even if there are limits to what may be questioned. In his methodology of scientific research program, Imre Lakatos (1971, 104) claimed that it was not irrational for an individual to pursue a so-called degenerating research program as long as one admitted the low odds of success; on the other hand, it was perfectly rational for funding agencies to deny support to such research—and, I would add, morally obligatory to not let degenerate science influence policy decisions.

Tolstoy said that happy families are all alike, while every unhappy family is unhappy in its own way. While not literally true (the happiest of families sometimes quarrel), perhaps something similar can be said about successful scientific projects. Any mature scientific theory gives unified explanations of disparate empirical phenomena and provides a basis for veridical predictions, while offering possibilities for further productive research. Pseudoscientific theories by contrast can exhibit any number of unhappy characteristics. Sometimes their entire focus is on phenomena that are neither reproducible nor intersubjectively observable—sightings of Loch Ness monsters and UFOs cannot be orchestrated. Ditto for near-death experiences and recovered memories. That does not necessarily mean that these experiences are not real. But it does mean that they are difficult to study. In other cases, the central causal conjectures of the pseudoscience are inconsistent with the most well-confirmed theories of contemporary science. Of course, fundamental revolutions in science sometimes occur, but not without the formation of a promising alternative paradigm. We have absolutely no reason to believe that pseudosciences such as homeopathy, astrology, or psychokinesis can even articulate an alternative theoretical system, let alone find evidence for it.

In this chapter, I have described another factor that inhibits researchers in pseudoscience: often their only colleagues are what I have called belief buddies, people who share a firm commitment to the stigmatized knowledge claims and who help collect supporting evidence and arguments but are very reluctant to encourage criticism. The result is that some pseudosciences are plagued with dubious reports from credulous amateurs or even charlatans looking for attention. We have a romantic image of the lone scientist working in isolation and, after many years, producing a system that overturns previous misconceptions. We forget that even the most reclusive of scientists these days is surrounded by peer-reviewed journals; and if our would-be genius does make a seemingly brilliant discovery, it is not enough to call a news conference or promote it on the web. Rather, it must survive the scrutiny and proposed amendments of the relevant critical scientific community.

The history of science reminds us of how difficult it is to establish and maintain these critical communities. There are always stresses from factors such as favoritism and funding, as well as possible distortions from ideology. Philosophers can join scientists in maintaining epistemic and empirical values within all branches of science and also in calling attention to the inadequacies of projects that we correctly label as pseudoscience.

NOTES

1. See Society for Scientific Exploration, "SSE Meetings," accessed October 7, 2012, http://www.scientificexploration.org/meetings/.

2. See International Society for Condensed Matter Nuclear Science, "JCMNS Publications," accessed October 7, 2012, http://www.iscmns.org/CMNS/publications.htm.

REFERENCES

Bauer, Henry H. 2001. *Science or Pseudoscience: Magnetic Healing, Psychic Phenomena, and Other Heterodoxies.* Urbana: University of Illinois Press.

Carey, Stephen S. 2011. *A Beginner's Guide to Scientific Method.* Boston: Wadsworth.

Dunning, Brian. 2008. "Here Be Dragons: An Introduction to Critical Thinking." Skeptoid Media, 41 mins. http://herebedragonsmovie.com/.

Goode, Erich. 2000. *Paranormal Beliefs: A Sociological Introduction.* Prospect Heights, IL: Waveland Press.

Grant, Edward. 2001. *God and Reason in the Middle Ages.* Cambridge: Cambridge University Press.

———. 2005. "What Was Natural Philosophy in the Late Middle Ages?" *History of Universities* 20 (2): 12–46.

Hull, David L. 1990. *Science as a Process: An Evolutionary Account of the Social and Conceptual Development of Science.* Chicago: University of Chicago Press.

Kahneman, Daniel, Paul Slovic, and Amos Tversky, eds. 1982. *Judgment Under Uncertainty: Heuristics and Biases.* Cambridge: Cambridge University Press.

Lakatos, Imre. 1971. "History of Science and Its Rational Reconstruction." In *Boston Studies in the Philosophy of Science*, vol. 8, edited by Roger C. Buck & Robert S. Cohen. Dordrecht: Reidel.

Newman, William R. 2006. *Atoms and Alchemy: Chymistry and the Experimental Origins of the Scientific Revolution.* Chicago: University of Chicago Press.

Newman, William R., and Lawrence M. Principe. 2002. *Alchemy Tried in the Fire: Starkey, Boyle, and the Fate of Helmontian Chymistry.* Chicago: University of Chicago Press.

Pigliucci, Massimo. 2002. *Denying Evolution: Creationism, Scientism, and the Nature of Science.* Sunderland: Sinauer Associates.

———. 2010. *Nonsense on Stilts: How to Tell Science from Bunk.* Chicago: University of Chicago Press.

Popper, Karl R. 1994. *In Search of a Better World: Lectures and Essays from Thirty Years.* London: Routledge.

Principe, Lawrence M. 1998. *The Aspiring Adept.* Princeton, NJ: Princeton University Press.

Sagan, Carl. 1995. *The Demon-Haunted World: Science as a Candle in the Dark.* New York: Random House.

Seethaler, Sherry. 2009. *Lies, Damned Lies, and Science: How to Sort Through the Noise Around Global Warming, the Latest Health Claims, and Other Scientific Controversies.* Upper Saddle River, NJ: FT Press Science.

Shermer, Michael. 2001. *The Borderlands of Science: Where Sense Meets Nonsense.* New York: Oxford University Press.

Singleton, Stephanie L. 2008. "According to Rumor, It's a Conspiracy: Conspiracy Theory as a Paradigmatic Construct." MA thesis, Indiana University, Dept. of Folklore and Ethnomusicology.

PART III

The Borderlands between Science and Pseudoscience

10

Science and the Messy, Uncontrollable World of Nature

CAROL E. CLELAND AND SHERALEE BRINDELL

Many public controversies over scientific claims involve cases in which scientists attempt to justify hypotheses and theories about complex systems occurring in the messy, uncontrollable world of nature. Examples include doubts about the relation between the vaccination of children and the rise of autism, doubts about climate change, and doubts about Darwinian evolution. In such cases, there appears to be a pronounced tendency to reject results that are widely accepted by specialists working within the pertinent area of science. And while this tendency is sometimes blamed on features of human psychology,[1] this explanation fails to make clear why such skepticism afflicts the findings of the field sciences to a much greater extent than those of the stereotypical experimental sciences. Moreover, a surprisingly large number of experimental scientists not only seem to share the skepticism of the public toward field science, but their skepticism sometimes reaches levels usually reserved for pseudosciences such as astrology and homeopathy![2] In this chapter, we argue that doubts about the scientific status of the field sciences, considered generally, commonly rest on mistaken preconceptions about the nature of the evaluative relation between empirical evidence and hypothesis or theory,[3] namely, that it is some sort of formal logical relation. We contend that there is a potentially more fruitful approach to understanding the nature of the support offered by empirical evidence to scientific hypotheses—an approach that promises a more accurate representation of the practices of scientists engaged in experimental as well as field research and reveals critical ways in which successful field sciences differ from pseudosciences such as astrology and homeopathy.

Philosophers have long sought a clear distinction between "proper" science and pseudoscience. This "demarcation problem," as Karl Popper famously dubbed it, was assumed to be solvable through clarification of the nature of the logical relation between hypothesis and empirical evidence. The Hypothetico-Deductive method, Hempel's (1962) Deductive Nomological and Inductive Statistical models of explanation, Carnap's (1928) *Aufbau*, and Popper's (1963) falsificationism are each exemplary illustrations. In the first part of this chapter, we briefly review the traditional philosophical take on the scientific method with an eye toward making clear its most serious problems. As will become apparent, these problems are greatly exacerbated when science moves from the artificially controlled environment of the laboratory to the messy uncontrollable world of nature. Like most contemporary philosophers of science, we reject the kind of crisp boundary between science and pseudoscience sought by Popper. But this does not mean that important differences between science and pseudoscience do not exist. We argue that failure to recognize that the core evaluative relations used by both experimental and field scientists in accepting and rejecting hypotheses involve appeals to extralogical, causal components is largely responsible for generalized doubts about the scientific status of the field sciences. Moreover, it is not at all clear that these critical causal components can be satisfactorily captured in purely structural (formal mathematical) causal analyses such as Bayesian causal networks,[4] which assume that the rationality of all scientific reasoning is secured by the axioms of the mathematical theory of probability. In a nutshell, it is our contention that an obsession with cashing out scientific reasoning in terms of formal logico-mathematical considerations lies behind much of the skepticism about the scientific status of work in the field sciences and has blinded philosophers to a potentially more satisfactory solution to the demarcation problem.

The second part of this chapter begins by revisiting Cleland's (2001, 2002, 2011) account of the methodology and justification of the historical field sciences. Her proposal departs from traditional philosophical accounts in that it grounds the acceptance and rejection of hypotheses in the historical natural sciences not on formal logico-mathematical considerations but on a physically pervasive, time asymmetry of causation. She contends that this "asymmetry of overdetermination," as it has come to be known, also sheds light on some aspects of evidential reasoning in the experimental sciences that have traditionally been viewed as problematic. As we discuss, this is not the only asymmetry of causation playing a critical role in the methodologies of historical and experimental science. The remainder of the chapter is devoted to ferreting out some highly general, causal components in the methodological reasoning of nonhistorical field scientists. We argue that differences in patterns of evidential reasoning in the experimental sciences versus the field sciences and in the historical versus nonhistorical field sciences seem tailored to pervasive causal differences in their epistemic situations. We close by suggesting that a philosophical search for additional, highly general, causal considerations keyed to the diverse epistemic situations in which scientists actually find themselves might lead to a philosophically more satisfactory understanding of the ways in which science differs from pseudoscience—for the evidential reasoning of scientists is far more complex than the logical empiricists and their successors portray. Good science exploits causal as well as logical constraints, and to the extent that an account of the difference between science and pseudoscience fails to take this into consideration, it is likely to misclassify some bona fide scientific practices as pseudoscientific.

The Logical Conception of Scientific Methodology

Consult almost any mid-twentieth-century textbook in the philosophy of science and the discussion about what constitutes proper evidential reasoning in science divides into two main camps: those who argue that some sort of formal inductive (probabilistic, or statistical) reasoning can be used to justify the acceptance and rejection of scientific hypotheses (and theories), and those who argue that deductive reasoning in the form of *modus tollens* can be used to justify the rejection but not the acceptance of scientific hypotheses. Both groups concur, however, that the evaluative relation between hypothesis and observation is primarily logico-mathematical. One group insists that this logical relation cannot be inductive, while the other group believes that it must be. The important point is that, whichever approach you endorse, the fundamental evaluative relation between hypothesis and empirical evidence is construed as formal and logical. But why imagine that only formal logico-mathematical relations have the power to protect scientific reasoning? After all, justificationist and falsificationist theories are riddled with well-known difficulties. As we argue here, some of these problems are exacerbated in the field sciences, but (as is well known but often downplayed) they also afflict experimental science.

The traditional philosophical approach to understanding scientific methodology goes something like this: scientists invent hypotheses in any number of ways; sometimes they observe certain regularities and begin to theorize about them, sometimes they have strange dreams, and sometimes they grasp at lucky straws. But however the hypothesis arises, what is important is that the scientist be able to *deduce* a test implication (in essence, a conditional prediction), which forms the basis for a search for confirming or disconfirming empirical evidence, whether in the lab or in the field. For example, suppose one has a hypothesis about cheese that it molds more quickly in a vacuum than it does when permitted to "breathe." It is fairly easy to see that the test one might deduce from this hypothesis requires that samples of cheese be stored in different containers, one of which is vacuum-sealed, to see which sample produces mold first. Let us also suppose that, sure enough, the cheese in the vacuum-sealed container produced signs of surface mold before the other sample did. On the justificationist model, we are permitted to infer that the truth of our hypothesis has just been rendered more probable. Had the test come out differently (either neutral or negative), we would have been required either to devise a new (and presumably better) test or to abandon our hypothesis altogether. But the important point is that, on this view, positive test results confer at least slightly higher degrees of probable truth on hypotheses.

The problems with traditional confirmation theories are well known. The most obvious and intractable is the problem of induction. One cannot justify a claim about unexamined instances by inference from examined ones, and one cannot infer universal laws of nature from the study of a limited number of instances. The famous paradoxes of confirmation are more subtle but of particular interest to us because they result directly from the attempt to characterize the relation between hypothesis and evidence in terms of first-order logic. The raven paradox asks us to consider the hypothesis "All ravens are black," which receives confirming support from the observation of a black raven. But the statement "All ravens are black" is logically equivalent to the statement "All non-black things are non-ravens," and the latter receives confirming support from the observation of a red rose since it is neither black nor a raven. Because these statements are logically equivalent, it seems that we must also conclude that the observation of a red rose confirms the hypothesis that all ravens are black as well. While a few philosophers are willing to bite the bullet and accept that the discovery of a red rose provides some empirical support for the hypothesis, most find such a claim deeply problematic.[5]

The infamous "gruesome paradoxes"—Goodman's (1955) "new riddle of induction"—also draw on a logical conception of the evidential relation between hypothesis and observation. Let the predicate "grue" express the property of being green before the year 2020 and blue afterward. The hypothesis "All emeralds are grue" is just as well supported by the available evidence to date as the hypothesis "All emeralds are green." Yet this seems counterintuitive. Various strategies for overcoming these and other paradoxes of confirmation have been well explored; for a succinct summary, see Trout (2000). But none of them has been fully successful, and the problem of induction continues to haunt even the most sophisticated of the probabilistic confirmation theories, including those grounded in Bayesian logic (Strevens 2004; Howson 2001). We suggest that the problem may lie in the sacrosanct assumption that the evaluative relation between hypothesis and empirical evidence must be fully characterized in terms of a formal logico-mathematical relation of some sort. This suspicion is reinforced when one considers the nature of the difficulties plaguing falsificationism.

Convinced that the problem of induction is unsolvable even for probabilistic theories of confirmation, Popper proposed a different approach, falsificationism. Like justificationism, the methodological ideal is to deduce a test implication from the hypothesis and design a test based on it. The differences between justificationism and falsificationism arise at this point. On a falsificationist approach, the "goal" is to attempt to disprove the hypothesis. Thus, the prediction deduced from the hypothesis ought to be such that one would not ordinarily expect it to be true if the hypothesis were not true; Popper famously called these "risky predictions" and contended that the riskier the prediction, the better the "test" of the hypothesis. The second important difference is that a "positive" test result does not count as confirmatory on this view, for that would be to implicitly accept some sort of principle of induction, the very thing that Popper is seeking to avoid. Instead, any failure to falsify one's hypothesis counts merely as "corroborating" it, which is not to be taken as raising its probability one iota. However, if the test result is negative—the prediction fails—the hypothesis is to be rejected. According to Popper, any temptation to make ad hoc adjustments to the theory in light of the negative results must be resisted, or one risks descending into pseudoscience. The logic of negative empirical results in bona fide science demands the ruthless rejection of the hypothesis being tested (Popper 1963). As is well known, it is this portion of the falsificationist position that is most problematic.

It is worth dwelling a little on the central problem with falsificationism because it highlights one of the major points we are trying to get across in this chapter: purely logico-mathematical characterizations of the evaluative relation between hypothesis and empirical evidence provide unsatisfactory accounts of the rejection and acceptance of scientific hypotheses and, as such, equally unsatisfactory solutions to the demarcation problem. The rationale behind Popper's demand that one ruthlessly reject a hypothesis in the face of a failed prediction is a rule of deductive logic, *modus tollens*. The problem is that predictions are treated as if they were self-contained, that is, involved no assumptions about a particular test situation other than those explicitly endorsed as boundary or initial conditions of the hypothesis under investigation or the theory in which it is embedded. But as Duhem (1954) pointed out some time ago, hypotheses and theories never stand alone when tested in real-world scenarios. A concrete test of a hypothesis, whether conducted in the lab or the field, involves an enormous number of auxiliary assumptions about the test situation, both theoretical and empirical, including assumptions about instrumentation and the absence of potentially interfering factors, many of which are well accepted, poorly understood, or simply unknown. The logical validity of the argument on which the rejection of a hypothesis in the face of a failed prediction is founded implicitly presupposes the truth of all of them. When they are explicitly conjoined to the target hypothesis H by means of placeholders $A_1, A_2 ..., A_n$ in the antecedent of the key premise ("If H and $\{A_i\}$ are both true, then P"), however, it becomes clear that logic is silent about whether one should reject the hypothesis in the face of a failed prediction P or search for a problematic auxiliary assumption(s). The most that logic can tell us is that either the hypothesis or one or more auxiliary assumptions about the test situation are false.

This helps to explain why scientists rarely if ever follow Popper and reject their hypotheses in the face of a failed prediction. Nineteenth-century astronomers, for example, had difficulty reconciling the orbit of Uranus with what Newtonian celestial mechanics suggested they should see. But rather than reject Newton's account, they elected to explain the anomaly by adjusting their background assumption regarding how many planets were in the solar system. Using the Newtonian formulae, they determined a location and mass that could account for the perturbations being observed and, training their telescopes on the location, found Neptune. Admittedly, from a Popperian perspective, such a move seems reasonable (as opposed to ad hoc) because astronomers were aware of the limitations of their telescopes and hence open to the possibility of undiscovered solar bodies gravitationally affecting the known solar bodies. Things did not go so well, however, when the same technique was applied to account for anomalies observed in the orbit of Mercury. In that case, adjustment of the same sort of auxiliary assumption—that there is no planet between Mercury and the sun—bore no fruit at all. Researchers searched in vain for the planet (dubbed "Vulcan" for its close proximity to the sun) and even explored the possibility of pinning the blame on other auxiliary assumptions (e.g., the sun's mass is not homogeneous). In hindsight, the latter efforts seem very ad hoc. Few astronomers saw it that way at the time, however. Indeed, up until the advent of Einstein's theory of general relativity, which solved the problem by dispensing with Newton's theory of celestial mechanics instead of an auxiliary assumption, astronomers persisted in trying to explain Mercury's anomalous orbit in terms of a defective auxiliary assumption. In retrospect, this hardly seems surprising. Popper and fellow travelers were never able to provide a principled (let alone logical) distinction between ad hoc and non–ad hoc modifications of auxiliary assumptions. In short, not only did astronomers not feel compelled to behave like falsificationists, logic did not compel them to do so, for logic was unable to determine whether the deviant orbits of Uranus and Mercury represented a defect in Newton's theory or an auxiliary assumption about the observational situations to which his theory was being applied.

The same difficulty about auxiliary assumptions also haunts successful predictions: was a prediction successful because the hypothesis is correct or was it a "false positive" for which success is due merely to the presence of extraneous conditions? Returning to our hypothesis about the mold on cheese, we might notice that even if we were to run the test with positive results several more times, there can be all kinds of alternative explanations about why the results obtain. Perhaps the samples differ in density or sugar (lactose) content, or perhaps the vacuum-sealing apparatus is accidentally introducing mold spores, or perhaps the sealed sample occupies a warmer pocket of the refrigeration unit. Any one of these conditions might be responsible for the successful prediction, and this is only the tip of the iceberg. A little ingenuity can uncover many more. While we might think that each of these alternative explanations is less plausible than our hypothesis, the point is that such thinking cannot be justified by reference to a *logical* relation between hypothesis and the evidence available.

Popper hoped to avoid the problems of justificationism by means of falsificationism, but it should be clear by now that the problem of false positives and false negatives afflicts both approaches and, most important, that logic cannot help us resolve it. Indeed, logic is an equal opportunity employer when it comes to deciding between auxiliary assumptions and the target hypothesis. And if that were not enough, these difficulties become exacerbated once science is practiced outside of the artificial confines of the lab. If identifying auxiliary assumptions is difficult in the ideal conditions of a controlled laboratory situation, then try doing so in the messy, uncontrollable world of nature.

Consider, for example, trying to control for every variable in the population during a drug trial. Imagine that—for whatever reason—the investigators end up with a large portion of their test population living in homes with unfinished basements in which they all spend considerable time. Let us further imagine that each of the basements has a level of radon present and that each member in the population with an unfinished basement happens to respond better to the medication being tested than other members. It would not be surprising if investigators never think to ask their subjects about their basements—who would imagine it to be relevant? False positives of this sort serve to undermine any claim a justificationist might make about having a viable method for doing science out "in the world."

More difficult still is the project of attempting to predict events within even larger and more complex, open systems. Despite the best modeling computers available, no one expects to be able to identify the precise strength or path of a hurricane; the system is simply too large and the number of variables incomprehensibly great. Even if we knew what all of them were (and we certainly do not!), it would not be possible to accommodate them all in the model. Thus, even when such unwieldy systems do not behave in exactly the way the model predicts, the presence of so many possible variables, many of which are not represented in the model because they are either unknown or not thought to be relevant, means that nothing about one's theory need follow from the negative result. There are many explanations for the result, and good scientific practice seems to be committed to identifying which of them is most plausible in light of the current state of our scientific knowledge.

From either a justificationist or falsificationist perspective, the field sciences seem even more hobbled in their dealings with nature than the experimental sciences. Because of their scale, complexity, and specificity, the physical systems targeted cannot be adequately represented in a laboratory setting for purposes of controlling for plausible interfering factors that could give rise to false positives and false negatives. Moreover, computer models of such systems cannot fully represent them and hence run the risk of failing to incorporate critical factors that are erroneously assumed not to be relevant. Under such circumstances, one would expect research in the field sciences to enjoy little in the way of successes. Yet in recent years, some of the field sciences have logged an impressive number of high-profile results. This is especially true of historical field sciences such as paleontology, paleoclimatology, epidemiology, evolutionary biology, historical geology, and cosmology: scientists believe that they have compelling evidence that the universe originated 13.7 billion years ago in a cosmic "Big Bang," that Earth is 4.5 billion years old, that there was life on Earth as early as 3.8 billion years ago, that all life on our planet descends from a last universal common ancestor, and that the end-Cretaceous mass extinction was caused by a gigantic meteorite. These successes strongly suggest that in some of the field sciences at least, the evaluative relation between hypothesis and empirical evidence rests on considerations in addition to those of formal logic—that the historical natural sciences are not as hobbled as traditional logico-mathematical accounts of justification and falsificationism seem to indicate.

Cleland (2001, 2002, 2011) has defended this claim in a number of papers. She argues that scientists investigating natural history exploit a pervasive time asymmetry of causation (aka the asymmetry of overdetermination) in their evidential reasoning, and that this physical relation provides the needed epistemic warrant for hypotheses about the remote past. She also argues that the asymmetry of overdetermination helps to explain the tendency of experimental scientists to ignore Popper's admonishments and search for defective auxiliary assumptions when faced with failed predictions; as will become apparent, experimentalists, who are somewhat disadvantaged by the asymmetry of overdetermination, exploit a different causal asymmetry in their evidential reasoning, namely, the asymmetry of manipulation. It is our contention that failure to recognize that the actual practices of experimental scientists and historical field scientists are tailored to deal with critical causal differences in their respective epistemic situations is largely responsible for the view that the latter borders on pseudoscience. The following section explains how historical and experimental scientists draw on these two asymmetries in their practices and evidential reasoning and, in so doing, points the way to a very different kind of solution to the problem of demarcation.

The Role of Causal Asymmetries in Evidential Reasoning

That there are prima facie differences in the ways in which hypotheses are typically tested and evaluated in experimental science and historical science has long been noted. What has been less well understood is that these differences track pervasive causal differences in the evidential situations in which these researchers typically find themselves. More specifically, localized events are causally connected in time in an asymmetric manner: they tend to *over*determine their causes (because the latter typically leave extensive and diverse effects) and *under*determine their effects (because they rarely constitute the total cause of an effect). As an illustration, consider an explosive volcanic eruption. Its effects include extensive deposits of ash, pyroclastic debris, masses of andesitic or rhyolitic magma, and a large crater. Only a small fraction of this material is required to infer the occurrence of the eruption. Indeed, any one of an enormous number of remarkably small subcollections of these effects will do. This helps to explain why geologists can confidently infer that a massive, caldera-forming eruption occurred 2.1 million years ago in what is now Yellowstone National Park. By contrast, predicting even the near-future eruption of a volcano such as Mt. Vesuvius is much more difficult. There are too many causally relevant conditions (known and unknown) in the absence of which an eruption simply will not occur. David Lewis (1991, 465–67) dubbed this time asymmetry of causation "the asymmetry of overdetermination."

The asymmetry of overdetermination is very familiar to physicists. One might even say that the asymmetry forms part of their "stock in trade." Examples such as an explosive volcanic eruption are amenable to explanation in terms of the second law of thermodynamics, which says that physical systems spontaneously move in the direction of increasing disorder but not vice versa. The natural processes that produce volcanoes are irreversible; one never sees a volcano literally swallow up the debris it produced and return the land around it to the condition it was in before the eruption occurred. The asymmetry of overdetermination also encompasses wave phenomena, which do not obviously admit of a thermodynamic explanation. Although traditionally associated with electromagnetic radiation (light, radio waves, etc.), the "radiative asymmetry" (as it is known) characterizes all wave-producing phenomena, including disturbances in water and air. By way of illustration, consider dropping a stone into a still pool of water.[6] Expanding concentric ripples spread outward from the point of impact. It is easy to explain these ripples in terms of a stone's entering the water at a small region of the surface of the pool. Indeed, one can pinpoint where the stone entered by examining a small segment of the pool's surface. But now consider eliminating all traces of the impact. An enormous number of separate and independent interventions are required all over the surface of the pool. Similarly, try explaining the time-reversed process. Ripples, which expanded outward from the point of impact, now contract inward to the point of impact. But there is no center of action to explain the simultaneous and coordinated behavior of the individual water molecules involved. From this time-reversed perspective, the contracting concentric waves seem to be a miracle; we can understand them causally only by running the process in the other direction (forward) in time.

Because localized events are causally connected in time in this asymmetric manner, historical scientists and experimental scientists typically (but not always) find themselves in quite different evidential situations. It is no surprise that their practices reflect these differences. The hypotheses tested in the lab by stereotypical ("classical") experimentalists usually have the form of generalizations expressing law-like regularities among generic events (event-types) (e.g., all copper expands when heated). The experimentalist proceeds roughly along the following (highly idealized) lines. She infers a test implication from the hypothesis specifying that an effect of a given sort E (expansion of a piece of copper) will be realized if a test condition of a certain kind C (the piece of copper is heated) is brought about. She then manipulates laboratory apparatus to bring C about and looks for the predicted effect E. In virtue of constituting only part of the total cause, localized conditions such as C typically underdetermine their effects, and the researcher is confronted with the very real possibility of false positives and false negatives. For this reason, she manipulates a host of variables (amounting to the denial of suspect auxiliary assumptions) while holding C constant. Sometimes an investigator will even remove C to evaluate whether something in the experimental situation other than C is giving rise to the successful or failed prediction. Depending on the ingenuity of her experimental design, further repetitions permit our investigator to build a body of evidence in support of her hypothesis.[7]

In contrast, many (but not all) of the hypotheses investigated by historical scientists designate particular events (event tokens). Consider the question of whether the end-Cretaceous mass extinction was caused by the impact of a gigantic meteorite. Obviously, there are no analogous manipulations that an investigator can perform to test a hypothesis about a particular event in the remote past. But this is not to say that he cannot procure empirical evidence for such hypotheses. Just as there are many different possible collections of evidence suitable for catching criminals, so there are many evidentiary traces for establishing what caused the extinction of the dinosaurs or the disappearance of the Clovis culture. Much like a good criminologist, historical scientists postulate a small handful of different causal etiologies for the traces they observe. Empirical research consists in searching for a "smoking gun," a trace (or traces) that when added to the prior body of evidence establishes that one (or more) of the hypotheses being entertained provides a better explanation for the total body of evidence now available than the others. The point is historical researchers are not concerned merely with discovering additional traces. They are searching for *telling* traces—traces capable of adjudicating among competing hypotheses for which the evidence available is ambiguous or neutral. The thesis of the asymmetry of overdetermination informs us that for any collection of rival historical hypotheses it is highly likely that such traces exist.[8] It thus underwrites the quest for a smoking gun; the research challenge is being clever enough to find traces of this sort. A quick examination of the history of the scientific debate over the meteor-impact hypothesis for the end-Cretaceous mass extinction nicely illustrates this distinctive pattern of evidential reasoning.

Prior to 1980, there were many different explanations for the extinction of the dinosaurs and other end-Cretaceous flora and fauna, including pandemic disease, evolutionary senescence, global climate change, volcanism, supernova, and meteorite impact. The father-and-son team of Luis and Walter Alvarez and others (1980) were surprised to discover[9] unusually high concentrations of the element iridium in the K-T boundary (a distinctive geological formation separating the rock record of the Cretaceous from that of the Tertiary). This serendipitous discovery focused attention on the volcanism and meteorite impact hypotheses since they are the only plausible mechanisms for the presence of so much iridium in such a thin layer of Earth's crust;[10] iridium is rare at the surface, but high concentrations exist in Earth's interior and in many meteors. Subsequent discoveries of large quantities of a special kind of shocked mineral (predominantly quartz) that forms only under enormous pressures and is found only in meteor craters and the sites of nuclear explosions (Alexopoulos et al. 1988) clinched the case in favor of meteorite impact for most geologists. In essence, the iridium anomaly coupled with extensive quantities of shocked quartz functioned as a smoking gun for the Alvarez hypothesis. While a crater of the right size and age was eventually identified off the coast of the Yucatan Peninsula, it was generally conceded that failure to find one would not count heavily against the hypothesis. The active geology of Earth could easily have obliterated all traces of the crater, particularly if the impact had occurred in the deep ocean. The Alvarez meteorite impact hypothesis became the widely accepted scientific explanation for the end-Cretaceous mass extinction because, of the available hypotheses, it provided the greatest *causal unity* to the diverse and puzzling body of traces (iridium anomaly, extensive quantities of shocked minerals, fossil records of end-Cretaceous plants and animals, etc.) that had been acquired through field investigations.[11]

What we are calling attention to is that both experimental and historical scientists exploit more than a formal logico-mathematical relation of induction (probabilistic, e.g., Bayesian, or statistical, e.g., frequentism) or deduction (falsification) in their testing and evaluation of hypotheses in light of empirical evidence. For the experimental investigator, working as she is in the direction from present to future, formal reasoning cannot fill in the gaps created by causal underdetermination. The experimenter is always threatened by the very real possibility of false positives in the case of successful predictions and false negatives in the case of failed predictions. She attempts to mitigate the problem by identifying and empirically testing suspect auxiliary assumptions. Her great investigative advantage—and the key to the genuinely scientific nature of her method—is that she is able to affect the future by exploiting her capacity to manipulate the present.[12] If she were a mere observer like Dummett's (1964) intelligent trees, she could not do this; she would be utterly at the mercy of the underdetermination of future events by whatever happens in the present.[13]

In some ways, the historical scientist is in the same position with respect to the past as Dummett's intelligent trees are with respect to the future: she cannot affect the past by manipulating the present. But she has an advantage that the experimentalist lacks, namely, the existence in the present of records (many and diverse collections of traces) of long-past events. Her research practices are designed to exploit these records; no such records of the future exist in the present for the experimentalist to exploit. The quest for a smoking gun is thus a search for additional evidential traces aimed at distinguishing which of several rival hypotheses provides the best explanation for the available body of traces. The overdetermination of the past by the localized present, a physically pervasive feature of our universe, ensures that such traces are likely to exist if the initial collection of traces shares a last common cause.[14]

For insofar as events typically leave numerous and diverse effects, only a small fraction of which are required to identify them, the contemporary environment is likely to contain many, as yet undiscovered, smoking guns for discriminating among rival common cause hypotheses. In this way, historical scientists are able to mitigate the epistemic threat of their own version of false positives and false negatives resulting from, among other things, the activities of information degrading processes on traces of the past (Cleland 2011).

In summary, experimental and historical scientists appeal to and cope with different extralogical considerations—respectively different sides of the asymmetries of manipulation and overdetermination—in the testing and evaluation of hypotheses. These considerations reflect pervasive and critical differences in their epistemic situations and, as such, demand different patterns of evidential reasoning. Experimentalists lack records of the future but, unlike Dummett's intelligent trees, have the capacity to affect it by manipulating the present. As a consequence, a key component of what comprises good experimental science is the requirement that they explore possibilities for false positives and false negatives by manipulating suspect conditions in controlled laboratory situations. Historical scientists, on the other hand, are unable to affect the past by manipulating the present but, in virtue of the asymmetry of overdetermination, have potential access to remarkably detailed records of it. Their central research task is to plumb these records for telltale traces of what actually happened, thereby allowing them to advance or exclude various possibilities. At the end of the day, any adequate account of the distinction between science and pseudoscience must recognize both the asymmetrical character of the causal reasoning patterns native to the historical and experimental sciences and, most important, their epistemic appropriateness. Indeed, it would be irrational for a historical scientist to behave like an experimentalist or vice versa, except in those rare cases in which experimentalists and historical scientists find themselves in evidentially analogous situations; see Cleland (2002) for a discussion.

Toward an Account of the Nonhistorical Field Sciences

The natural sciences may be crudely divided into two broad categories, experimental and field science, and the latter may be further divided into historical and nonhistorical field science. We have shown how differences in the practices of experimental and historical scientists can be explained in terms of pervasive causal features of the situations in which they typically find them-selves. In this section, we hope to motivate the idea that similar considerations can be used to shed light on some central but poorly understood features of the methodology of the nonhistorical field sciences—features that seem to underlie much of the general skepticism directed toward them by educated, well-intentioned people.[15]

The epistemic situation of many field scientists engaged in nonhistorical research is more complex than that of either historical scientists or experi-mentalists. Like experimentalists, field scientists have the burden of forecast-ing future events. But unlike experimentalists, they have little to no recourse with respect to the experimental manipulation of key parameters and suspect auxiliary assumptions; the systems in which they operate are physically open, large in scale, and highly complex. Consider, for example, the difficulty of try-ing to manipulate any of the variables in the epidemiology of AIDS research: even if—contrary to fact—there were no strictures on research conducted on human beings, the natural mechanisms of disease transmission simply are not amenable to this sort of manipulation. Similarly, seismologists con-cerned with forecasting earthquakes cannot manipulate (selectively rupture) the responsible faults. But while this makes their task more difficult, it should not be taken to mean that their findings are no more reliable than those of pseudosciences such as astrology and homeopathy. Rather, we suggest that they compensate for their epistemic predicament by extracting information from the historical record and using it to emulate controlled experiments via model simulations.[16]

The use of computer models and simulations is pervasive among scientists attempting to understand phenomena outside the sterile, artificial environs of the lab. Models amount to theoretical hypotheses about complex natural phenomena. Computer simulations, in which the values of key variables of the model are "manipulated" and suspect auxiliary assumptions explored by introducing and omitting various parameters, function as proxies for con-trolled experiments. Studies of the historical record play important roles at all stages of the design, calibration, and evaluation of models. Climate sci-ence provides a good illustration. Because of the asymmetry of overdeter-mination, the present-day environment contains varied and rich sources of information about paleoclimate that would be relevant to performing con-trolled experiments on future climate *if* we could manipulate the conditions concerned.[17] Data are extracted from the historical record in a variety of ways, including gross observation, ice core samples, measurements of cosmogenic isotope exposure, and radiocarbon dating of marine and lake sediment core samples.[18] Climate modelers use this information, along with pertinent information drawn from chemistry and physics (thermodynamics, fluid dynamics, etc.), to develop and refine climate models. By adjusting a wide range of fairly well-understood parameters such as atmospheric and thermohaline circulation, or fluctuations in albedo, they can implement a diversity of model simulations. The more accurately these simulations describe past climate events for which there exist sufficient paleoclimate data, and the more tightly they converge on forecasts of future climatic events, the more confident the researcher is that his projection does not suffer from falsely positive or falsely negative results.

The use of paleoclimate data to couple Arctic amplification (the phenomenon whereby temperature changes observed at the Arctic tend to exceed those observed in the rest of the Northern Hemisphere) more tightly to global average temperature changes provides a good illustration. Long believed to play an important role in global warming and cooling events, Arctic amplification is incorporated into most climate models and is said to "force" events by increasing concentrations of greenhouse gases. Recent field observations indicate that it has played a significant role in the warming trend of the past century, suggesting that it will play an important role in the future. The problem is calibrating its magnitude in light of different forcing and feedback mechanisms. To this purpose, Miller and colleagues (2010) have run model simulations on paleoclimate data from four intervals in the past 3 million years that were both warmer and cooler than today. They describe these simulations as "the natural experiments of the past."

The guiding maxim in the use of the historical record in non-historical field research is that *in situ* causal relations extending from present to future will resemble those extending from past to present in critical (but not all) respects. From a traditional justificationist perspective this maxim is deeply problematic. Its plausibility seems to rest on a principle of induction, yet one cannot rule out the possibility that the future will someday cease to resemble the past. But suppose there *were* a change in the course of nature and the asymmetry of overdetermination ceased to apply? In that case, one could no longer do historical science. Similarly, one could not do experimental science if one were, like Dummett's unfortunate intelligent trees, unable to manipulate the present. In short, we concede that our account cannot escape the problem of induction. But we are hardly alone in being unable to do this. No inductivist or falsificationist account founded on logical considerations has been able to escape the hoary problem of induction, and this includes Bayesian confirmation theory (Howson 2001). Nonetheless, it is an empirical fact that the future *does* have a pronounced tendency to resemble the past, especially in ultimately general ways such as the asymmetry of overdetermination. One of the central purposes of this chapter is to point out that the evidential reasoning of scientists is designed to exploit not only logical but also pervasive causal features of the universe, and moreover that it is rational for them to do so. To put a new twist on an old saw: those who refuse to learn from the past cannot hope to forecast the future.

In conclusion, as we have seen, experimental and historical scientists draw on extralogical, causal considerations—the asymmetries of manipulation and overdetermination, respectively—in the testing and evaluation of their hypotheses. These considerations reflect critical differences in their epistemic situations. Nonhistorical field scientists frequently find themselves in yet another epistemic situation, one (so to speak) straddling that of the historical scientist and the experimental scientist. It is thus hardly surprising that their research practices often draw on the methodologies of both, exploiting the asymmetry of overdetermination for information potentially relevant to manipulation and control and using this information to conduct proxy experiments in the form of model simulations. To the extent that the core evidential relation involved in accepting and rejecting scientific hypotheses has traditionally been taken to be logico-mathematical and the relevant causal and epistemic asymmetries that are also involved have gone unrecognized, it is hardly surprising that the practices of field scientists have been viewed with skepticism. It is important, however, to keep in mind that it is a myth that the evidential relation exploited in classical experimental science is purely logico-mathematical. Experimental scientists also exploit causal features of their epistemic situations. In this context one cannot help but wonder whether the asymmetries of overdetermination and manipulation are the only extralogical features of the universe that could be exploited for rationally (but not logically) justifying hypotheses on the basis of observation. A philosophical search for additional such factors in the context of analyses of the diverse epistemic situations in which scientists find themselves might lead to a philosophically more satisfying understanding of scientific reasoning, and hence of the distinction between genuine science and pseudoscience.

NOTES

1. See, e.g., Mooney (2011); Kahan, Jenkins-Smith, and Braman (2011); Kunda (1990); or Lord, Ross, and Lepper (1979) for psychological accounts of skepticism about scientific findings.

2. Most experimentalists holding this view do not express it in public. But there are some revealing exceptions. A salient illustration is Henry Gee, at the time an editor of the prestigious science journal *Nature,* who explicitly attacked the scientific status of all the historical field sciences on the grounds that their hypotheses "can never be tested by experiment" (1999, 5). Another good illustration is "A Scientific Dissent from Darwinism," a privately funded statement signed by approximately one hundred scientists and published in the *New York Review of Books* in 2001 (23). The vast majority of signatories were experimentalists; they listed their fields after their names.

3. We use the latter terms interchangeably in this chapter; nothing of importance rides on this.

4. See Pearl (2000) and Sprites et al. (2000) for more detail on these formal Bayesian structures.

5. Few scientists would be willing to accept a red rose as evidence for the hypothesis that all ravens are black; indeed, a red rose seems irrelevant. For an analysis of the paradox and the assumptions on which it rests, as well as some possible resolutions, see Maher (1999).

6. We owe this example to Popper (1956, 538).

7. A caveat is in order: philosophical investigations—for example, Hacking (1983) and Franklin (1999)—into the methodology of experimental science have established that much of the work that goes on there is more exploratory and lacks the character of classical experimental science. Nonetheless, the latter is traditionally held up as the ideal form of experimentation.

8. It is important to keep in mind that a trace has the status of a smoking gun only in the context of a given body of evidence and a collection of competing hypotheses. That is to say, the concept of a smoking gun is inherently comparative; it does not provide absolute support for a hypothesis considered in isolation.

9. As Cleland (2011) argues, the iridium anomaly was not predicted and indeed could not be predicted from a meteorite impact even today because, among other things, not all meteorites are rich in iridium, only those left over from the formation of the solar system.

10. See Clemens et al. (1981) for the original paper in defense of the volcanism hypothesis for the iridium anomaly.

11. Many paleontologists, while conceding that a gigantic meteorite impact occurred at the end of the Cretaceous, nonetheless remained unconvinced that the second prong of the Alvarez hypothesis (that the impact explained the extinctions) was true. The story of their field investigations and eventual conversion to the Alvarez hypothesis is fascinating but beyond the scope of our discussion. As Cleland (2002) points out, studies of the fossil records of end-Cretaceous ammonites, flowering plant pollen, and fern spores demonstrated that the extinction was worldwide and geologically instantaneous, providing a smoking gun for the second prong of the hypothesis.

12. Turner (2007) also emphasizes this point in a discussion of the methodology of experimental and historical science, although he downplays the epistemic threat posed to experimental science by the asymmetry of overdetermination as well as the epistemic ad-

vantage that the asymmetry of overdetermination confers on historical science; see Cleland (2011) for a discussion.

13. Woodward (2003, 11), perhaps the most influential contemporary advocate of a manipulability theory of causation, contends that passive observers such as Dummett's intelligent trees could not even develop a concept of causation. But he treats causal relations (potential manipulations) as purely formal structures, more specifically, as directed graphs closely resembling Bayesian networks. It seems clear that there are important features of causal interactions and manipulations that cannot be captured in formal networks; one could, for instance, easily construct purely spatial analogues involving no causal interactions whatsoever.

14. See Cleland (2011) for a discussion of the justification of the principle of the common cause, and hence common cause explanation, in terms of the thesis of the asymmetry of overdetermination.

15. We are not denying, of course, that some work in the field sciences is shoddy or otherwise inadequate. It is important to keep in mind, however, that the same is true of the (classical, or hypothesis testing) experimental sciences; consider, for instance, the much maligned experimental work of electrochemists Stanley Pons and Martin Fleischmann on cold fusion in the late 1980s.

16. Model simulations are also pervasive in contemporary historical research. But alas, we do not have the space here to compare and contrast the ways in which they are used in historical and nonhistorical field science; unsurprisingly some of the differences track important differences in the epistemic situations of their practitioners.

17. Woodward (2003) also emphasizes the use of information that is potentially relevant for manipulation and control in circumstances where actual manipulation is impossible. But he fails to recognize the critical role of the asymmetry of overdetermination in supplying such information via records of the past.

18. See Steig et al. (1998) and Miller et al. (2010) for examples of the use of proxies for paleoclimate data.

REFERENCES

Alexopoulos, J. S., R. A. F. Grieve, and P. B. Robertson. 1988. "Microscopic Lamellar. Deformation Features in Quartz: Discriminative Characteristics of Shock-Generated Varieties." *Geology* 16:796–99.

Alvarez, Luis W., Walter Alvarez, F. Asaro, and H. V. Michel. 1980. "Extraterrestrial Cause for the Cretaceous-Tertiary Extinction." *Science* 208:1095–1108.

Carnap, Rudolf. 1928. *Der Logische Aufbau der Welt.* Leipzig: Meiner.

Clemens, W. A., J. D. Archibald, and L. J. Hickey. 1981. "Out with a Whimper Not a Bang." *Paleobiology* 7:293–98.

Cleland, Carol E. 2001. "Historical Science, Experimental Science, and the Scientific Method." *Geology* 29:987–990.

———. 2002. "Methodological and Epistemic Differences between Historical Science and Experimental Science." *Philosophy of Science* 69:474–96.

———. 2011. "Prediction and Explanation in Historical Natural Science." *British Journal of Philosophy of Science* 62 (3): 551–82.

Duhem, Pierre. 1954. *The Aim and Structure of Physical Theory.* Translated by Philip P. Wiener. Princeton, NJ: Princeton University Press.

Dummett, Michael. 1964. "Bringing About the Past." *Philosophical Review* 73:338–59.

Franklin, Allan. 1999. *Can That Be Right?* Dordrecht: Kluwer Academic.

Gee, Henry. 1999. *In Search of Deep Time.* New York: Free Press.

Goodman, Nelson. 1955. *Fact, Fiction, and Forecast.* Cambridge, MA: Harvard University Press.

Hacking, Ian. 1983. *Representing and Intervening.* Cambridge: Cambridge University Press.

Hempel, Carl G. 1962. "Deductive-Nomological vs. Statistical Explanation." In *Minnesota Studies in the Philosophy of Science*, vol. 3, edited by Herbert Feigl and Grover Maxwell. Minneapolis: University of Minnesota Press.

Howson, Colin. 2001. *Hume's Problem: Induction and the Justification of Belief.* Oxford: Oxford University Press.

Kahan, Dan, Hank Jenkins-Smith, and Donald Braman. 2011. "Cultural Cognition of Scientific Consensus." *Journal of Risk Research* 14:147–74.

Kunda, Ziva. 1990. "The Case for Motivated Reasoning." *Psychological Bulletin* 108 (3): 480–98.

Lewis, David. 1991. "Counterfactual Dependence and Time's Arrow." *Nous* 13 (4): 455–76.

Lord, Charles G., Lee Ross, and Mark R. Lepper. 1979. "Biased Assimilation and Attitude Polarization: The Effects of Prior Theories on Subsequently Considered Evidence." *Journal of Personality and Social Psychology* 37 (11): 2098–2109.

Maher, Patrick. 1999. "Inductive Logic and the Ravens Paradox." *Philosophy of Science* 66:50–70.

Miller, Gifford H., R. B. Alley, J. Brigham-Grette, J. J. Fitzpatrick, L. Polyak, M. C. Serreze, and J. W. C. White. 2010. "Arctic Amplification: Can the Past Constrain the Future?" *Quarternary Science Reviews* 29:1779–90.

Mooney, Chris. 2011. "Rapture Ready: The Science of Self Delusion." *Mother Jones*, May/June.

Pearl, Judea. 2000. *Causality: Models, Reasoning and Inference.* Cambridge: Cambridge University Press.

Popper, Karl 1956. Letter. *Nature* 177:538.

———. 1963. "Science: Conjectures and Refutations." In *Philosophy of Science,* edited by Martin Curd and Jan Cover, 3–10. New York: Norton.

Spirtes, Peter, Clark Glymour, and Richard Scheines. 2000. *Causation, Prediction, and Search.* Cambridge, MA: MIT Press.

Steig, Eric J., Alexander P. Wolfe, and Gifford H. Miller. 1998. "Wisconsinan refugia and the Glacial History of Eastern Baffin Island, Arctic Canada: Coupled Evidence from Cosmogenic Isotopes and Lake Sediments." *Geology* 26 (9): 835–38.

Strevens, Michael. 2004. "Bayesian Confirmation Theory: Inductive Logic, or Mere Inductive Framework?" *Synthese* 141:365–79.

Trout, J. D. 2000. "Confirmation, Paradoxes of." In *A Companion to the Philosophy of Science,* edited by W. H. Newton-Smith, 53–55. Oxford: Oxford University Press.

Turner, Derek. 2007. *Making Prehistory.* Cambridge: Cambridge University Press.

Woodward, James. 2003. *Making Things Happen: A Theory of Causal Explanation.* Oxford: Oxford University Press.

11

Science and Pseudoscience

The Difference in Practice and the Difference It Makes

MICHAEL SHERMER

Pseudoscience is necessarily defined by its relation to science and typically involves subjects that are either on the margins or borderlands of science and are not yet proven, or have been disproven, or make claims that sound scientific but in fact have no relationship to science. Pseudonyms for pseudoscience include "bad science," "junk science," "voodoo science," "crackpot science," "pathological science," and, most pejoratively, "nonsense." In this chapter, I consider the popularity of pseudoscience among members of the general public and the numerous problems in finding agreement among scientists, philosophers, and historians of science on how best to demarcate science from pseudoscience. I also examine how science is defined as a way of distinguishing it from pseudoscience; some examples of science, pseudoscience, and in-between claims; as well as how the legal system deals with the demarcation problem in court cases that require such a determination to adjudicate a legal dispute. Moreover, I discuss some possible lines of demarcation between science and pseudoscience, examples of scientists and pseudoscientists, a tale of two fringe scientists and why one may succeed while the other may not, and finally a pragmatic solution to the problem of distinguishing between science and pseudoscience.

Some Demographics of Pseudoscientific Beliefs

When 2303 adult Americans were asked in a 2009 Harris poll to "indicate for each one if you believe in it, or not," the following results were revealing (Harris Interactive 2009):

God	82%
Miracles	76%
Heaven	75%
Jesus is God or the son of God	73%
Angels	72%
Survival of the soul after death	71%
The resurrection of Jesus Christ	70%
Hell	61%
The virgin birth (of Jesus)	61%
The devil	60%
Darwin's theory of evolution	45%
Ghosts	42%
Creationism	40%
UFOs	32%
Astrology	26%
Witches	23%
Reincarnation	20%

More Americans believe in angels and the devil than believe in the theory of evolution. To scientists, this is a disturbing finding. And yet, such results match similar survey findings for belief in pseudoscience and the paranormal conducted over the past several decades (Moore 2005), including internationally (Lyons 2005). For example, a 2006 *Readers Digest* survey of 1006 adult Britons reported that 43 percent said that they can read other people's thoughts or have their thoughts read, more than half said that they have had a dream or premonition of an event that then occurred, more than two-thirds said they could feel when someone was looking at them, 26 percent said they had sensed when a loved one was ill or in trouble, and 62 percent said that they could tell who was calling before they picked up the phone. A fifth said they had seen a ghost, and nearly a third said they believe that near-death experiences are evidence for an afterlife ("Britons Report" 2006). Less common but still interesting, more than 10 percent thought they could influence machinery or electronic equipment using their minds, and another 10 percent said something bad had happened to another person after they had wished for it to happen. A 2005 Gallup Poll found these levels of belief in subjects that most scientists would consider to be pseudoscience or nonsense (Moore 2005):

Psychic or spiritual healing	55%
Demon possession	42%
ESP	41%
Haunted houses	37%
Telepathy	31%
Clairvoyance (know past/predict future)	26%
Astrology	25%
Psychics are able to talk to the dead	21%
Reincarnation	20%
Channeling spirits from the other side	9%

Although most scientists reject such beliefs, from the perspective of those making the unaccepted claims, what is being presented is something more like a new aspect of science, an alternative science, a prescience, or a revolutionary science. In a culture in which science is given high status (indeed, we are often said to be living in the age of science), one would expect political theories (scientific socialism), religions (Christian science, scientology, creation science), and even literature (science fiction) to try to associate themselves with science. And as in these examples, they often do. Precisely for this reason, the boundaries between science and pseudoscience must be explored to distinguish pseudoscience from mistaken science or not fully accepted science.

The Demarcation Problem

This distinction between science and pseudoscience is known as the demarcation problem—where do we draw the boundaries between science and pseudoscience, or between science and nonscience? The problem is that it is not always, or even usually, clear where to draw such a line. Whether a particular claim should be put into the set labeled science or pseudoscience will depend not only on the claim per se, but on other factors as well, such as the proponent of the claim, the methodology, the history of the claim, attempts to test it, the coherence of the theory with other theories, and the like. Here it is useful to expand our heuristic into three categories: normal science, pseudo-science, and borderlands science. The following are examples of claims that might best be placed in each of these three bins:

- *Normal science*: heliocentrism, evolution, quantum mechanics, Big Bang cosmology, plate tectonics, neurophysiology of brain functions, economic models, chaos and complexity theory, intelligence testing.
- *Pseudoscience*: creationism, Holocaust revisionism, remote viewing, astrology, Bible code, alien abductions, unidentified flying objects (UFOs), Bigfoot, Freudian psychoanalytic theory, reincarnation, angels, ghosts, extrasensory perception (ESP), recovered memories.
- *Borderlands science*: string theory, inflationary cosmology, theories of consciousness, grand theories of economics (objectivism, socialism, etc.), Search for Extraterrestrial Intelligence (SETI), hypnosis, chiropractic, acupuncture, other alternative medical practices as yet untested by medical researchers.

Since membership in these categories is provisional, it is possible for theories to be moved and reevaluated with changing evidence. Indeed, many normal science claims were at one time either pseudoscience or borderland science. SETI, for example, is not pseudoscience because it is not claiming to have found anything (or anyone) yet; it is conducted by professional scientists who publish their findings about factors that could lead to the evolution of life elsewhere in the cosmos (extrasolar planets, atmospheres surrounding moons in our solar system, and even the atmospheres of planets revolving around other stars) in peer-reviewed journals; it polices its own claims and does not hesitate to debunk the occasional signals found in the data; and it fits well within our understanding of the history and structure of the cosmos and the evolution of life. But SETI is not normal science either because its central theme has yet to surface as reality. Ufology, by contrast, is pseudoscience. Proponents do not play by the rules of science, do not publish in peer-reviewed journals, ignore the 90 to 95 percent of sightings that are fully explicable, focus on anomalies, are not self-policing, and depend heavily on conspiratorial theorizing about government cover-ups, hidden spacecraft, and aliens holed up in secret Nevada sites (Michaud 2007; Shermer 2001; Shostak 2009; Webb 2003; Achenbach 1999; Sagan 1996).

Likewise, string theory and inflationary cosmology are at the top of borderlands science, soon to be either bumped up into full-scale normal science or abandoned altogether, depending on the evidence that is now starting to come in for these previously untested ideas. What makes them borderlands science instead of pseudoscience (or nonscience) is that the practitioners in the field are professional scientists who publish in peer-reviewed journals and are trying to devise ways to test their theories and falsify their hypotheses. By contrast, creationists who devise cosmologies that they think will best fit biblical myths are typically not professional scientists, do not publish in peer-reviewed journals, and have no interest in testing their theories except against what they believe to be the divine words of God (Pigliucci 2010; Kuhn 1962, 1977; Bynum, Browne, and Porter 1981; Gardner 1981; Taubes 1993; Randi 1982; Olson 1982, 1991).

Theories of consciousness grounded in neuroscience are borderlands science that are progressing toward acceptance as mainstream science as their models become more specific and testable, whereas psychoanalytic theories are pseudoscience because they are either untestable or have failed the tests repeatedly and are grounded in discredited nineteenth-century theories of the mind. Similarly, recovered memory theory is bunk because we now understand that memory is not like a videotape that one can rewind and play back, and that the very process of "recovering" a memory contaminates it. But hypnosis, by contrast, is tapping into something else in the brain, and sound scientific evidence may very well support some of its claims, so it remains in the borderlands of science.

From a pragmatic perspective, science is what scientists do. So if we want to know what science is and how it differs from pseudoscience, we should ask those who practice it. Consider the wisdom of one of the greatest scientists of the twentieth century, the Nobel prize-winning Caltech physicist Richard Feynman. In a 1964 lecture at Cornell University, Feynman explained to an audience of would-be scientists the three steps in discovering a new law in nature:

> How do we look for a new law? First, we guess it. Don't laugh. That's really true. Then we compute the consequences of the guess to see what it would imply. Then we compare those computation results to nature—or to experiment, or to experience, or to observation—to see if it works. If it disagrees with experiment, it's wrong. In that simple statement is the key to science. It doesn't

make any difference how beautiful your guess is, it doesn't make any difference
how smart you are, who made the guess, or what his name is. If it disagrees
with experiment, it's wrong. That's all there is to it. (NOVA 1993)

Science Defined

Is that all there is to it? Well, not exactly. If your name is Feynman—or Einstein,
Hawking, Diamond, or Pinker—you may initially receive a more favorable
hearing. But as Hollywood pundits say about the extensive studio promotion
of a film, a big name or a big promotion will buy you only a week—after that
it stands or falls on its own merits. That is, you have to have the goods, which
in the case of science means you need evidence. This leads to a more precise
definition of science, as follows:

> Science is a set of methods designed to describe and interpret observed or
> inferred phenomena, past or present, aimed at building a testable body of
> knowledge, which means that it is open to rejection or confirmation. (Shermer
> 1997, 18–19)

Even more succinctly:

> Science is a testable body of knowledge open to rejection or confirmation.

The description of methods is important because it shows how science actually works. Included in the methods are such cognitive products as hunches,
guesses, ideas, hypotheses, theories, and paradigms, and testing them involves background research, experiments, data collection and organization,
colleague collaboration and communication, correlation of findings, statistical analyses, conference presentations, and publications.

Although philosophers and historians of science debate passionately about
what science is, they agree generally that science involves what is known for-
mally as the "hypothetico-deductive method," which involves: (1) formulating a hypothesis, (2) making a prediction based on the hypothesis, and
(3) testing whether the prediction is accurate. In formulating hypotheses and
theories, science employs natural explanations for natural phenomena. These
characteristics of science were adopted by the legal system in two important
evolution-creationism trials in the 1980s, one in Arkansas and the other in
Louisiana, the latter of which on appeal went to the US Supreme Court.

Science and Pseudoscience in the Court

Although the law does not determine the nature of science, trials have a way of narrowing our focus and clarifying our definitions because trials do not allow for the kind of highly technical debates often practiced by philosophers of science about the nature of science.

The 1981 Arkansas trial was over the constitutionality of Act 590, which required equal time in public school science classes for "creation science" and "evolution science." The federal judge in that case, William R. Overton, ruled against the creationists on the following grounds: first, he said, creation science conveys "an inescapable religiosity" and is therefore unconstitutional: "Every theologian who testified, including defense witnesses, expressed the opinion that the statement referred to a supernatural creation which was performed by God" (Overton 1985, 280). Second, Overton said that the creationists employed a "two model approach" in a "contrived dualism" that "assumes only two explanations for the origins of life and existence of man, plants and animals: it was either the work of a creator or it was not." In this either-or paradigm, the creationists claim that any evidence "which fails to support the theory of evolution is necessarily scientific evidence in support of creationism." Overton slapped down this tactic: "evolution does not presuppose the absence of a creator or God and the plain inference conveyed by Section 4 [of Act 590] is erroneous" (280).

More important, Judge Overton summarized why creation science is not science. His opinion in this case was rendered important enough to be republished in the prestigious journal *Science*. Overton explained what science is:

1. It is guided by natural law.
2. It has to be explanatory by reference to natural law.
3. It is testable against the empirical world.
4. Its conclusions are tentative.
5. It is falsifiable. (281)

Overton concluded: "Creation science as described in Section 4(a) fails to meet these essential characteristics," adding the "obvious implication" that "knowledge does not require the imprimatur of legislation in order to become science" (281).

The 1987 Louisiana case reinforced the description of science even more because this case was appealed all the way to the US Supreme Court, thereby fulfilling the American Civil Liberty Union's original intent for the 1925 Scopes's Tennessee trial. For this case, seventy-two Nobel laureates, seventeen state academies of science, and seven other scientific organizations submitted an amicus curiae brief to the Supreme Court justices in support of the appellees in *Edwards v. Aguillard*, the case testing the constitutionality of Louisiana's "Balanced Treatment for Creation-Science and Evolution-Science Act," an equal-time law passed in 1982 and subsequently challenged by the ACLU. The brief is one of the most important documents in the history of the evolution-creation debate, and I have written extensively about its import (Shermer 1997). For our purposes here, the brief presents the best short statement on the central tenets of science endorsed by the world's leading scientists and science organizations.

The amicus curiae brief is concise, well documented, and demonstrates first that "creation science" is just a new label for the old religious creationism of decades past. It then moves toward a discussion of why creation science does not meet the criteria of "science" as defined in the brief by the amici: "Science is devoted to formulating and testing naturalistic explanations for natural phenomena. It is a process for systematically collecting and recording data about the physical world, then categorizing and studying the collected data in an effort to infer the principles of nature that best explain the observed phenomena." The Nobelists then explain the scientific method, as follows:

- *Facts.* "The grist for the mill of scientific inquiry is an ever increasing body of observations that give information about underlying 'facts.' Facts are the properties of natural phenomena. The scientific method involves the rigorous, methodical testing of principles that might present a naturalistic explanation for those facts."
- *Hypotheses.* Based on well-established facts, testable hypotheses are formed. The process of testing "leads scientists to accord a special dignity to those hypotheses that accumulate substantial observational or experimental support."
- *Theories.* This "special dignity" language refers to "theory." When it "explains a large and diverse body of facts," it is considered robust. When it "consistently predicts new phenomena that are subsequently observed," it is considered to be reliable. Facts and theories are not to be used interchangeably or in relation to one another as more or less true. Facts are the world's data, and because we cannot interpret those facts without

some theory, theories become the explanatory ideas about those data. By contrast, nontestable statements are not a part of science. "An explanatory principle that by its nature cannot be tested is outside the realm of science."

- *Conclusions*. It follows from this process that no explanatory principles in science are final. "Even the most robust and reliable theory . . . is tentative. A scientific theory is forever subject to reexamination and—as in the case of Ptolemaic astronomy—may ultimately be rejected after centuries of viability. In an ideal world, every science course would include repeated reminders that each theory presented to explain our observations of the universe carries this qualification: 'as far as we know now, from examining the evidence available to us today.'"

- *Explanations*. Science also seeks only *naturalistic explanations* for phenomena. "Science is not equipped to evaluate supernatural explanations for our observations; without passing judgment on the truth or falsity of supernatural explanations, science leaves their consideration to the domain of religious faith." (It should be noted that many scientists do not agree with this demarcation between religion and science, holding that such epistemological walls protects religion from science when it makes empirical claims about the world, such as the age of the earth.) Any body of knowledge accumulated within these guidelines is considered "scientific" and suitable for public school education; and any body of knowledge not accumulated within these guidelines is not considered scientific. "Because the scope of scientific inquiry is consciously limited to the search for naturalistic principles, science remains free of religious dogma and is thus an appropriate subject for public-school instruction."

This case was decided on June 19, 1987, with the Court voting seven to two in favor of the appellees, holding that "the Act is facially invalid as violative of the Establishment Clause of the First Amendment, because it lacks a clear secular purpose" and that "the Act impermissibly endorses religion by advancing the religious belief that a supernatural being created humankind." Since the creationists are not practicing science, as defined by these *ne plus ultra* practitioners of science—the Nobel laureates—they have been disallowed to teach their doctrines as science. (See Shermer 1997 for a complete review and analysis of the amicus curae brief and the primary participants involved.)

Intelligent Design Pseudoscience in Dover, Pennsylvania

In late 2005, Judge John E. Jones III issued a decision in *Kitzmiller v. Dover Area School District* that settled yet another case in which the pseudoscience of creationism under the guise of "Intelligent Design theory" was challenged. *Kitzmiller* was an exceptional court case both for what it revealed about the motives of the ID creationists, and the clarity and severity of the conservative judge's decision against the ID proponents. The Thomas More Law Center (TMLC), founded by conservative Catholic businessman Tom Monaghan and former Kevorkian prosecutor Richard Thompson, was itching for a fight with the ACLU from the time of its formation in 1999. Declaring themselves the "sword and shield for people of faith" and the "Christian Answer to the ACLU," TMLC sought out confrontations with the ACLU on a number of fronts, from public nativity and Ten Commandment displays to gay marriage and pornography. But the fight they really wanted, it seems, was over evolution in public school science classrooms, a fight that would take five years to occur.

TMLC representatives traveled the country from at least early 2000, encouraging school boards to teach ID in science classrooms. From Virginia to Minnesota, TMLC recommended the textbook *Of Pandas and People* as a supplement to regular biology textbooks, promising to defend the schools free of charge when the ACLU filed the inevitable lawsuit. Finally, in summer 2004, they found a willing school board in Dover, Pennsylvania, a board known to have been searching for a way to get creationism inserted into its science classrooms for years.

On October 18, 2004, the Dover school board voted six to three to add the following statement to their biology curriculum: "Students will be made aware of the gaps/problems in Darwin's theory and of other theories of evolution including, but not limited to, intelligent design. Note: Origins of life is not taught." The next month, the board added a statement to be read to all ninth-grade biology classes at Dover High:

> The Pennsylvania Academic Standards require students to learn about Darwin's theory of evolution and eventually to take a standardized test of which evolution is a part.
>
> Because Darwin's Theory is a theory, it is still being tested as new evidence is discovered. The Theory is not a fact. Gaps in the Theory exist for which there is no evidence. A theory is defined as a well-tested explanation that unifies a broad range of observations.

> Intelligent design is an explanation of the origin of life that differs from Darwin's view. The reference book, *Of Pandas and People*, is available for students to see if they would like to explore this view in an effort to gain an understanding of what intelligent design actually involves.

> As is true with any theory, students are encouraged to keep an open mind. The school leaves the discussion of the origins of life to individual students and their families. As a standards-driven district, class instruction focuses upon preparing students to achieve proficiency on standards-based assessments.

Copies of the book *Of Pandas and People* were made available to the school by William Buckingham, the chair of the curriculum committee, who raised $850 from his church to purchase copies of the book for the school. As he told a Fox affiliate in an interview the week after the school board meeting, "My opinion, it's okay to teach Darwin, but you have to balance it with something else such as creationism." Eleven parents of students enrolled in Dover High would have none of this, and on December 14, 2004, they filed suit against the district with the legal backing of the ACLU and Americans United for Separation of Church and State. The TMLC had the fight it was aching for, but it was not going to get the outcome it wanted. The suit was brought in the US District Court for the Middle District of Pennsylvania, and a bench trial was held from September 26 to November 4, 2005, presided over by Judge John E. Jones III, a conservative Christian appointed to the bench in 2002 by President Bush.

The primary task of the prosecution was to show not only that ID is not science, but that it is just another name for creationism, which the US Supreme Court had already decided in *Edwards v. Aguillard*—the Louisiana case—could not be taught in public schools. Expert scientific witnesses testified on behalf of the prosecution, such as Brown University molecular biologist Kenneth Miller and University of California at Berkeley paleontologist Kevin Padian, both of whom rebutted specific ID claims. More important were the expert testimonies of the philosophers Robert Pennock, from Michigan State University, and Barbara Forrest from Southeastern Louisiana University, both of whom had authored definitive histories of the ID movement. Pennock and Forrest presented overwhelming evidence that ID is, in the memorable phrase of one observer, nothing more than "creationism in a cheap tuxedo."

It was revealed, for example, that the lead author of the book *Of Pandas and People*, Dean Kenyon, had also written the foreword to the classic creationism textbook *What Is Creation Science?* by Henry Morris and Gary Parker. The second author of *Pandas*, Percival Davis, was the coauthor of a young-earth creationism book called *A Case for Creation*. But the most damning evidence was in the book itself. Documents provided to the prosecution by the National Center for Science Education revealed that *Of Pandas and People* was originally titled *Creation Biology* when it was conceived in 1983, then *Biology and Creation* in a 1986 version, which was retitled yet again a year later to *Biology and Origins*. Since this was before the rise of the ID movement in the early 1990s, the manuscripts referred to "creation," and fundraising letters associated with the publishing project noted that it supported "creationism." The final version, by now titled *Of Pandas and People*, was released in 1989, with a revised edition published in 1993. Interestingly, in the 1986 draft, *Biology and Creation*, the authors presented this definition of the central theme of the book, creation:

> Creation means that the various forms of life began abruptly through the agency of an intelligent creator with their distinctive features already intact. Fish with fins and scales, birds with feathers, beaks, and wings, etc.

Yet, in *Of Pandas and People*, published after *Edwards v. Aguillard*, the definition of creation mutated to this:

> Intelligent design means that various forms of life began abruptly through an intelligent agency, with their distinctive features already intact. Fish with fins and scales, birds with feathers, beaks, wings, etc.

So there it was, the smoking gun that the textbook recommended to students as the definitive statement of ID began life as a creationist tract. If all this were not enough to indict the true motives of the creationists, the prosecution punctuated the point by highlighting a statement made by the purchaser of the school's copies of *Pandas*, William Buckingham, who told a local newspaper that the teaching of evolution should be balanced with the teaching of creationism because "two thousand years ago, someone died on a cross. Can't someone take a stand for him?"

This was all too much even for the ultraconservative Judge Jones. On the morning of December 20, 2005, he released his decision—a ringing indictment of both ID and religious insularity (Kitzmiller 2005, 136):

> The proper application of both the endorsement and Lemon tests to the facts of this case makes it abundantly clear that the Board's ID Policy violates the

Establishment Clause. In making this determination, we have addressed the seminal question of whether ID is science. We have concluded that it is not, and moreover that ID cannot uncouple itself from its creationist, and thus religious, antecedents.

Judge Jones went even further, excoriating the board members for their insistence that evolutionary theory contradicts religious faith (136):

> Both Defendants and many of the leading proponents of ID make a bedrock assumption which is utterly false. Their presupposition is that evolutionary theory is antithetical to a belief in the existence of a supreme being and to religion in general. Repeatedly in this trial, Plaintiffs' scientific experts testified that the theory of evolution represents good science, is overwhelmingly accepted by the scientific community, and that it in no way conflicts with, nor does it deny, the existence of a divine creator.

Demonstrating his understanding of the provisional nature of science, Judge Jones added that uncertainties in science do not translate into evidence for a nonscientific belief (136–37):

> To be sure, Darwin's theory of evolution is imperfect. However, the fact that a scientific theory cannot yet render an explanation on every point should not be used as a pretext to thrust an untestable alternative hypothesis grounded in religion into the science classroom or to misrepresent well-established scientific propositions.

The judge pulled no punches in his opinion about the board member's actions and especially their motives, going so far as to call them liars (137):

> The citizens of the Dover area were poorly served by the members of the Board who voted for the ID Policy. It is ironic that several of these individuals, who so staunchly and proudly touted their religious convictions in public, would time and again lie to cover their tracks and disguise the real purpose behind the ID Policy.

Finally, knowing how his decision would be treated in the press, Judge Jones forestalled any accusations of him being an activist judge, and in the process took one more shot at the "breathtaking inanity" of the Dover school board (137–38):

Those who disagree with our holding will likely mark it as the product of an activist judge. If so, they will have erred as this is manifestly not an activist Court. Rather, this case came to us as the result of the activism of an ill-informed faction on a school board, aided by a national public interest law firm eager to find a constitutional test case on ID, who in combination drove the Board to adopt an imprudent and ultimately unconstitutional policy. The breathtaking inanity of the Board's decision is evident when considered against the factual backdrop which has now been fully revealed through this trial. The students, parents, and teachers of the Dover Area School District deserved better than to be dragged into this legal maelstrom, with its resulting utter waste of monetary and personal resources. (See also Humburg and Brayton 2005.)

Some Lines of Demarcation between Science and Pseudoscience

Creation science (and its most recent hybrid, Intelligent Design theory) is just one of many beliefs that most mainstream scientists reject as pseudoscience. But what about those claims to scientific knowledge that are not so obviously classified as pseudoscience? When encountering a claim, how can one determine whether it constitutes a legitimate assertion as scientific? What follows is a list of ten questions that get to the heart of delimiting the boundaries between science and pseudoscience.

1. *How reliable is the source of the claim?* All scientists make mistakes, but are the mistakes random, as one might expect from a normally reliable source, or are they directed toward supporting the claimant's preferred belief? Ideally, scientists' mistakes are random; pseudoscientists' mistakes tend to be directional and systematic, and this is, in fact, how scientific fraud has been uncovered by searching for intentional bias.

2. *Does this source often make similar claims?* Pseudoscientists have a habit of going well beyond the facts, and so when individuals make many extraordinary claims, they may be more than iconoclasts; for example, those who believe in one form of paranormal belief tend to believe most other paranormal claims as well. What one is looking for here is a pattern of fringe thinking that consistently ignores or distorts data.

3. *Have the claims been verified by another source?* Typically, pseudoscientists make statements that are unverified or are verified by a source within their own belief circle. One must ask who is checking the claims and even who is checking the checkers.

4. *How does the claim fit with what is known about how the world works?* An extraordinary claim must be placed in a larger context to see how it fits. When people claim that the pyramids and the Sphinx were built over 10,000 years ago by an advanced race of humans, they are not presenting any context for that earlier civilization. Where are its works of art, weapons, clothing, tools, and trash?

5. *Has anyone made an effort to disprove the claim, or has only confirmatory evidence been sought?* This is the confirmation bias or the tendency to seek confirmatory evidence and reject or ignore disconfirmatory evidence. The confirmation bias is powerful and pervasive. This is why the scientific method—which emphasizes checking and rechecking, verification and replication, and especially attempts to falsify a claim—is critical.

6. *Does the preponderance of evidence converge on the claimant's conclusion or a different one?* The theory of evolution, for example, is proved through a convergence of evidence from a number of independent lines of inquiry. No single fossil or piece of biological or paleontological evidence has the word "evolution" written on it; instead, there is a convergence from tens of thousands of evidentiary bits that adds up to a story of the evolution of life. Creationists conveniently ignore this convergence, focusing instead on trivial anomalies or currently unexplained phenomena in the history of life.

7. *Is the claimant employing the accepted rules of science and tools of research, or have those rules and tools been abandoned in favor of others that lead to the desired conclusion?* Ufologists, for example, exhibit this fallacy in their continued focus on a handful of unexplained atmospheric anomalies and visual misperceptions by eyewitnesses while ignoring that the vast majority of UFO sightings are fully explicable. This is an example of data mining or cherry-picking examples to fit one's belief.

8. *Has the claimant provided a different explanation for the observed phenomena, or is it strictly a matter of denying the existing explanation?* This is a classic debate strategy to avoid criticism: criticize your opponent and never affirm what you believe. This strategy is unacceptable in science.

9. *If the claimant has proffered a new explanation, does it account for as many phenomena as does the old explanation?* For a new theory to displace an old theory, it must explain what the old theory did and then some.

10. *Do the claimants' personal beliefs and biases drive the conclusions or vice versa?* All scientists have social, political, and ideological beliefs that potentially could slant their interpretations of the data, but at some point,

usually during the peer-review process, those biases and beliefs are rooted out or the paper or book is rejected for publication.

The primary difference between science and pseudoscience is the presumption made about a claim before going into the research protocol to test it. Science begins with the null hypothesis, which assumes that the claim under investigation is not true until proven otherwise. Of course, most scientists have some degree of confidence that their hypothesis will be supported in their experiment or else they would likely not pursue a given line of research, but other scientists acting as skeptics will demand evidence strong enough to reject the null hypothesis. The statistical standards of proof needed to reject the latter are substantial. Ideally, in a controlled experiment, we would like to be 95 to 99 percent confident that the results were not due to chance before we offer our provisional assent that the effect may be real. Failure to reject the null hypothesis does not make the claim false; conversely, rejecting the null hypothesis is not a warranty on truth. Nevertheless, from a pragmatic perspective, the scientific method is the best tool ever devised to discriminate between true and false patterns, to distinguish between reality and fantasy, and to demarcate science from pseudoscience.

The concept of the null hypothesis makes clear that the burden of proof is on the person asserting a positive claim, not on the skeptics to disprove it. I once appeared on *Larry King Live* to discuss UFOs (a perennial favorite of his), along with a table full of ufologists. Larry's questions for me and other skeptics typically miss this central tenet of science. It is not up to the skeptics to disprove UFOs. Although we cannot run a controlled experiment that would yield an exact probability figure for the likelihood of a sighting being an example of aliens visiting Earth, proof would be simple: show us an alien spacecraft or an extraterrestrial body. Until then, keep searching and get back to us when you have something.

Unfortunately for ufologists, scientists cannot accept as definitive proof of alien visitation such evidence as blurry photographs, grainy videos, and anecdotes about spooky lights in the sky. Photographs and videos are often misperceived and can be easily doctored, and lights in the sky have many prosaic explanations (aerial flares, lighted balloons, experimental aircraft, even the planet Venus). Nor will government documents with redacted paragraphs count as evidence for ET contact because we know that governments keep secrets for a host of reasons having to do with military intelligence and national security. Terrestrial secrets do not equate to extraterrestrial cover-ups.

So many claims of this nature are based on negative evidence. They typically take the form: if science cannot explain X, then my explanation for X is necessarily true. Not so. In science, lots of mysteries are left unexplained until further evidence arises, and problems are often left unsolved until another day. I recall a mystery in cosmology in the early 1990s whereby it appeared that there were stars older than the universe itself—the daughter was older than the mother! Thinking that I might have a hot story to write about that would reveal something deeply wrong with current cosmological models, I first queried the Caltech cosmologist Kip Thorne, who assured me that the discrepancy was merely a problem in the current estimates of the age of the universe and that it would resolve itself in time with more data and better dating techniques. It did, as so many problems in science eventually do. In the meantime, it is okay to say "I don't know," "I'm not sure," and "Let's wait and see."

Such provisional qualifications are at the very heart of science and are what, at least in part, help to distinguish it from pseudoscience.

A Tale of Two Heretics: A Case Study in Demarcation

Consider the following quotations, written by two different authors from two self-published books purporting to revolutionize science:

> This book is the culmination of nearly twenty years of work that I have done to develop that new kind of science. I had never expected it would take anything like as long, but I have discovered vastly more than I ever thought possible, and in fact what I have done now touches almost every existing area of science, and quite a bit besides. I have come to view [my discovery] as one of the more important single discoveries in the whole history of theoretical science.

> The development of this work has been a completely solitary effort during the past thirty years. As you will realize as you read through this book, these ideas had to be developed by an outsider. They are such a complete reversal of contemporary thinking that it would have been very difficult for any one part of this integrated theoretical system to be developed within the rigid structure of institutional science.

Both authors worked in isolation for decades. Both make equally extravagant claims about overturning the foundations of physics in particular and science in general. Both shunned the traditional route of peer-reviewed scientific papers and instead chose to take their ideas straight to the public through popular tomes. And both texts are filled with hundreds of self-produced diagrams and illustrations alleging to reveal the fundamental structures of nature. There is one distinct difference between the two authors: one was featured in *Time, Newsweek*, and *Wired*, and his book was reviewed in the *New York Times*. The other author has been completely ignored, with the exception of being featured in an exhibition in a tiny Southern California art museum. Their bios help clarify these rather different receptions.

One of the authors earned his PhD in physics at age twenty at Caltech, where Richard Feynman called him "astonishing," and he was the youngest to ever win a prestigious MacArthur genius award. He founded an institute for the study of complexity at a major university and then quit to start his own software company, where he produced a wildly successful computer program used by millions of scientists and engineers. The other author is a former abalone diver, gold miner, filmmaker, cave digger, repairman, inventor, proprietor of a company that designs and builds underwater lift bags, and owner-operator of a trailer park.

The first quotation comes from Stephen Wolfram (2002), the Caltech whiz and author of *A New Kind of Science*, in which the fundamental structure of the universe and everything in it is reduced to computational rules and algorithms that produce complexity in the form of cellular automata. The second comes from James Carter (2000), the abalone diver and author of *The Other Theory of Physics*, proffering a "circlon" theory of the universe, where all matter is founded on these hollow, ring-shaped tubes that link everything together, from atoms to galaxies.

Whether Wolfram is right remains to be seen (although at the time of this writing it doesn't look good), but eventually we will find out because his ideas will be tested in the competitive marketplace of science. We will never know the veracity of Carter's ideas because they will never be taken seriously by scientists; although he is the subject of a book-length treatment of fringe theories in physics by the science writer Margaret Wertheim (2011) entitled *Physics on the Fringe* and is featured in her art exhibition that is a part of her Institute for Figuring. Why? Because, like it or not, in science, as in most human intellectual endeavors, who is doing the saying matters as much as what is being said, at least in terms of getting an initial hearing. (If Wolfram is wrong, his theory will go the way of phlogiston, aether, and, well, the circlon.)

Science is, in this sense, conservative and sometimes elitist. It has to be to survive in a surfeit of would-be revolutionaries. Given limited time and resources, and the fact that for every Stephen Wolfram there are a hundred James Carters, one has to be selective. There needs to be some screening process whereby true revolutionary ideas are weeded out from ersatz ones. Enter the skeptics. We are interested in the James Carters of the world, in part because in studying how science goes wrong, we learn how it can go right. But we also explore the interstices between science and pseudoscience because it is here where the next great revolution in science may arise. Although most of these ideas will join phlogiston on the junk heap of science, you never know until you look more closely.

A Pragmatic Solution to the Demarcation Problem

When discussing pseudoscience, we should bear in mind that those whom scientists and skeptics label as "pseudoscientists" and their practice as "pseudoscience" naturally do not consider themselves or their work as such. In their minds (to the extent we have access to them), they are cutting-edge scientists on the verge of a scientific breakthrough. As the Princeton University historian of science Michael D. Gordin (2012, 1) observes in his book *The Pseudoscience Wars*, "No one in the history of the world has ever self-identified as a pseudoscientist. There is no person who wakes up in the morning and thinks to himself, 'I'll just head into my pseudolaboratory and perform some pseudoexperiments to try to confirm my pseudotheories with pseudofacts.'" As Grodin documents with detailed examples, "individual scientists (as distinct from the monolithic 'scientific community') designate a doctrine a 'pseudoscience' only when they perceive themselves to be threatened—not necessarily by the new ideas themselves, but by what those ideas represent about the authority of science, science's access to resources, or some other broader social trend. If one is not threatened, there is no need to lash out at the perceived pseudoscience; instead, one continues with one's work and happily ignores the cranks" (Gordin 2012, 2–3).

Indeed, most scientists consider creationism to be pseudoscience not because its proponents are doing bad science—they are not doing science at all—but because they threaten science education in America, they breach the wall separating church and state, and they confuse the public about the nature of evolutionary theory and how science is conducted in this and other sciences.

Here, perhaps, is a practical criterion for resolving the demarcation problem: the conduct of scientists as reflected in the pragmatic usefulness of an idea. That is, does the revolutionary new idea generate any interest on the part of working scientists for adoption in their research programs, produce any new lines of research, lead to any new discoveries, or influence any existing hypotheses, theories, models, paradigms, or worldviews? If not, chances are it is pseudoscience or not science at all.

In other words, we can demarcate science from pseudoscience less by what science is and more by what scientists do. As noted earlier, science is a set of methods to describe and interpret observed or inferred phenomena aimed at testing hypotheses and building theories. If a community of scientists actively adopts a new idea, and if that idea then spreads through the field and is incorporated into research that produces useful knowledge reflected in presentations, publications, and especially new lines of inquiry and research, chances are it is science.

This demarcation criterion of usefulness has the advantage of being bottom up instead of top down, egalitarian instead of elitist, nondiscriminatory instead of prejudicial. Let science consumers in the marketplace of ideas determine what constitutes good science, starting with working scientists themselves and filtering through science editors, educators, and readers. As for potential consumers of pseudoscience, that's what skeptics are for, but as always, caveat emptor.

REFERENCES

Achenbach, Joel. 1999. *Captured by Aliens: The Search for Life and Truth in a Very Large Universe.* New York: Simon and Schuster.

"Britons Report 'Psychic Powers.'" 2006. BBC News, May 26. http://news.bbc.co.uk/2/hi/uk_news/5017910.stm.

Bynum, William F., E. Janet Browne, and Roy Porter. 1981. *Dictionary of the History of Science.* Princeton, NJ: Princeton University Press.

Carter, James. 2000. *The Other Theory of Physics.* Enumclaw, WA: Absolute Motion Institute.

Gardner, Martin. 1981. *Science: Good, Bad, and Bogus.* Amherst, NY: Prometheus Books.

Gordin, Michael D. 2012. *The Pseudoscience Wars.* Chicago: University of Chicago Press.

Harris Interactive. 2009. "Harris Poll 2009." Accessed December 7, 2012. http://www.harris interactive.com/vault/Harris_Poll_2009_12_15.pdf.

Humburg, Burt, and Ed Brayton. 2005. "Dover Decision—Design Denied: Report on *Kitzmiller et al. v. Dover Area School District.*" *Skeptic* 12 (2): 44–50.

Kitzmiller v. Dover School District. 2005. 400 F. Supp. 2d 707 (*M.D. Pa.*). http://www.pamd
.uscourts.gov/kitzmiller/decision.htm.

Kuhn, Thomas. 1962. *The Structure of Scientific Revolutions*. Chicago: University of Chicago
Press.

———. 1977. *The Essential Tension: Selected Studies in Scientific Tradition and Change*. Chicago: University of Chicago Press

Lyons, Linda. 2005. "Paranormal Beliefs Come (Super) Naturally to Some." Gallup News
Service, November 1. http://www.gallup.com/poll/19558/Paranormal-Beliefs-Come
-SuperNaturally-Some.aspx.

Michaud, Michael. 2007. *Contact with Alien Civilizations: Our Hopes and Fears about Encountering Extraterrestrials*. New York: Copernicus Books.

Moore, David W. 2005. "Three in Four Americans Believe in Paranormal." Gallup News Service, June 16. http://www.gallup.com/poll/16915/Three-Four-Americans-Believe
-Paranormal.aspx.

NOVA: The Best Mind Since Einstein. 1993. Videocassette. Boston, Massachusetts: WGBH.

Olson, Richard. 1982. *Science Deified and Science Defied: The Historical Significance of Science
in Western Culture from the Bronze Age to the Beginnings of the Modern Era, ca. 3500 B.C.
to A.D. 1640*. Berkeley: University of California Press.

———. 1991. *Science Deified and Science Defied: The Historical Significance of Science in Western Culture from the Early Modern Age Through the Early Romantic Era, ca. 1640 to 1820*.
Berkeley: University of California Press.

Overton, William R. 1985. "Memorandum Opinion of United States District Judge William R.
Overton in *McLean v. Arkansas*, 5 January 1982." In *Creationism on Trial*, edited by Lang-
don Gilkey, 280–83. New York: Harper and Row.

Pigliucci, Massimo. 2010. *Nonsense on Stilts*. Chicago: University of Chicago Press.

Randi, James. 1982. *Flim-Flam!* Amherst, NY: Prometheus Books.

Sagan, Carl. 1996. *The Demon-Haunted World: Science as a Candle in the Dark*. New York:
Ballantine Books.

Shermer, Michael. 1997. *Why People Believe Weird Things*. New York: W. H. Freeman.

———. 2001. *The Borderlands of Science: Where Sense Meets Nonsense*. New York: Oxford
University Press

Shostak, Seth. 2009. *Confessions of an Alien Hunter: A Scientist's Search for Extraterrestrial
Intelligence*. Washington, DC: National Geographic.

Taubes, Gary. 1993. *Bad Science*. New York: Random House.

Webb, Stephen. 2003. *If the Universe Is Teeming with Aliens . . . Where Is Everybody? Fifty Solutions to Fermi's Paradox and the Problem of Extraterrestrial Life*. New York: Springer.

Wertheim, Margaret. 2011. *Physics on the Fringe: Smoke Rings, Circlons, and Alternative Theories of Everything*. New York: Bloomsbury Books.

Wolfram, Stephen. 2002. *A New Kind of Science*. Champaign, IL: Wolfram Media.

12

Evolution

From Pseudoscience to Popular Science, from
Popular Science to Professional Science

MICHAEL RUSE

The Greeks did not have the idea of evolution. When Empedocles suggested that life might have started naturally, he was criticized savagely by Aristotle (Sedley 2007). It was not that Aristotle and the other great philosophers were prejudiced against some kind of developmental origin for organisms. It was rather that they saw organisms as functioning, as having ends—what Aristotle was to call exhibiting "final causes"—and they could not see how final cause could be brought about by the workings of blind law (Ruse 2003). It was at the beginning of the seventeenth century, the opening of the time known as the Age of Enlightenment, that evolutionary ideas first appeared and began to win some acceptance (Ruse 1996). This was because a new ideology had appeared on the scene and was thought by enthusiasts to trump the problem of final cause. We have therefore a three-hundred-year history for evolution: a history that (as we shall see) falls into three parts, divided first by the arrival of Charles Darwin's *Origin of Species* in 1859 and then by the incorporation of Mendelian genetics into the picture by a number of mathematically gifted scientists around 1930.

To understand our history, I want to make a threefold distinction, between what I call "professional science," what I call "popular science," and what I call "pseudoscience" (Hanen, Osler, and Weyant 1980; Ruse 1996). Although temporarily I am working backward, it is easiest conceptually to begin with professional science. Here I am not offering any surprises, for I mean the science that is done today in university science departments and commercial laboratories and so forth. I mean science that, speaking epistemologically, is based on empirical experience and testing, that is law-like, that takes the search for causes very seriously, and that leads to explanations and predictions. I take molecular genetics to be a paradigmatic example of a professional science.

In a way, popular science is parasitic on professional science, although it may prepare the way for professional science or be created after the professional science is underway. As its name suggests, it is the science of the public domain, aimed for the nontechnical reader. As such, it probably drops a lot of the tough reasoning, especially the mathematics. It may well be a lot more visual. It is the science that you find in the public sides of great science and natural history museums. It is the science that features in major newspapers— the Tuesday section of the *New York Times* being a good example. It is the science that comes in books written by science journalists and the like. But although it may be somewhat watered down and not particularly causal, it does aim to give a view of reality that would be accepted and understood by the professional. Some professionals, like the late Stephen Jay Gould, are very good at writing popular science.

Pseudoscience is a very different kettle of fish. It is science of the fringe and beyond. It has proven very difficult to give a precise definition of the notion, but I take it as an absolute that it does exist and that people do recognize it. For instance, about ten years ago, Florida State University started a medical school. One of the state's influential legislators, a chiropractor, passed a bill giving a great deal of money to the university to start a department of chiropractic within the medical school. The faculty of the new school rose up and condemned chiropractic as a pseudoscience with no legitimate place on campus. They did not argue about the notion of pseudoscience. Nor did they argue about so describing chiropractic. They just argued about whether it was appropriate to have it on campus and in the medical school. And that, as it happens, was that. The money was rejected.

This particular story is instructive. Epistemologically, it is clear that when we talk of pseudoscience, we tend to talk of something that plays fast and loose with the evidence. That precisely is the problem with fringe medicine in the eyes of conventional practitioners. The claims lack genuine causal underpinnings, statistical surveys are not carried out or are ignored, enthusiasts cherry-pick significant cures and failures, and much more. Also, pseudosciences generally, and this is particularly true in the medical field, tend to be fueled by strong value considerations, favored ideologies, and the like. There is perhaps a view of human nature—of nature generally—that is holistic and spiritual and at odds with the mechanistic reductionism of modern science.

Having said all of this, however, the trouble is that in significant respects we have been speaking as much of professional science as of pseudoscience (Laudan 1983)! Thomas Kuhn (1962) sensitized us to the extent to which regular science refuses to take no for an answer. If something does not turn out as predicted, the first move is not to reject the theory but to devise or modify auxiliary hypotheses to escape the refutation. It is clear, therefore, that although epistemology is not irrelevant—professional scientists do give up hypotheses in a way that pseudoscientists do not—more psychological and sociological factors are also very relevant. Pseudosciences are those areas of inquiry so labeled by professional scientists! This seems almost vacuous, but it is not really. It points to the ways in which an area can move in and out of focus with respect to pseudoscience status. It also points to the very important fact that professional scientists may be calling something a pseudoscience less from a commitment to epistemic purity and more from insecurity, trying to bolster their own position and status (see Thurs and Numbers, chapter 7, in this volume; Ruse 2013b; Gordin 2012).

In the Florida State medical school instance, it is obvious that not everyone thought of chiropractic as a pseudoscience: the legislators who voted the money, for instance, and a good number of the university's higher administrators. So the term, or rather the application of the term, was very much contested. Being a pseudoscience was not something waiting out there to be found. Additionally, although perhaps this was not a major factor in the medical school case, often the people committed to what is being labeled a pseudoscience are not entirely adverse to their minority and despised position and reputation. It is not so much that they have a persecution or martyr complex, but that they do revel in having esoteric knowledge unknown to or rejected by others, and they have the sorts of personalities that rather enjoy being on the fringe or outside. Followers of Rudolf Steiner's biodynamic agriculture are particularly prone to this syndrome. They just know they are right and get a big kick out of their opposition to genetically modified foods and so forth. There is great satisfaction when people from more central positions agree that they were forerunners in offering warnings and advocating alternative strategies.[1]

Picking up now on the particular story that I want to tell, let me stress that even if the notion of pseudoscience had no history before the writing of the paragraphs above, it would still be appropriate to use it to understand our history. But, as it happens, even though the term apparently was first used around the middle of the nineteenth century, the idea of a pseudoscience was well understood back in the eighteenth century (Ruse 2010a; Thurs and Numbers, chapter 7, in this volume). The French king Louis XVI asked Benjamin Franklin to chair a committee looking at the then-popular craze for mesmerism, a therapy for various diseases supposed to work through a kind of body magnetism. Franklin's committee ruled firmly that mesmerism is about as clear-cut a case of a pseudoscience as it is possible for something to be—they found that it was riddled with ideology and failed properly controlled tests, that it made no pretense at explaining through natural law, and that it therefore failed in the key scientific functions of explanation and prediction.

Franklin's committee rather took for granted what constitutes the opposite of a pseudoscience. For them, that was a good, functioning area of empirical inquiry in some sense. Since the great Antoine Lavoisier was a member of the committee, presumably the new chemistry for which he was responsible would have been taken as the paradigm of something that is very much not a pseudoscience. However, in the context of our discussion, I shall use the trichotomy given above, distinguishing against pseudoscience between "professional science" and "popular science," where a professional science is mature science done by trained experts in the field, and a popular science is something created more for the layperson, and as such may well be much more explicit about values that, even if they underlie professional science, the practitioners probably try to avoid making explicit mention in their work.

The Age of Pseudoscience

With this trichotomy in mind, let us turn now to the history of evolutionary theory. The all-powerful new ideology in the Enlightenment was the idea of progress (Bury 1920). In the cultural realm, this refers to the belief that humans unaided can improve their lot. It is possible through human reason and effort to develop our understanding of the natural world (i.e., to develop *scientific* understanding of the natural world); it is possible to improve education and health care; one can with effort make for a better political society; and thus, overall, we ourselves can create a better world for all humankind. Progress is not in itself antireligious, and indeed most of the early enthusiasts believed in a god of one kind or another. However, progress is opposed to the Christian notion of Providence, where it is believed that only through the intervention of God—the death of Jesus on the cross—is ultimate improvement or salvation possible (Ruse 2005). Hence, whereas Christians are theists—meaning they believe in a god who intervenes in the creation—progressionists tend to be deists—meaning they believe in a god who set things in motion and then allowed items and events to follow through unbroken law.

It was a very easy move in the minds of most early progressionists to go from the cultural world to the world of organisms. As one sees or hopes for improvement and development in the world of humans, so also one sees or hopes for improvement in the world of organisms. This idea of organic change—what we today would call "evolution" (back then terms like "transmutation" were more common)—was therefore always read as showing upward development, from the simple to the complex, from the less worthwhile to the more worthwhile, from what was generally spoken of as the "monad" to the "man." Usually, in a happy circular fashion, once having claimed that progress must be a major feature of development in the organic world, people then used their findings to justify their commitment to progress in the cultural or social world.

Erasmus Darwin, a British physician who was a good friend of industrialists as well as of a variety of advanced political thinkers (for a while he was close to Benjamin Franklin), was a paradigmatic example of an eighteenth-century progressionist who took his beliefs in the social world and applied them directly to the organic world. Fond of expressing his ideas in poetry, Darwin wrote:

> Organic Life beneath the shoreless waves
> Was born and nurs'd in Ocean's pearly caves;
> First forms minute, unseen by spheric glass,
> Move on the mud, or pierce the watery mass;
> These, as successive generations bloom,
> New powers acquire, and larger limbs assume;
> Whence countless groups of vegetation spring,
> And breathing realms of fin, and feet, and wing.
>
> Thus the tall Oak, the giant of the wood,
> Which bears Britannia's thunders on the flood;
> The Whale, unmeasured monster of the main,
> The lordly Lion, monarch of the plain,
> The Eagle soaring in the realms of air,

Whose eye undazzled drinks the solar glare,
Imperious man, who rules the bestial crowd,
Of language, reason, and reflection proud,
With brow erect who scorns this earthy sod,
And styles himself the image of his God;
Arose from rudiments of form and sense,
An embryon point, or microscopic ens!
(E. Darwin 1803, 1, 11, 295–314)

Explicitly, he tied in this vision of the organic world with his hopes and beliefs about the social world: the idea of organic progressive evolution "is analogous to the improving excellence observable in every part of the creation; . . . such as the progressive increase of the wisdom and happiness of its inhabitants" (E. Darwin 1794–96, 509).

Note that evolution back then was not in any sense an empirically rooted enterprise. There was a good reason for this, namely that no one really had that much pertinent information. It was only then that the fossil record was being unearthed, and it was not until the nineteenth century that this was interpreted in any systematic fashion. Little was known about the geographical distribution of organisms: it was not without good reason that Africa, for instance, was known as the Dark Continent. And in other areas also, for instance embryology, researchers were only then slowly coming to an adequate understanding of the nature of things. In short, the idea of organic development—transmutation or evolution—was purely and simply an epiphenomenon on the back of the cultural notion of progress: the belief that humans unaided can improve their lot. It was a pseudoscience. Moreover, evolution was seen to be a pseudoscience, especially by its many critics. When it was thought necessary to destroy the reputation of Erasmus Darwin—support of the American Revolution and then of the French Revolution (at least before things got out of hand) was, with reason, considered politically highly dangerous—his conservative opponents did not attempt to do so on empirical or other scientific grounds. Instead they parodied his poetry, making particular fun of the underlying commitment to progress. A system founded on an ideology had to be challenged through that ideology and in no other way. Erasmus Darwin (inspired by the work of the Swedish taxonomist Linnaeus) had written a work, "The Love of the Plants." The conservatives—led by the politician George Canning—wrote "The Love of the Triangles." In case anyone doubted their intention—to destroy progress as a viable idea—in the introduction to the poem (published in their magazine the *Anti-Jacobin*),[2] they firmly linked Darwin's thinking to other progressionists whose ideas they parodied in another poem, "The Progress of Man." Here is a sample: having killed a pig, the savage turns to bigger game:

> Ah, hapless porker! what can now avail
> Thy back's stiff bristles, or thy curly tail?
> Ah! what avail those eyes so small and round,
> Long pendant ears, and snout that loves the ground?

> Not unreveng'd thou diest!—in after times
> From thy spilt blood shall spring unnumber'd crimes.
> Soon shall the slaught'rous arms that wrought thy woe,
> Improved by malice, deal a deadlier blow;
> When social man shall pant for nobler game,
> And 'gainst his fellow man the vengeful weapon aim.
> (Canning et al. 1798)

Not much progress here. In this poem, Erasmus Darwin is picked out explicitly as a target of their scorn.

It is indeed true that, as the nineteenth century got under way, more and more pertinent empirical information was uncovered. Nevertheless, it was still the belief in progress that drove people to evolutionary speculations, and it was the opposition to progress that underlay much of the opposition. The French biologist Jean-Baptist de Lamarck was a very highly qualified taxonomist, but his excursions into evolutionary thinking—notably in his *Philosophie Zoologique* of 1809—were very much grounded in his beliefs about the possibility of the upward improvement of humankind.[3] Lamarck's great critic was the so-called father of comparative anatomy, Georges Cuvier. It is true indeed that Cuvier brought empirical arguments to bear against Lamarck's evolutionism: for instance, it was Cuvier who first started to explore the fossil record in some detail, and he made much of the gaps between different forms. Also Cuvier (who was much influenced by the German philosopher Immanuel Kant) stressed that evolutionists had no explanation of final causes. Very much in the tradition of Aristotle (whom Cuvier venerated greatly), he argued that blind law cannot possibly have brought about the intricate nature of functioning organisms (Cuvier 1817). But Cuvier's main objections to evolution were cultural and political. He was ever a servant of the state, both during the reign of Napoleon and after, and with good reason he saw evolutionary ideas as having contributed to political unrest and upheaval. Therefore, he opposed them with every ounce of his being (Coleman 1964).

We see here also an exemplification of a point made in the introduction, about the way in which giving something pseudoscience status is often bound up with the agenda of the person (especially a professional scientist) making the charge. Cuvier—who had a hugely powerful role in French science as one of the two permanent secretaries of the French Academy of Sciences—was trying desperately to upgrade the status of biology as a science. He wanted to make it as well regarded as something like astronomy or (a very important area of inquiry in France at this time) optics. It was for this reason he introduced his notion of the "conditions of existence," the design-like constraints under which any organism must operate:

> Natural history nevertheless has a rational principle that is exclusive to it and which it employs with great advantage on many occasions; it is the conditions of existence or, popularly, final causes. As nothing may exist which does not include the conditions which made it possible, the different parts of each creature must be coordinated in such a way as to make possible the whole organism, not only in itself but in relationship to those which surround it, and the analysis of these conditions often leads to general laws as well founded as those of calculation or experiment. (Cuvier 1817, 1, 6)

Notice the aim for laws, and the use of calculation and experiment. The sort of stuff that Lamarck was producing was just the very sort of work that would be scorned by the professional science that was Cuvier's aim.

Bringing the story quickly up to the middle of the nineteenth century and returning to Britain, the most notorious proponent of evolutionary thinking was the anonymous author of the smash-hit, best-seller *Vestiges of the Natural History of Creation* (1844). The author, now known to be the Scottish publisher Robert Chambers, revealingly started out writing a book on the most notorious pseudoscience of them all, phrenology, the belief that the shape of the skull offers insight into intelligence. It then morphed into a tract on evolution, offering a veritable mishmash of fact and fiction, going from the belief that primitive organisms are spontaneously generated from the frost patterns on windows in winter days, to the possibility that the newly crowned Queen Victoria might represent a more highly evolved type of being. Through and through his work, Chambers showed his deep commitment to the idea of progress. It was this and nothing else that underlay his world picture:

> A progression resembling development may be traced in human nature, both in the individual and in large groups of men. . . . Now all of this is in conformity with what we have seen of the progress of organic creation. It seems but the minute hand of a watch, of which the hour hand is the transition from species to species. Knowing what we do of that latter transition, the possibility of a decided and general retrogression of the highest species towards a meaner type is scarce admissible, but a forward movement seems anything but unlikely. (Chambers 1846, 400–402)

Those who liked *Vestiges*—and there were many, including the poet Alfred Tennyson—liked it because of this message of progress. The final stanzas of *In Memoriam*, Tennyson's great poem to the memory of his dead friend Arthur Hallam, are taken almost directly from *Vestiges* with its suggestions that evolution continues to evolve upward. Perhaps, suggested Tennyson (1850), it is Hallam who truly represents the future higher type, unfortunately having come before his time.

> A soul shall strike from out the vast
> And strike his being into bounds,
>
> And moved thro' life of lower phase,
>   Result in man, be born and think,
>   And act and love, a closer link
> Betwixt us and the crowning race
>
>   . . . . .
> Whereof the man, that with me trod
>   This planet, was a noble type
>   Appearing ere the times were ripe,
> That friend of mine who lives in God.

Fascinatingly, and picking up on a theme mentioned in the introduction, we find support for evolution coming from those who had precisely the personalities of dissent and rejection of the majority view. Notable is Alfred Russel Wallace, the codiscoverer of natural selection and he who pushed Darwin into publishing. Let no one detract from Wallace's genius and achievements, but he had a lifelong affinity for minority—many would say crackpot—positions. Evolution (before it was acceptable), spiritualism, socialism, feminism, land reformism, vegetarianism (for health reasons he was forced to eat raw chopped liver), and more. Progress was meat and drink to him (Ruse 2008).

The great critics of *Vestiges* opposed it because they loathed and detested the message of progress. The Cambridge professor of geology, Adam Sedgwick (1845), was nigh apocalyptic. On the one hand, as a member of an institution that got its financial support from rents on land and property and who therefore had a vested interest in the status quo, he feared the possible social upheavals that would follow from a commitment to the ideology of progress. On the other hand, as an ordained clergyman in the Church of England, Sedgwick made clear his personal commitment to the idea of Providence. He wanted no truck with thoughts that humans unaided can create a new Jerusalem here on Earth. Sedgwick speculated that so vile a work must have been written by a woman; he then pulled back and opined that no woman could be so far fallen as to write such a dreadful book: "the ascent up the hill of science is rugged and thorny, and ill suited for the drapery of the petticoat" (Sedgwick 1845, 4). David Brewster, general man of Scottish science, likewise picked on this sad aspect of the work: "It would auger ill for the rising generation, if the mothers of England were infected with the errors of Phrenology: it would auger worse were they tainted with Materialism." The problem, Brewster gloomily concluded, was with the slackness of the curriculum in present-day schools and universities. "Prophetic of infidel times, and indicating the unsoundness of our general education, 'The Vestiges . . .' has started into public favour with a fair chance of poisoning the fountains of science, and of sapping the foundations of religion" (Brewster 1844, 503).

Apart from noting the linking of evolution and phrenology, and identifying the materialism that supposedly lay behind evolutionary speculations (my suspicion is that most early evolutionists were more deists or pantheists than outright materialists), we should pick up on the tensions and insecurities of the likes of Sedgwick and Brewster. These men were treading a very fine line, between pushing for the establishment of professional science in Britain—they were articulating the norms of good science, they were looking out for good junior prospects (like Charles Darwin!), they were founding societies (notably the British Association for the Advancement of Science)—and making sure that the religious establishment did not fault them for infidelity. Times were tense in the 1840s, not just socially with the Chartist movement stirring working men to strikes and rebellion, but also within the church (the Anglican Church, of which Sedgwick was an ordained minister). John Henry Newman and his followers were moving over to Rome, *über*-Protestant literalists (like Dean Cockburn of York) were on the prowl for all other signs of unorthodoxy, and so people like Sedgwick and Brewster simply had to screech their opposition to something like evolution. It threatened them personally (Ruse 1979).

The Age of Popular Science

It is against this background that Charles Darwin published his great work, the *Origin of Species*, in 1859. As is well known, in the *Origin* Darwin set out to do two things. First, he wanted to convince his readers of the fact of evolution. This he did by detailed coverage of all of the then-known facts in the various life sciences. Thus, he went systematically through our understanding of instinct (covering what we today would more broadly speak of as "social behavior"); next to the fossil record and our knowledge of paleontology; he looked in detail at geographical distributions (including the reptiles and birds of the Galapagos Archipelago, the nature of which there is reason to think was a vital component in his becoming an evolutionist); and so on to classification or taxonomy, anatomy, and embryological development. Second, Darwin presented his mechanism of natural selection: a differential reproduction brought about by the struggle for existence. What is all important about natural selection is that at long last there was someone who was speaking to the problem of final cause. Darwin argued that those organisms that are naturally selected are so because they have characteristics not possessed by those who fail, and over time this leads not just to new characteristics, but new characteristics that help their possessors—so-called adaptations.

Strong evidence suggests that Darwin hoped very much to create a fully functioning branch of science, a branch of that area that was now known as biology—something that would be underpinned by evolution through natural selection. The *Origin* was published just at the time that the ancient universities were introducing degree programs in the life sciences. Additionally, in the increasing number of newly founded universities, the sciences generally (including biology) figured high in the curricula. Darwin hoped and had some reason to think that the *Origin* might be the start of a functioning, mature science. The year 1859 would mark the point where evolutionary thinking moved directly from the status of pseudoscience to that of a professional science. This does not mean that Darwin himself had no beliefs in progress.

There is every reason to think that he was committed to progressivist ideas both culturally and biologically. However, he made it clear that, inasmuch as there is progress in biology, it is not something simply founded on a cultural ideology. Indeed, he was adamantly opposed to this kind of thinking. Rather, Darwin introduced the notion of what today is known as an "arms race," arguing that lines of organisms compete against each other—the predators gets faster and in tandem the prey gets faster—and thus adaptations are improved. Darwin speculated that intelligence would emerge and improve, and so the highest forms of organisms, namely human beings, would emerge. But this was intended to be a purely mechanistic explanation without reference to value-impregnated social concepts.

Darwin did not leave things purely to chance. Although he suffered from a never-ending illness and had rather isolated himself in the village of Downe in Kent, through the late 1840s and all of the 1850s he had carefully cultivated a group of friends and supporters, notably the botanist Joseph Hooker and the anatomist Thomas Henry Huxley in Britain, as well as the botanist Asa Gray over in the United States of America. It was his hope that these men would carry the banner and that through them a professional science of evolution, based on the thinking of the *Origin of Species*, would come into being. In other words, in moving evolution from pseudoscience status, Darwin worked hard on the social side of things as well as on the epistemological support—bearing out incidentally the point made earlier about the notion of pseudoscience being not something existing objectively and eternally in a Platonic heaven but very much a matter of negotiation and labeling. Unfortunately, in his hope of pushing evolution right into the professional science category, Darwin was to be sadly disappointed. The key person was Thomas Henry Huxley. Although as a young man he had been extremely critical of *Vestiges*, Huxley was completely won over by Darwin's evolutionism. He became known as Darwin's "Bulldog," a nickname he was happy to carry. He talked about evolution and pushed the idea at every possible opportunity. However, he was never very committed to natural selection—he always thought it would need to be supplemented by other mechanisms, and more important, it was not something of great value to him as a morphologist or as a paleontologist, an area of inquiry to which he turned increasingly. Huxley's organisms were always dead! Hence, he was basically indifferent to adaptation. Greek worries about final cause meant nothing to him, and he was unappreciative of Darwin's solution to this problem (Desmond 1994, 1997).

Huxley therefore felt no great commitment to evolution as a professional science. The almost indifference to causes shows this very clearly. Moreover, there were practical factors that pointed him away from promoting evolution to such a status. Huxley was greatly involved both in secondary education (he became a founding member of the London school board) and in tertiary education (he became science dean of the newly founded, London-based science university, now known as Imperial College). As an educator, and increasingly as an education administrator, Huxley knew that the key to success was finding support for his teaching establishments. Most particularly, he had to sell people on the worth of science degrees. This he did very successfully, persuading the teaching profession that anatomy was ideal training for youngsters in the new industrial age, and persuading the medical profession that physiology and embryology are absolutely crucial foundations for those about to embark on clinical training. Evolution had no place in this vision. However, Huxley could see a rather different role for the belief in the natural origins of organisms. He saw that the greatest opposition to the reforms that he and his fellows were trying to bring about in mid-Victorian Britain came from the established church, the Church of England. With good reason, it was known as "the Tory party at prayer." Huxley therefore saw that inasmuch as Christianity was the foundation of an unreformed Britain, he had to attack it. And there is no better method of attack than by proposing an alternative. Huxley therefore took over evolution as a kind of secular religion, one firmly based on notions of progress, and on the podium and in print pushed it as an alternative to the existing spiritual religion (Ruse 2005).

We have a move, but not a complete move. Thanks to Darwin, evolution now was more than just a pseudoscience. Enough empirical evidence existed to convince people of its truth, and it had been Darwin's genius in the *Origin* to show precisely why the empirical evidence is so convincing. Marshalling the facts very much like a skilled lawyer, Darwin had shown how evolution throws light on so many problems in the life sciences: why in the fossil record are so many early forms seemingly a cross between different later forms? Because of evolution! Why are the inhabitants of the Galapagos like the inhabitants of South America and not like the inhabitants of Africa? Because of evolution! Why are the embryos of different forms like human and dog so very similar? Because of evolution! Conversely, Darwin had argued that these explanations should convince one of the truth of what he called "descent with modification." (This feedback method of argumentation, known as a "consilience of inductions," he got from the philosopher and historian William Whewell (1840).)

However, thanks primarily to Huxley, evolution did not gain the general status of a professional science. It was not just a question of evidence and theory, the worries about causation for instance, but a positive, sociological urge to keep evolution as a popular science, one that could be used in the public domain as a tool to push an overall metaphysical and social vision. This said, obviously, there was some professional work done on evolution. In Germany, thanks particularly to the inspiration and leadership of Ernst Haeckel, there was much interest in tracing phylogenies (Richards 2008). In England, there were always those who actually tried to use selective explanations to account for adaptations. This group was generally drawn from those interested in fast-breeding organisms like butterflies and moths. Some did work that by any standard merits the label "professional." At the end of the nineteenth century, Raphael Weldon (1898) particularly should be mentioned as one who did groundbreaking experiments on marine organisms, showing how selective pressures can operate in nature. But, by and large, these were exceptions. Most of the effort expended on evolution, even by those who worked as full-time, professional biologists, fell on the side of popularization. Articles and books were written explicitly aimed at the general public, and no one had any hesitation about using evolution as a vehicle for social and political ideas. The natural home of evolution was the museum rather than the laboratory. (It is worth mentioning how many of the natural history museums then being erected, the leader being the museum built in London next door to Huxley's science college, were architecturally explicitly modeled on medieval cathedrals. Instead of communion on Sunday morning, the intent was fossil-based panoramas on Sunday afternoon.)

Pertinent to the point being made now is that, in respects, far more popular than Charles Darwin by the end of the nineteenth century was his fellow Englishman Herbert Spencer (Ruse 2013a). The so-called Synthetic Philosophy of Spencer enjoyed incredible success, not only in England but through the empire and most particularly in the United States. Spencer (1857) was an ardent progressionist and never lost the opportunity to use evolutionary speculations to promote his social and cultural beliefs. His disciples were many, and just as different things are claimed in the names of leaders of conventional religions—for every warmonger turning to Jesus for support, you can find a pacifist turning to the same source—so different things were claimed in the name of evolution. Some American industrialists, for instance, used Spencer as justification for their extreme laissez-faire, cutthroat business tactics. Other industrialists used Spencer as justification for widespread benevolence. The steel magnate

Andrew Carnegie, for instance, explicitly appealed to Spencerian ideals when he started to finance public libraries. He argued that a public library would allow poor but gifted children to better themselves. Hence, they would rise up through the struggle for existence and prove themselves, thanks to evolution, to be morally and educationally qualified as tomorrow's leaders (Russett 1976).

The Age of Professional Science

Well into the twentieth century, evolution was the triumph of a popular science. One could be an evolutionist and enjoy respect. One was not dabbling in pseudoactivities like mesmerism or phrenology. And yet one could use evolution as a way of promoting various cultural or social causes. But overall it did not have the status of something like physics or chemistry, or even the professional branches of biology like physiology and embryology. Things started to change again with the rediscovery, at the beginning of the twentieth century, of the true principles of heredity (Provine 1971). First uncovered around the time of Darwin by the obscure Moravian monk Gregor Mendel, these principles were ignored for nearly forty years. Then, once underway, the early geneticists (as they became known) saw their science as a substitute for Darwinian natural selection. Thanks to a focus primarily on large variations, it was believed that all significant change takes place in single jumps ("saltations"). Selection was relegated to a kind of cleaning up role. Only slowly and with much effort was it realized that there is a range of variation and that much evolution might result from slowly acting, almost invisible increments.

This opened the way for the full-blown operation of natural selection, and by around 1930 a number of mathematically gifted thinkers were creating systems that incorporated natural selection operating effectively on small-scale Mendelian changes. Then, following the work of the theoreticians—notably R. A. Fisher (1930) and J. B. S. Haldane (1932) in England and Sewall Wright (1931, 1932) in America—the empiricists moved in and (as it were) put experimental and naturalistic flesh on the mathematical skeleton. In Britain, the most important figure was the Oxford University–based E. B. Ford (1931, 1964), who founded what he called the school of "ecological genetics." His group included the gifted evolutionists Philip Sheppard (1958), A. J. Cain (1954), and H. B. D. Kettlewell (1973). In America, the most important figure was the Columbia University–based, Russian-born Theodosius Dobzhansky. He was the author of the influential work *Genetics and the Origin of Species* (1937). A group builder like Darwin, Dobzhansky encouraged the German-born ornithologist Ernst Mayr, author of *Systematics and the Origin of Species* (1942); the paleontologist George Gaylord Simpson, author of *Tempo and Mode in Evolution* (1944); and then, somewhat later, the botanist and geneticist G. Ledyard Stebbins, author of *Variation and Evolution in Plants* (1950).

Finally, evolution had become a professional science. This did not happen by chance. Not only were articles and books being produced, but the key players were working on a social level to upgrade evolutionary studies. Symptomatic was the founding in the late 1940s of the journal *Evolution*, first editor Ernst Mayr. Money was scrounged from the American Philosophical Society (in Philadelphia), key proselytizing articles were commissioned, grandiose claims were made for the importance of the work being produced. And it worked! By around 1959, the hundredth anniversary of the publication of the *Origin of Species*, there were evolutionary biologists based in universities, getting grants, directing students, and doing full-time research. It was as professional as anything to be found in the academy. Evolutionists explained through unbroken law, they formulated predictions and then went out to test them, and always the basic tool of research was Darwinian selection working in the context of modern genetics, formerly Mendelian genetics and now smoothly giving way to molecular genetics. What of ideology? What of progress? The paradox was that virtually every one of the new breed of professional evolutionists was deeply committed to thoughts of social and cultural progress, and most of them also had a hankering in the direction of biological progress. Simpson, to mention just one example, never made any secret of his belief that evolution overall has been an upwardly directed process and that humans are the end point of nearly four billion years of success (Simpson 1949). However, the new breed of evolutionists realized that promoting progress explicitly in their professional work would be fatal to the status of the discipline. Hence, they made a conscious decision to remove any and all trace of ideology from the professional science. This did not mean, however, that they no longer believed in progress or that they were unwilling to write about it. Instead, having completed the professional work, many of the new evolutionists then turned to the popular, public domain and penned volumes that were brimming with beliefs about progress and prescriptions for social change. (I discuss all of this in great detail in my *Monad to Man: The Concept of Progress in Evolutionary Biology* [1996].)

This is the pattern that has continued down to the present, although to a certain extent the thinking of evolutionists reflects changes in the culture in which they are part. Today, there are few people who subscribe to quite such a happy view of social and cultural progress as was held by people in the eighteenth and nineteenth centuries. It is realized that unbroken, upward change is a mirage, and that for every step forward there tends to be a step backward in the direction of war, corruption, poverty, illness, and the other disasters periodically striking the human condition. In an age of nuclear weaponry, religious fanaticism, worldwide hunger, uncontrollable diseases, who dare today speak of progress? One finds that those evolutionists who venture into the popular domain reflect these concerns. This was very much the case for the late Stephen Jay Gould, a fully professional paleontologist, who (as noted) became one of the greatest popular writers of his time. He thought progress to be a pernicious and unrealizable doctrine, something that accepted uncritically could lead to grave moral and social consequences (Gould 1989). He therefore used his popular writing about evolution to argue as vigorously against progress as two centuries earlier Erasmus Darwin had used his writing about evolution to argue for progress!

Conclusion

Our story is told. The history of evolutionary thinking over the past three centuries falls naturally into three parts. For the first one hundred and fifty years evolution was—and was seen to be—a pseudoscience. It was a vision of the organic world that emerged simply because living things were viewed through the lens of an ideology about the cultural and social world. It was an epiphenomenon on the back of hopes of cultural and social progress. It made little or no pretense that it was doing the things that one expects of good quality, empirical enquiry. Charles Darwin's *Origin of Species* raised the status of evolutionary thinking. However, it did not do everything that the great naturalist had intended. Darwin had wanted his mechanism of evolution through natural selection to be the foundation of a new branch of professional science, the new branch of the life sciences devoted to organic change. This did not come about, primarily because Darwin's supporters—notably Thomas Henry Huxley—had other ends in mind. Thanks particularly to Huxley, evolution was used as a kind of secular religion or what one might call a popular science. It was interesting and respectable. It was, however, not an area of inquiry like physics and chemistry, or the university-based areas of the life sciences like physiology and embryology. Its natural home was the museum rather than the laboratory. It was something on which its enthusiasts could hang their beliefs about cultural and social norms, most particularly their aspirations about progress.

The second major change in the status of evolutionary thinking, that which opened the third and final part of the history of the idea, came around 1930 when Darwinian selection was brought together fruitfully with the newly developed Mendelian (later molecular) genetics. Now finally there was a professional science of evolution. It was one that did what mature science is supposed to do. It was explanatory and predictive, based on extensive empirical study in the laboratory and in nature. It was also one that eschewed social and cultural theorizing. Progress had been expelled. However, a popular side to evolutionary thinking exists still. One finds it in print, in film and particularly on television, and increasingly on the Internet thanks especially to the many enthusiastic evolutionary bloggers. Expectedly, as social and cultural norms have changed generally, so the social and cultural thinking of evolutionists reflects these changes. This is what one expects of a popular science. It is not and does not aspire to be "knowledge without a knower," to use the felicitous phrase of Karl Popper (1972) speaking of what here is called professional science.

From pseudoscience to popular science; from popular science to professional science; the history of a great idea through the past three hundred years! And at the same time, this demonstrates that, although notions like pseudoscience are notoriously slippery and hard to catch, the effort pays major dividends. In thinking about something as a pseudoscience as opposed to other levels of acceptability, one reveals much about the motivations of all concerned. Almost paradoxically, this applies as much, if not more, to the regular scientists, to the professionals, as well as to the outliers. A mask of confidence is an absolute prerequisite for a successful science—for success in anything really—as is a willingness to take on fringe players and those who do not really play the game at all. But all that glitters is not gold, and all that seems powerful and confident is not necessarily so. In a very deep way, the story of evolutionary thinking is a very human story, and the categories used in this essay help to show why this is so.

NOTES

1. This is truly a case where "personal experience" is the appropriate reference. All of my family, with me as the exception, are deeply committed to Steiner's philosophy, so-called

anthroposophy. If there is one thing to which I can claim deep insight, it is into the minds and personalities of enthusiasts for what the world labels "pseudoscience." There is no experience to equal turning on the television idly, on a Sunday morning, to find that the news item features stark naked relatives in an English meadow, spelling out with their bodies "No GM Foods."

2. The title reflected their opposition to the English supporters of the French Revolutionaries, the Jacobins.

3. Significantly, although Lamarck was from the minor nobility, during the French Revolution he had thrived and much improved his personal status.

REFERENCES

Brewster, David. 1844. "Vestiges." *North British Review* 3: 470–515.

Bury, John B. 1920. *The Idea of Progress; An Inquiry into its Origin and Growth*. London: MacMillan.

Cain, Arthur J. 1954. *Animal Species and Their Evolution*. London: Hutchinson.

Canning, George, John Hookham Frere, and George Ellis. 1798. "The Loves of the Triangles." *Anti-Jacobin*, April 16, April 23, and May 17.

Chambers, Robert. 1844. *Vestiges of the Natural History of Creation*. London: Churchill.

———. 1846. *Vestiges of the Natural History of Creation*. 5th ed. London: Churchill.

Coleman, William. 1964. *Georges Cuvier Zoologist: A Study in the History of Evolution Theory*. Cambridge, MA: Harvard University Press.

Cuvier, Georges. 1817. *Le regne animal distribué d'aprés son organisation, pour servir de base q l'histoire naturelle des animaux et d'introduction q l'anatomie comparée*. Paris: Deterville.

Darwin, Charles. 1859. *On the Origin of Species by Means of Natural Selection, or the Preservation of Favoured Races in the Struggle for Life*. London: John Murray.

Darwin, Erasmus. 1794–96. *Zoonomia; or, The Laws of Organic Life*. London: J. Johnson.

———. 1803. *The Temple of Nature*. London: J. Johnson.

Desmond, Adrian. 1994. *Huxley, the Devil's Disciple*. London: Michael Joseph.

———. 1997. *Huxley, Evolution's High Priest*. London: Michael Joseph.

Dobzhansky, Theodosius. 1937. *Genetics and the Origin of Species*. New York: Columbia University Press.

Fisher, Ronald A. 1930. *The Genetical Theory of Natural Selection*. Oxford: Oxford University Press.

Ford, Edmund B. 1931. *Mendelism and Evolution*. London: Methuen.

———. 1964. *Ecological Genetics*. London: Methuen.

Gordin, Michael D. 2013. *The Pseudoscience Wars: Immanuel Velikovsky and the Birth of the Modern Fringe*. Chicago: University of Chicago Press.

Gould, Stephen Jay. 1989. *Wonderful Life: The Burgess Shale and the Nature of History*. New York: W. W. Norton.

Haldane, John B. S. 1932. *The Causes of Evolution*. Ithaca, NY: Cornell University Press.

Hanen, Marsha P., Margaret J. Osler, and Robert G. Weyant. 1980. *Science, Pseudo-Science and Society*. Waterloo, Ontario: Wilfred Laurier Press.

Kettlewell, H. Bernard D. 1973. *The Evolution of Melanism*. Oxford: Clarendon Press.

Kuhn, Thomas. 1962. *The Structure of Scientific Revolutions*. Chicago: University of Chicago Press.

Lamarck, Jean-Baptiste. 1809. *Philosophie zoologique*. Paris: Dentu.

Laudan, Laurens. 1983. "The Demise of the Demarcation Problem." In *Physics, Philosophy, and Psychoanalysis: Essays in Honor of Adolf Grübaum,* edited by R. S. Laudan and L. Cohen, 111–27. Dordrecht: Reidel.

Mayr, Ernst. 1942. *Systematics and the Origin of Species.* New York: Columbia University Press.

Popper, Karl R. 1972. *Objective Knowledge.* Oxford: Oxford University Press.

Provine, William B. 1971. *The Origins of Theoretical Population Genetics.* Chicago: University of Chicago Press.

Richards, Robert J. 2008. *The Tragic Sense of Life: Ernst Haeckel and the Struggle over Evolutionary Thought.* Chicago: University of Chicago Press.

Ruse, Michael. 1979. *The Darwinian Revolution: Science Red in Tooth and Claw.* Chicago: University of Chicago Press.

———. 1996. *Monad to Man: The Concept of Progress in Evolutionary Biology.* Cambridge, MA: Harvard University Press.

———. 2003. *Darwin and Design: Does Evolution have a Purpose?* Cambridge, MA: Harvard University Press.

———. 2005. *The Evolution-Creation Struggle.* Cambridge, MA: Harvard University Press.

———. 2008. "Alfred Russel Wallace, the Discovery of Natural Selection and the Origins of Humankind." In *Rebels, Mavericks, and Heretics in Biology,* edited by O. Harman and M. R. Dietrich, 20–36. New Haven, CT: Yale University Press.

———. 2010. "Evolution and Progress." In *Biology and Ideology from Descartes to Dawkins,* edited by D. Alexander and R. N. Numbers. Chicago: University of Chicago Press.

———, ed. 2013a. *The Cambridge Encyclopedia of Darwin and Evolutionary Thought.* Cambridge: Cambridge University Press.

———. 2013b. *The Gaia Hypothesis: Science on a Pagan Planet.* Chicago: University of Chicago Press.

Russett, Cynthia E. 1976. *Darwin in America: The Intellectual Response, 1865–1912.* San Francisco: Freeman.

Sedgwick, Adam. 1845. "Vestiges." *Edinburgh Review* 82:1–85.

Sedley, David. 2007. *Creationism and Its Critics in Antiquity.* Berkeley: University of California Press.

Sheppard, P. M. 1958. *Natural Selection and Heredity.* London: Hutchinson.

Simpson, George Gaylord. 1944. *Tempo and Mode in Evolution.* New York: Columbia University Press.

———. 1949. *The Meaning of Evolution.* New Haven, CT: Yale University Press.

Spencer, Herbert. 1857. "Progress: Its Law and Cause." *Westminster Review* 67:244–67.

Stebbins, G. Ledyard. 1950. *Variation and Evolution in Plants.* New York: Columbia University Press.

Tennyson, Alfred. 1850. "In Memoriam." *In Memoriam: An Authoritative Text Backgrounds and Sources Criticism,* edited by R. H. Ross, 3–90. New York: W. W. Norton.

Weldon, W. F. Raphael. 1898. "Presidential Address to the Zoological Section of the British Association." In *Transactions of the British Association (Bristol)*, 887–902. London: John Murray.

Whewell, William. 1840. *The Philosophy of the Inductive Sciences.* London: Parker.

Wright, Sewall. 1931. "Evolution in Mendelian Populations." *Genetics* 16:97–159.

———. 1932. "The Roles of Mutation, Inbreeding, Crossbreeding and Selection in Evolution." *Proceedings of the Sixth International Congress of Genetics* 1:356–66.

PART IV

Science and the Supernatural

13

Is a Science of the Supernatural Possible?

EVAN FALES

Protocol

The search for a criterion (or, more plausibly, a set of criteria) of demarcation between science and other supposed ways to gain knowledge of the world around us has not, on the whole, led to very encouraging results. Perhaps the best we can hope for is a set of standards that can be met in varying degrees and that can justify rough-and-ready judgments that a given line of research is excellent, or passable, or marginal science—or more or less beyond the pale. Nevertheless, it would be of significant interest to be able to show that certain sorts of claims, and certain ways of asking and answering questions about the world, were in principle "beyond the pale." Such claims and methods, then, whatever independent justification they might be able to muster, would be properly excluded from scientific purview.

In one fashion or another, exactly this sort of status has been attributed to claims about the supernatural. Such claims include claims both about what supernatural beings there are (or are not), and about how such beings interact with the world. It would follow from a successful argument to this effect that such appeals to the supernatural are not scientific and that any research program that invokes supernatural beings or forces is pseudoscientific.

Here, I want to examine arguments for the view that any science of the supernatural must be a pseudoscience (see also Boudry, chapter 5, in this volume). I shall try to show that many of these arguments are not good arguments. I shall then offer an argument that—if sound—does succeed in ruling out scientific appeals to the supernatural (because it denies supernatural beings any possibility of interacting causally with the world). I shall also defend a weaker conclusion, viz. that even if the argument is not sound, it still does not follow that supernaturalistic theories, such as have been offered, are spelled out in such as way as to qualify as legitimate guides to scientific research or as good science.

However, before we can begin to discuss these issues, it is essential to say something about how to draw the distinction between the natural and the supernatural. This is by no means a trivial matter.[1] It might be tempting to say that the natural consists just of everything that is material (and of the space-time framework that provide the arena for the interactions of matter)—where by "material" I mean all of the manifestations of matter, including energy, fields, and their properties. This suggestion runs afoul of three difficulties. First, there may be (material) universes that are causally isolated from ours, so that we have no possibility of observing or investigating them; and on some multiverse cosmologies, the laws of nature may differ sharply from those obtaining in our universe. That may raise questions about the identity conditions for matter, and it suggests as well that being "material" is not a sufficient condition for being open to scientific investigation. (Possibly we can have very indirect evidence of their existence or of the possibility that they exist, but nothing more.)

A second difficulty is presented by the question of the status of minds. We cannot seriously doubt that minds exist, and we certainly consider them suitable targets for scientific investigation (they are in the domains of psychology and the social sciences). But what if they, or at least many of their properties, are immaterial (however that might be understood)? Of course, if one is a reductive materialist, minds will not pose an in-principle problem. If one is a materialist of a nonreductive stripe, matters become more complex, but we may still accept a happy naturalism that confines itself to the material. But it would be a mistake, I think, in the present context, to bind naturalism to a commitment that minds are material. That is arguably not something that science alone can settle, but however it is settled, we should not hold psychology and the social sciences hostage to the outcome. Thus I propose that the right sort of gerrymander here, to give us what matters, is one that rules out disembodied minds.[2] Naturalism, then, is committed to there being none of those.

But (third) what about abstracta (e.g., propositions, universals, sets, numbers, perhaps possibilia, and the like)? Some naturalists want to banish these from their ontology, and perhaps they are right (though I think not). But if such things do exist and are essential as truth makers for, for example, mathematical truths, then surely science need have no quarrel with realism respecting abstracta.

So I adopt the following way of framing the issue I wish to discuss. Take ontological naturalism to be just the minimal thesis that there are no disembodied minds, and methodological naturalism to be the thesis that science should eschew appeal to such minds by way of explaining the "empirical data"—both what we experience by means of our senses and what John Locke called "reflection"—that is, by means of introspection. The phrase "should eschew appeal" is itself ambiguous, and I will have more to say about that presently.

With this much stage setting, let me pose the question I mean to address. Should (or must) a scientist, in his or her role as a scientist, adopt methodological naturalism? And I shall argue that the answer to that question is only a very weak form of methodological naturalism, unless the following ontological claim can be ruled out—viz., that (a) there are disembodied minds, and (b) these minds can and do causally interact (either uni-directionally or bi-directionally) with material bodies and/or embodied minds. Let me call this two-part claim (or simply (b), which presupposes (a)) "theism." I use the scare quotes because theism is ordinarily the claim that there is exactly one god, and this god satisfies condition (b). I mean "theism" to be the broader thesis that there is at least one disembodied mind that satisfies condition (b).

I shall proceed as follows. After considering some other arguments for ruling out scientific consideration of "theism," I shall offer what I take to be the best anti-"theistic" argument, one that would sustain a strong form of methodological naturalism. It is an argument whose conclusion allows that there may be disembodied minds—gods and the like—but denies that they can causally affect the world, including us, in any way. Finally, I shall consider what sort of methodological naturalism survives if that argument is unsound.

Origins and Unrepeatables

Let me first dispose, rather summarily, of two related arguments that are sometimes offered—usually by theists—for the view that acts of God are beyond scientific investigation. The first is that science cannot discover origins—in particular, it cannot explain the origin of life or of the universe itself. These—

since they must have a cause—must be the handiwork of the divine. The second is that science cannot hope to account for miracles, for miracles are by nature unique and unrepeatable, and science can deal only with events that are repeatable. The first claim—the one about beginnings—appears to be just a special case of the second, for both are, at heart, motivated by the thought that science requires repeatability. And both fall prey to similar objections.

Miracles, to be sure, are not repeatable by us; but then, neither are supernovae. That does not prevent astronomers from explaining the latter. And miracles are surely repeatable by God. Indeed, biblical literalists are committed to there being various miracles (e.g., resurrections) that God has repeated (and will repeat). So that cannot be a reason to exclude miracles from scientific purview. In fact, if David Hume is correct and miracles are to be thought of as "violations" of laws of nature, then a precise understanding of those laws is a prerequisite for identifying an event as a miracle. (I shall say more about this presently.)

But repeatability (and actual repetitions) is a red herring in any case. The underlying rationale for an insistence on repeatability seems to be either that such repetition is necessary for a discovery of the laws of nature or that the replicability of experiments is a necessary feature of the scientific method. Both those claims contain a kernel of truth, but they cannot be wielded to yield the desired conclusion. Here is the kernel of truth. Concrete, complex events (certainly macroscopic ones) are probably never repeated in every detail in the entire history of the universe. How, then, are laws discovered? Through systematic application of J. S. Mill's methods, scientists (and, less systematically, ordinary folk) manage to isolate causally relevant features of causally related complex events, and abstract from these generalizable laws. To fully (causally) explain a complex event requires identifying those distinct features that can be subsumed under such laws, and then identifying prior events certain of whose features satisfy the antecedents of those laws.[3] Repeatability in experimental contexts is usually a matter of establishing the proper control of confounds.

One-off originating events or processes can be scientifically understood if they can be seen to have features or constituents that, because the latter are occasioned elsewhere, are understood to be subsumable under known laws.[4] The possibility of explaining a miracle—even if it is unique and unprecedented—will depend inter alia on our having sufficient knowledge of the antecedent natural circumstances and relevant natural laws to rule out the sufficiency of natural causes. Absent that, we cannot even identify the event as a miracle.

But if we can ever establish with reasonable certainty that an event has no suf-ficient natural cause, and if it can be shown to be the sort of thing that a god could and quite possibly would cause, then what is wrong with the suggestion that the best science has pointed toward a supernatural explanation?

Consider "skeptical theists," who hold that no amount of apparently gra-tuitous evil can serve as evidence against the existence of God, as God's pur-poses, and the goods God may achieve by way of those evils, may be beyond our ken. Atheists who similarly insist that no event, no matter how apparently miraculous, can serve as evidence of the supernatural because the operations of Mother Nature are so far beyond our ken as to preclude ruling out natural causes, deserve the moniker "skeptical naturalists." Neither position is plau-sible in my view. Skeptical theists claim knowledge of God's purposes when it suits them and ignorance when that suits them. Skeptical naturalists claim scientific understanding when dialectically helpful and are prepared to invoke the open-endedness of science to the bitter end when (or if) they should be confronted by anything that defies current understanding. But *imagining* clear evidence for the miraculous is not that hard (see, e.g., Deitl 1968).

The Wayward Ways of Deity

It is generally assumed in debates about methodological naturalism that what is to be ruled out of bounds for science is not simply disembodied "some-things," but more particularly disembodied minds or persons. But what is wrong about appealing to such persons as agents whose agency can be the subject of scientific scrutiny? After all, there has been a good deal of scientific work done to investigate paranormal phenomena, and at least some of those phenomena have been chalked up, by some investigators, to the doings of disembodied minds/persons (or spooks, as I shall henceforth call them). Of course, one might protest that much of this work has been bad science, but it is one thing for the methodology to be bad because of, for example, poor experimental controls or bad statistical analyses, and another thing to dismiss it simply because it sought to establish, or to disprove, the activities of this or that spook.

What is often said to be wrong about admitting spooky action to explain phenomena scientifically is that persons are essentially capricious. So, for ex-ample, one argument for a commitment to naturalism in science centers on the idea that our very ability to get empirical information about the world requires reliance on the consistent and universal operation of laws of nature—

laws that are put in jeopardy with the admission of a deity who can override them. A chief defender of this line of argument is Robert Pennock (2001, 83), who takes "the Naturalist view of the world [to have] become coincident with the scientific view of the world, whatever that may turn out to be."

At first blush, this looks like an ontological thesis: to be a naturalist is to accept the ontology of science. Of course, if that is what it is to be a naturalist, it appears an open possibility that the naturalist might come to accept the existence of supernatural beings: it all depends on whether "science" comes to accept the existence of spooks.[5] But in any case, Pennock clearly means the principle to be understood as a methodological principle—something like an imperative: believe about the world whatever science says. But why should one's noetic structure be ruled by such a principle as that? Clearly, Pennock thinks there is something about the methods of scientific investigation that recommends the principle. And those methods, it allegedly turns out, rule out appeal to supernatural causes—or at least appeal to a spook like Yahweh. The offending characteristic of God, in this case, is that he is not merely able to meddle in the affairs of the world, but that he can do things that involve overriding the laws of nature—that is, he can do miracles, and can do them at will.[6]

The supposition that there is such a spook as that wreaks havoc on scientific method; specifically, it undermines the reliability of our methods of observing the world and making measurements of any sort. So Pennock (2001, 88–89) claims that

> without the constraint of lawful regularity, inductive evidential inference cannot get off the ground. Controlled, repeatable experimentation, for example, . . . would not be possible without the methodological assumption that supernatural entities do not intervene to negate lawful natural regularities. . . .
>
> Supernatural theories . . . can give no guidance about what follows or does not follow from their supernatural components. For instance, nothing definite can be said about the process that would connect a given effect with the will of the supernatural agent—God may simply say the word and zap anything into or out of existence. . . . Science assumes Methodological Naturalism because to do otherwise would be to abandon its empirical evidential touchstone.

Pennock's worry raises, I think, two distinguishable questions. One is whether a sufficiently whimsical god could throw into disarray our efforts to get a reliable "fix" on those features of the world that are of scientific interest. The other is whether the mere introduction of divine agency (where that agent exercises his will freely) into our picture of the world infirms scientific knowledge claims.

Now in the first place, it must be observed that if there is a deity who tinkers from time to time in significant ways with physical processes, then adopting the sort of methodological naturalism recommended by Pennock looks like bad advice, advice that will, at best, blind us to something important about our world. To be sure, a sufficiently willful god could, no doubt, make our world so capricious that all bets would be off: inductive reasoning frustrated (if it tries to infer anything other than randomness), predictions going regularly astray, and so on. But this is to raise the specter of radical inductive skepticism—of a world ruled by a divinely evil demon. If that sort of skepticism is in play, then why worry about the occasional machinations of a god? The skeptic, in any case, will demand an answer to the much more fundamental question by what right we assume the operation, ever, of lawful natural regularities.

Short of our world being one that visibly forces skepticism on us, we should note that experimental science is full of wayward factors that frustrate reliable and accurate measurement. Much as philosophers of science puzzle over the problems posed by Pierre Duhem's thesis and the underdetermination of theory by data, working scientists do regularly debug their experiments and work around confounds. Again: an insistent divine "bug" might be ineliminable, but then we were never promised a scientific rose garden. But the mere fact that experimentation sometimes goes off the rails does not show that the entire scientific enterprise is under mortal threat. As Plantinga (e.g., 2000, 405–7) laconically observes, an occasional miracle is not the stuff of which science-destroying skepticism is made. For all that, this much of Pennock's counsel seems right: faced with an anomaly, we should look very hard for natural causes. More of that anon.

But what of Pennock's assertion that supernatural theories "can give no guidance" concerning "what follows from their supernatural components"? I assume that what he means by this is that such theories can tell us nothing about what we should expect God to do, or how. Why should we think that? The fact is that theologies regularly offer all sorts of claims about the character and dispositions of divine beings. Presumably, then, they tell us something— perhaps a great deal—about when and how we should expect those beings to act on our world. Perhaps Pennock thinks that none of the free actions of an agent can be scientifically understood, predicted, or controlled. But that does not seem right—unless Pennock means to be disqualifying all of the human sciences, the sciences that deal with those free and rational agents about whose existence we are not in doubt.[7] So if God is, like us, a free, rational agent with certain hypothesized enduring character traits, why should we not have some basis for making predictions about his ways of influencing the world? Pennock evidently thinks that there would be no way at all to test such a theory. But that seems precisely wrong: almost all of natural theology (with the exception of such a priori arguments as the ontological argument) are concerned with assessing the empirical evidence for and against the existence of such a being, on the basis of his presumed effects on the world. There is thus, on Pennock's showing, no reason in principle why supernaturalistic hypotheses could not figure as eligible for scientific investigation.

Divine Transcendence

It is often said that God transcends the material world. What does that mean? Well, one thing that is often meant, or entailed, by that claim is that God is not a material being—not only does he not have a material body, but he does not exist in space at all, and (according to a common theological opinion) not in time either. That makes God quite peculiar since, unlike abstracta that also have no spatiotemporal location, he is a concrete particular. And one might think that this very fact makes God inaccessible to scientific investigation and therefore beyond the purview of scientific consideration. How can scientists observe, or measure, a being like that? How could one hope to capture something of such a being's nature in a laboratory or in the eyepiece of a telescope? Some, to be sure, have claimed to have seen God or to have been contacted by him, but their experiences are of a subjective kind not open to the rest of us to confirm and so beyond the proper scope of anything that can be accepted as a scientific datum. Science—so this line of reasoning goes—must therefore confine itself to observing and explaining the behaviors of objects locatable within space and time, in terms of the powers and properties of other objects also so locatable.

Perhaps there is something to this. It may be rooted in the mechanistic conception of causal influence that, it seems to me, we all acquire as a central part of our most primitive experience of causation—viz., our experience of pushes and pulls.[8] But even though a force is exerted at a time and is experienced as a vector possessing spatial location and direction, it cannot be assumed straightaway that the source of a force must have spatiotemporal location. Our naïve conception of causal processes has already been forced to undergo fundamental revision in order to accommodate action at a distance and quantum processes. Why refuse to allow that there might be supernatural causes—causes whose source lies outside the spatiotemporal order?[9] If the notion of a-spatial and/or a-temporal causes is not incoherent (as it may be; see below), why can't our concept of causation be stretched to accommodate that possibility? And if the possibility is admitted, then, surely, science ought in principle to be open to it. After all, if the fundamental "missions" of science are prediction and explanation, and if there are events whose explanation involves supernatural agents, why should science arbitrarily allow its hands to be tied with respect to the pursuit of such explanations?

Immaterial Causes?

However, there is an argument for the exclusion of the supernatural that, in my opinion, merits serious consideration. It is an argument for the conclusion that immaterial beings cannot causally influence material things. To properly discuss this argument, we must do three things. First, we must become clear what supernatural (or, as I shall just say, divine) intervention—for example, a miracle—amounts to. Second, we must see in what sense miracles might be impossible. And third, we must consider the objection that, perhaps evidently or in any case for all we know, we have abundant instances of immaterial beings—namely human minds—that are able to causally influence material things (e.g., human bodies): so the argument must be unsound.

Hume describes miracles as violations of the laws of nature, but also as events outside the order of nature. The latter characterization is more cogent: a miracle is an event whose natural causes, if any, are insufficient to produce it, without the superadded "help" of an immaterial cause—viz. God or some other spirit. That immaterial cause, we must suppose, involves a divinely generated physical force, directed on some portion of the material universe, that causes its matter to behave in ways it would not have if, all else being equal, that force had been absent.[10] The question before us then is: could there be such immaterial causes, causes that generate forces that permit control over, or influence on, the denizens of the material world?

Much of the relevant debate over the central issue of immaterial causes has appeared in the context of modern defenses of Cartesian dualism. For here, too, the question is whether (human) minds, conceived as immaterial substances, can direct the movements of (human) bodies.[11] If such a feat is

possible for immaterial finite minds, perhaps little stands in the way of supposing it possible for an "infinite" mind.[12] This debate has, unfortunately, been
handicapped by a rather poor understanding of the physical principles involved. An early contribution was C. D. Broad's suggestion that an immaterial
mind could affect the brain without violating the principle of conservation of
energy (COE) by affecting the brain's energy distribution, with no change in
the total energy (Broad 1925, chap. 3, sec. 2). But while this is theoretically
possible, it avails little. Such an energy-preserving shift in the motions of particles requires that the force be applied always in a direction perpendicular to
their motions; and this will in any event produce changes in their (linear and/
or angular) momenta.[13]

A second strategy is to look for room to maneuver in the fact that en-
ergy/momentum need not be conserved over very short intervals of time/
space because of the Heisenberg Uncertainty Principle. But, as David Wilson
(1999) has shown, neural processes (and, for our purposes, macroscopic systems generally) are far too large to permit the deviations from conservation
required to explain them.

Much more common are arguments that allege that the conservation laws
are, in effect, defeasible. The motivating idea is that physical systems obey
these laws only if "closed" or isolated from external sources of energy/mo-
mentum, and that the mind (or God) can supply precisely such a source, even
if all external physical influences are barred (e.g., Larmer 1986; von Wachter
2006; Averill and Keating 1981).Thus, for example, Larmer (1986) distinguishes between a weak and a strong form of COE:

(a)   Weak COE: The total amount of energy in an isolated system remains
      constant.
(b)   Strong COE: Energy can be neither created nor destroyed.

Larmer argues that (a) is consistent with immaterial causes, as material systems affected by such causes are not (causally) "closed," whereas the scientific evidence and the requirements of scientific explanation warrant only
(a) and not (b). But this seems misguided. Of course, as Larmer concedes,
there is nothing in principle that prevents the detection of "surplus" energy in
the human body (presumably, in the brain), not attributable to any physical
source. This energy might be small; it might be required only to operate neural "switches" that amplify its effects by controlling much larger energy flows.
Of course, no such "un-provenanced" energy has been detected. But, were

it found, such surplus energy would indeed call for explanation and would provide evidence for either the creation of energy ex nihilo or the conclusion that the mind (or God) lost some energy. Larmer is, in other words, playing a mind-of-the-gaps game here. (Similarly, God could hide his influence on the mundane world by making the additions of energy/momentum sufficiently small or hidden away in remote places or times to be unobservable by us.) What are the chances of this? If immaterial minds and gods do not—cannot—have any energy to gain or lose, then we must suppose that energy/momentum conservation laws apply only to material causes and effects. But as we shall see, this way of admitting immaterial causes by limiting the scope of the conservation laws runs up against a fundamental difficulty.

Still, one might argue that the issue should be settled by asking whether conservation laws that forbid immaterial causes provide the best explanation of all the phenomena. So, for example, Bricke (1975) argues that when we take account of the psychological evidence that our action intentions cause appropriate bodily movements, we have (in the case of finite minds) evidence that trumps whatever reasons we have for denying that minds can create energy (and thus, a fortiori, trumps a reason for thinking that God can't). But this won't serve. As Hume pointed out, we have no acquaintance with the causal particulars of how our mental states produce bodily motions: we understand (nowadays, but not by introspection) many of the neural pathways involved, but we have no evidence—introspective or otherwise—that the causal chains can be traced back to immaterial mental states.

It is Averill and Keating (1981), arguing in the same vein as Larmer, who eventually put their finger on the decisive issue. The law of conservation of momentum follows from the Newtonian principle that, as it is often put, there is for every action an equal and opposite reaction. In more proper terminology,

> F: Whenever something exerts a force upon an object, the object exerts an equal and opposite force upon the original thing.

This entails that the vector sum of the forces—the rate of change of the total momentum p_T (of the system comprising those two things or of any n interacting things)—be zero: $\partial p_T / \partial t = 0$.[14] It is precisely this law that immaterial causes appear to violate. And there are, so far as I can see, only two ways to avoid this conclusion. One—the one Averill and Keating adopt—is to restrict the law to interactions between physical objects. But this flies in the face of the fact that the "resistive," or counteracting, force exerted by an object on whatever pushes it is a consequence of its inertia—that is, the fact that it has mass. Nothing in this requires that the source of the force on it be itself a material object. So, if a material object were pushed by an immaterial object, it would still exert a counteracting force—but on what?

The only move remaining to the immaterialist is to claim that an immaterial source—for example, God—influences the behavior of material objects without exerting on them any sort of force. And that—since $F = ma$ is indefeasible—appears equally out of the question, at least if the changes in the matter involve any change in motion (or potential energy). For a change of motion is an acceleration—and thus implies the exertion of a force whose magnitude is given by this law of nature.[15] In short, we may conclude that if not even God can violate the laws of nature, then he cannot influence the world in any way that involves pushing matter around. And unless there are other measurable changes that God can effect, this, in turn, provides a decisive reason to exclude divine influence from the sphere of scientific explanation—indeed, from the sphere of explanation of anything at all in the physical world.

Scientific Method and Supernatural Meddling

But suppose, for the sake of argument, that the conclusion just reached is mistaken: suppose, that is, that God is (somehow) able to meddle in the meanderings of matter. What would be the implications of this for scientific method? Perhaps nothing much—so long as God does not meddle much. We might even be able to discover where and how God intervenes—by discovering events whose energy and momentum budgets simply are not fully accounted for by natural causes. That seems possible, at least in principle.[16] But in the absence of really dramatic evidence,[17] this should not give supernaturalists much hope, for two reasons.

The first reason is that supernaturalistic hypotheses are typically devoid of the kind of explanatory detail that we expect of scientific hypotheses.[18] This asymmetry was quite dramatically on display in the testimony of Intelligent Design (ID) defender Michael Behe in the notorious *Kitzmiller v. Dover School Board* trial, in which the plaintiffs' lawyer, Eric Rothschild, elicited from Behe the following two contrasting claims. On the one hand, Behe criticized the neo-Darwinian theory of evolution on the grounds that it had not provided detailed mechanisms to explain how various subcellular structures (so-called irreducibly complex systems) evolved. But, on the other hand, when asked about the Intelligent Designer and how it might have gone about fashioning its biological designs, Behe both denied having any knowledge of this and denied that ID was under any obligation to provide answers to these questions.[19] But such eschewing of basic questions about explanatory processes and mechanisms is not only surprising (surely normal scientific curiosity would be aroused concerning the nature and modus operandi of such a remarkable being!) but scientifically irresponsible. It is a reflection of the explanatory poverty typical of supernaturalistic hypotheses.

The second reason is the inductive grounds we have for expecting the supernatural to be evicted from the gaps that remain in naturalistic explanations by the long history of such evictions that naturalistic science has achieved in the past (note, too, that the evictions are one-sided; it seems never to happen that a naturalistic explanation is chased from the field by a triumphant supernaturalistic one). But this yields at most a pragmatic counsel to look preferentially for naturalistic explanations of puzzling phenomena (for, odds on, that is where you will find the explanation), not anything like a principled dictum (let alone a prejudice) to the effect that the supernatural cannot possibly explain anything.

NOTES

1. Some folks think that science is just by definition limited in its purview to naturalistic explanations. That would settle the matter posthaste: if only they would now tell us exactly what they mean by "naturalistic." Note that I did not say: limited to natural phenomena. For it is standard theistic fare that God can and does produce natural phenomena; if so, then divine action is (at least part of) the (true) explanation of those phenomena.

2. And spirits/souls, if there can be such things and they are not minds. St. Paul, for example, distinguished between the soul (*psyche*), the spirit (*pneuma*), and our cognitive faculty (*nous*)—see, e.g., 1 Cor. 15—whatever he meant by the former two.

3. This is highly simplified. It leaves out of account statistical laws, and it leaves out the fact that (so I maintain) causal relations in special cases can be perceived in single causal sequences (see Fales 1990, chap. 1). But it is good enough for present purposes. I should be clear that I am not presupposing a regularity account of causation or of laws of nature—a view that I in fact reject.

4. A possible special case is the origin of the universe, where theists invoke an antecedent that allegedly cannot be investigated by Mill's methods. I shall have more to say about that presently.

5. Pennock (2001, 84) in fact allows for such a possibility—provided that what is being proposed is certain kinds of deism, pantheism, or identification of God as a transcendent "ordering principle." But he clearly intends to rule out, for example, Krishna or Allah or Yahweh.

6. Pennock (2001, 88) takes this power to be definitive of supernatural beings. I cannot see why; if there are angels, I suppose we ought to think of them as supernatural beings, but we are not under obligation—at least not as a matter of definition—to think of them as having the power to contravene laws of nature.

7. At stake here, to be sure, are issues concerning the nature of free agency and the role of covering laws in explaining or predicting the actions of such agents. There is no space here to discuss these large matters (concerning which see Fales (1994) and Fales (2011)). Here I shall only record my view that human beings have freedom (as libertarians view freedom) and that their actions are to be understood by appeal to their reasons, not to causal laws. Nevertheless, that seems to me in no way to undermine the credentials of the human sciences.

8. Indeed, a popular argument against immaterial causes (whether supernatural or natural minds) is that, as they are not located in space, they cannot be spatially contiguous to their (material) effects (see, e.g., Sober 2000, 24). This argument seems far from decisive, inasmuch as I know of no decisive demonstration of the claim that causes must be spatially related to their effects.

9. The question whether an a-temporal god could bear causal relations to the temporal world—for example, perceiving what happens in this world at given times and causing things to happen here at given times—is vexed. A well-known attempt to make sense of this is Stump and Kretzmann (1981). For criticisms, see Craig (1994) and Fales (1997); see also Leftow (1991) and Helm (1988).

10. This way of thinking about miracles avoids the obvious objection that (since laws of nature presumably entail universal generalizations), a violation miracle would be logically impossible. Of course, many laws are defeasible, but some laws—in particular certain conservation laws—are in a relevant sense indefeasible. As we shall see, much of the debate hinges on whether miracles would violate two of these laws, the laws of conservation of energy and momentum.

11. My thanks to Keith Augustine (private communication) for a very helpful survey of major contributions to this literature. See also, for the analogy, Vallicella (2009). His *modus ponens* is my *modus tollens*.

12. The cautionary "perhaps" signals one crucial difference—viz. that human minds, but not the divine mind, are embodied. But how that could be relevant depends entirely on how embodiment is to be understood, of which more anon.

13. Could it happen that some physical systems, in virtue of having emergent nonphysical properties, exert independent causal influences on their subvenient physical states, without exerting forces on their physical constituents? I very much doubt that this sort of "downward causation" is possible; but it would avail the theist nothing in any case, requiring its source to have a physical body. Richardson (1982) argues that Descartes plausibly held that psychophysical causation is unique and quite distinct from physical causation—concerning which see below. Lund (2009, 66–67) argues that mind/body causation may be just a brute, irreducible fact—as, allegedly, physical/physical causation is as well. But this ignores the fact that causation as we know it is mediated by forces, whatever their source (see Fales 1990, chap. 1); it would apparently involve a mind's moving matter about without exerting any forces on it.

14. That is, the partial derivative of the total momentum with respect to time.

15. I know of no change in physical properties that involves neither of these.

16. I have met atheists who insist, as a methodological rule, on the principle that natural causes are to be sought no matter what. But that seems misguided: what reason can be given for insisting that, no matter what the evidence, supernatural causes are always to be ruled out on the grounds that such a cause could never provide the best explanation?

17. Such as all the stars blinking on and off in synchrony so as to spell out, in Morse code and in multiple human languages, some eschatologically significant message. It is telling that the most heralded recent evidence for paranormal phenomena tends to be produced by randomizing targets and finding slightly better-than-chance selectivity on the part of subjects in very large trial sizes. The fact is that tiny confounds will be amplified by such large trial numbers into statistically significant anomalies.

18. Theists sometimes contrast personal explanation with physical or mechanical causation, and so will appeal to God's purposes and reasons to explain why he acts in certain alleged ways (see, e.g., Swinburne 1968), but almost never discuss how the divine will is put into effect, other than an appeal to omnipotence and an implicit claim that divine volitions are proximate causes of certain physical events; see Fales 2010.

19. See the trial testimony for the morning of day 12 of *Kitzmiller v. Dover School Board* (2005). For the former claim, see about one-third of the way through part 1 of that testimony; for the latter, see, for example, the exchange about halfway through part 2, concerning the bacterial flagellum, and part 2 of the afternoon testimony on day 11. It was likely also a reflection of Behe's efforts to avoid entangling ID in religious commitments, though he allowed that he thought the evidence pointed to the likelihood of a supernatural creator. But it is also certain, I think, that Behe has little or nothing to offer by way of independently testable hypotheses regarding the nature and powers of the Intelligent Designer.

REFERENCES

Averill, Edward, and B. F. Keating. 1981. "Does Interactionism Violate a Law of Classical Physics?" *Mind* 90:102–7.

Bricke, John. 1975. "Interaction and Physiology." *Mind* 84:255–59.

Broad, C. D. 1925. *The Mind and Its Place in Nature*. London: Keegan Paul.

Craig, William Lane. 1994. "The Special Theory of Relativity and Theories of Divine Eternity." *Faith and Philosophy* 11:19–37.

Dietl, Paul. 1968. "On Miracles." *American Philosophical Quarterly* 5:130–34.

Fales, Evan. 1990. *Causation and Universals*. New York: Routledge.

———. 1994. "Divine Freedom and the Choice of a World." *International Journal for Philosophy of Religion* 35:65–88.

———. 1997. "Divine Intervention." *Faith and Philosophy* 14:170–94.

———. 2010. *Divine Intervention: Metaphysical and Epistemological Puzzles*. New York: Routledge.

———. 2011 "Is Middle Knowledge Possible? Almost." *Sophia* 50 (1): 1–9.

Helm, Paul. 1988. *Eternal God: A Study of God Without Time*. Oxford: Clarendon Press.

Kitzmiller v. Dover School Board. 2005. "Trial Transcripts, Day 11 (October 18, 2005) and Day 12 (October 19, 2005)." TalkOrigins Archive. Last updated September 28, 2006. http://www.talkorigins.org/faqs/dover/kitzmiller_v_dover.html.

Larmer, Robert A. 1986. "Mind-Body Interactionism and the Conservation of Energy." *International Philosophical Quarterly* 26:277–85.

Leftow, Brian. 1991. *Time and Eternity*. Ithaca, NY: Cornell University Press.

Lund, David H. 2009. *Persons, Souls, and Death: A Philosophical Investigation of the Afterlife*. Jefferson, NC: McFarland.

Pennock, Robert T. 2001. "Naturalism, Evidence, and Creationism: The Case of Phillip Johnson." In *Intelligent Design and Its Critics: Philosophical, Theological, and Scientific Perspectives*, edited by Robert T. Pennock, 77–97. Cambridge, MA: Bradford, MIT Press.

Plantinga, Alvin. 2000. *Warranted Christian Belief.* New York: Oxford University Press.

Richardson, R. C. 1982. "The Scandal of Cartesian Dualism." *Mind* 91:20–37.

Sober, Elliot. 2000. *Philosophy of Biology.* 2nd ed. Boulder, CO: Westview Press.

Stump, Eleonore, and Norman Kretzmann. 1981. "Eternity." *Journal of Philosophy* 78:429–58.

Swinburne, Richard. 1968. "The Argument from Design." *Philosophy* 43:199–212.

Vallicella, William. 2009. "Is the Problem of Miracles a Special Case of the Interaction Problem?" *Maverick Philosopher* (blog). November 22. http://maverickphilosopher.typepad .com/maverick_philosopher/2009/11/is-the-problem-of-miracles-a-special-case-of-the -interaction-problem.html.

Von Wachter, Daniel. 2006. "Why the Argument from Causal Closure Against the Existence of Immaterial Things Is Bad." In *Science—A Challenge to Philosophy?*, edited by H. J. Koskinen, R. Vilkko & S. Philström, 113–24. Frankfurt: Peter Lang.

Wilson, David L. 1999. "Mind-Brain Interactionism and the Violation of Physical Laws." *Journal of Consciousness Studies* 6:185–200.

14

Navigating the Landscape between Science and Religious Pseudoscience

Can Hume Help?

BARBARA FORREST

David Hume, whose empiricist epistemology and trenchant critique of supernaturalism helped lay the groundwork for modern science, was under no illusion that his religious skepticism would become popular, as indeed it did not. He would probably not be surprised that three centuries after his birth, religious pseudoscience is among the modern world's most stubborn problems. In the United States, the most tenacious form is creationism. The sheer doggedness of its proponents, who seek political sanction for personal belief, has weakened the teaching and public understanding of science.

Creationists fit neatly among the disputants against which Hume directed his criticisms of supernaturalist religion.[1] Young-earth creationists unabashedly invoke the supernatural, while Intelligent Design (ID) creationists attempt to create a more sophisticated scientific façade. Yet ID's supernaturalism is now so well established that reiterating its purveyors' self-incriminating statements is unnecessary (see Forrest and Gross 2007). Creationists recognize neither a methodological nor a metaphysical boundary between the natural world and the supernatural, and therefore none between science and their religious pseudoscience. Hume's insights, supplemented by modern cognitive science, can help locate this boundary by defining the limits of cognition, although one must look beneath methodology and metaphysics to the most fundamental issues of epistemology to find it. Epistemology is fundamental to understanding both the parameters within which workable methodologies can be developed and the kinds of metaphysical views for which evidential justification is possible. *How* the mind acquires knowledge determines *what* humans can know about the world. In fact, the *concept* of metaphysics is itself the product of our cognitive capability. We cannot justifiably claim to know anything for which our cognitive faculties are insufficient; a particular metaphysics may transcend what is epistemically accessible, thereby necessitating the reliance on faith, scripture, and religious authority.

My central contention is that the boundary between the naturalism of science and the supernaturalism of religion—and, by extension, between science and religious pseudoscience—is set by the cognitive faculties that humans have and the corresponding kinds of knowledge of which we are capable. Recognizing this boundary is crucial to properly understanding science. Although one should certainly not teach Hume uncritically, his work can help students (at the university level) and the public to see the distinction between science and supernaturalist religion underlying that between science and religious pseudoscience. His still-relevant insights also serve as an entré to an area of empirical research that he presciently foresaw: cognitive science, particularly the cognitive science of religion.[2]

Brief Comments about Demarcation

To distinguish between science and religious pseudoscience, students and the public need guidelines for the more basic distinction between science and religion, an issue that has generated long-standing debate. General criteria for distinguishing science from pseudoscience, although helpful (see Bunge 1984), are insufficient for religious pseudoscience such as creationism. The cultural power and respectability of religion make creationism appealing even to people who reject more mundane forms of pseudoscience involving paranormal phenomena. Surveys show that evangelical Protestants, creationism's most aggressive constituency, nonetheless reject most paranormal beliefs for religious reasons that they do not apply to creationism (Bader et al. 2006; Pew Forum 2009). This suggests that the distinction between science and religion is a more basic one for which separate criteria are needed.

Hume's clarification of the limits of cognition facilitates this distinction. Aiming his work at both scholars and the literate public, he attempted "to bridge the gap between the learned world of the academy, and the world of 'polite' civil society and the literary market" (Copley and Edgar 2008, ix). Like Hume, Robert Pennock today urges philosophers to enunciate a workable distinction between science and pseudoscience for a broader audience than just other philosophers (Pennock 2011, 195). Pennock argues that we need not "an ahistorical formal definition [of demarcation] but . . . a *ballpark* demarcation" that underscores science's inability to incorporate the supernatural (Pennock 2011, 183–84). Noting that the demarcation issue encompasses both religious and paranormal claims, Pennock invokes Hume's insight about the limits of cognition, which is directly relevant to creationism: "As Hume pointed out, we have no experience and thus no knowledge of divine attributes" (Pennock 2011, 189). Consequently, we can draw no conclusions about a supernatural designer.

Unless they can identify special cognitive faculties for the supernatural—an epistemic challenge no one has met—religious believers, including creationists, are paradoxically forced to rely on their natural faculties when they invoke supernatural explanations. So Pennock is right: we do not need a one-size-fits-all demarcation criterion in order to say what science *is* because we can confidently say what science is *not*: it is not an enterprise in which supernatural explanations can be invoked in any workable, intersubjective way. And Hume is right: an epistemic line does mark the inaccessibility of the supernatural.

We can now survey aspects of Hume's work that clarify the distinction between science and religion and, by extension, between science and religious pseudoscience. His insights provide a conceptual introduction to cognitive science research that helps to explain not only the origin of supernatural belief but also the tenacity of creationism.

Survey of Hume's Relevant Work

As early as *A Treatise of Human Nature*, Hume understood (1) that cognition must be studied empirically, and (2) that it is constrained by sense experience and the natural world. He realized that we can properly understand the mind only through "careful and exact experiments" and that "we cannot go beyond experience" (Hume [1739] 1968, xxi).[3]

These epistemological insights are preserved in *An Enquiry Concerning Human Understanding*, in which Hume's analysis of religious belief includes both a recognizable description of pseudoscience and its remedy. Most metaphysics, he says, is not "properly a science" but the product of either "fruitless efforts . . . [to] penetrate into subjects utterly inaccessible to the understanding" or "the craft of popular superstitions, which . . . overwhelm [the mind] with religious fears and prejudices" (Hume [1772] 2000, 9).[4] The only remedy is "an exact analysis of [the] powers and capacity" of the mind in order to map out a "geography" of the cognitive landscape in which the "distinct parts and powers of the mind" are clearly delineated (Hume [1772] 2000, 9–10).

Hume was well acquainted with religious pseudoscience; it was rampant in the early Royal Society, whose 1663 charter incorporated the goal of illuminating "the providential glory of God" (Force 1984, 517). According to James E. Force in "Hume and the Relation of Science to Religion Among Certain Members of the Royal Society," some members were "apologist-scientists" who attempted to "institutionalize the design argument," according to which a "celestial watchmaker" God could miraculously contravene natural laws, a position that required them "to balance naturalism and supernaturalism" (Force 1984, 519–20). Although there is no evidence that Hume aimed the *Enquiry*'s arguments specifically at the Society, Force contends that Hume's critique of religion helped erode the Society's "scientific theism," which had developed over "some eighty years of religious propagandizing" (Force 1984, 517–18). At Hume's death in 1776, the Society was largely secularized, reflecting an evolving administrative philosophy that solidified when Martin Folkes became president in 1741. Folkes and his followers had staged a "palace revolt" against the apologists, ridiculing "any mention . . . of Moses, of the deluge, of religion, scriptures, &c." (Force 1984, 527).

Some modern scholars such as Philip Kitcher interpret the creationism of scientists of this period as science that simply had to be "discarded, consigned to the large vault of dead science" (Kitcher 2007, 22). Even after the Society's change of leadership, the design argument enjoyed a measure of scientific respectability among English scientists, whose faith often inspired their work (Kitcher 2007, 12). However, as I point out elsewhere, "the historical entanglement of science and religion does not make the religious inspiration of scientific discoveries itself scientific" (Forrest 2010, 431). Society apologists' inability to disentangle their science from their religion meant that their effort to explain natural phenomena by appeals to the supernatural was doomed to fail for epistemological reasons that Hume recognized, even if many of his contemporaries did not.

The Society's chartering of its religious apologetics remarkably prefigures the current aims of ID creationists at the Discovery Institute. Their strategy document, "The Wedge," incorporates the antiquated goal of replacing the "materialist worldview" with "a science consonant with Christian and theistic convictions," which they propose to advance through "apologetics seminars" (Discovery Institute 1998; Forrest and Gross 2007, chap. 2). ID exhibits hallmarks of pseudoscience as discussed by Maarten Boudry and Johan Braeckman, the most relevant being the invoking of "invisible or imponderable causes"—a supernatural designer—to "account for a range of phenomena," the workings of which "can only be inferred *ex post facto* from their effects" (Boudry and Braeckman 2011, 151; see also Boudry, chapter 5, in this volume). Leading ID scientist Michael Behe rejects the "rule . . . disbarring the supernatural" from science (Behe 1996, 240). His ID colleague William Dembski, arguing "that empirical evidence fails to establish the reduction of intelligent agency to natural causes," asserts that God created the world miraculously through a divine speech act: "God speaks and things happen" (Dembski 1999, 224). Both aim their work at students and the popular audience (Forrest and Gross 2007, 69, 116).

Whereas science's naturalistic methodology is well established nowadays, leaving ID creationists no excuse for promoting pseudoscience, seventeenth-century scientists' religious apologetics reflected the hazy understanding of scientific reasoning that marked their historical context. Yet Hume presciently helped lay the groundwork for science's epistemological disconnection from the supernatural by recognizing that cognition is bounded by the natural world and enunciating this insight's implications for supernaturalism. In the *Enquiry*, he combined such insights with incisive analyses of supernaturalism such as miracle claims and the design argument. He continued this scrutiny in *The Natural History of Religion* (*NHR*) (Hume [1757] 2008), strongly prefiguring current research in the cognitive science of religion.[5]

Relevant Points from the *Enquiry*

Anticipating scientific advances by a secularized Royal Society, Hume "envisioned a philosophy that employed the experimental method to make the study of the mind more a science than a speculative metaphysics" (Beauchamp 2000, xxvi). Although in the *Treatise* he noted philosophers' attribution of puzzling phenomena to "a faculty or an occult quality" (Hume [1739] 1968, 224), the *Enquiry* yields more mature insights upon which a distinction between science and pseudoscience can still be built.

The standard undergraduate introduction to Hume begins, appropriately, with his explanation in the *Enquiry* of the "origin of ideas" in sense perception and his distinction between "impressions" and "ideas" (Hume [1772] 2000, 13–14). However, concerning the ontological difference between the natural and the supernatural that underlies the distinction between science and religious pseudoscience, Hume's most useful insight is his recognition of the cognitive limitations intrinsic to sense experience, which constrains what even the imagination can conceive. Although we can imagine "the most distant regions of the universe; or even beyond," the mind's cognitive creativity "amounts to no more than the faculty of compounding, transposing, augmenting, or diminishing the materials afforded us by the senses and experience" (Hume [1772] 2000, 14). Yet, with these minimal, basic cognitive mechanisms, the mind can construct a "golden mountain," a "virtuous horse"—or a supernatural entity, "the idea of God" (Hume [1772] 2000, 14). Hume thus explains plausibly how we produce supernatural ideas via natural cognitive processes, foreshadowing the cognitive science of religion. Moreover, his explanation of how the mind can easily form the concept of God comes not as a shock but flows intuitively from the uncontroversial examples that precede it: "The idea of God, as meaning *an infinitely intelligent, wise, and good Being*, arises from reflecting on the operations of our own mind, and augmenting, without limit, those qualities of goodness and wisdom (Hume [1772] 2000, 14).

Hume next introduces the "association of ideas," the mind's mechanism of producing spontaneous order and unity among its ideas. This mechanism is "a principle of connexion between the different thoughts or ideas of the mind . . . that . . . introduce each other with a certain degree of method and regularity," differentiating into three forms: "*Resemblance, Contiguity* in time or place, and *Cause* or *Effect*" (Hume [1772] 2000, 17). My point here is not to arbitrarily assert the correctness of this aspect of Hume's epistemology but to highlight his recognition of what Todd Tremlin, in *Minds and Gods*, a survey of the cognitive science of religion, calls the "mental tools" through which humans generate basic and higher-order concepts (Tremlin 2006, 75).

Through the association of ideas, one idea automatically activates another, as when "a picture naturally leads our thoughts to the original" (Hume [1772] 2000, 17), generating the narrative unity that enables us both to communicate coherently and to compose fictitious, including supernatural, stories that nonetheless make sense. For example, Ovid's stories of "fabulous [fictitious] transformation[s], produced by the miraculous power of the gods" make sense because of the unity created by the "resemblance" of events, that is, their intuitively recognizable common feature of supernaturalness (Hume [1772] 2000, 18). Hume applies the same logic to John Milton's biblically inspired *Paradise Lost*. Although "the rebellion of the angels, the creation of the world, and the fall of man, *resemble* each other, in being miraculous and out of the common course of nature," they "naturally recall each other to the thought or imagination" with "sufficient unity to make them be comprehended in one fable or narration" (Hume [1772] 2000, 22). His recognition that such fictions make intuitive sense to humans is, as I show later, strikingly similar to Tremlin's in *Minds and Gods*.

However, recognizing danger in the mind's construction of intuitively credible but false narratives, Hume saw the need for some distinction between fantasy and reality, or "fiction and belief": "the difference between *fiction* and *belief* lies in some sentiment or feeling . . . which depends not on the will. . . . It must be excited by nature . . . and must arise from the particular situation, in which the mind is placed" (Hume [1772] 2000, 40). He views belief formation as a basic, involuntary process springing from our unavoidable interaction with the natural world, whereas fiction lacks this natural involuntariness. Belief unavoidably tracks experience in ways that fiction does not. But this raises an important question: since the structure and coherence of supernatural religious narratives give them a "natural," intuitive plausibility (especially when one has heard them since birth), is there any line of demarcation enabling us to classify them as fiction? Given the mind's imaginative ability to escape the natural world, what enables us to consciously and accurately distinguish between the natural and the supernatural?

Hume provides a workable "ballpark" criterion (Pennock 2011, 184) into which he builds the open-endedness of modern scientific reasoning by appealing to the intrinsic limitations of sense experience. This means that the cumulative, empirical knowledge of the physical world, to which human cognitive faculties are naturally receptive, never yields logical certainty; it does, however, provide a reasonably reliable epistemic map of that world. Consequently, science, or "natural philosophy," despite enabling us to successfully navigate the world, remains forever incomplete. Moreover, the mind's search for "ultimate"—which in Western culture usually means supernatural—causes runs smack into the involuntary boundary of our sensory faculties. The farther an idea recedes from its traceable sensory origin, the less likely we are to find a counterpart in the natural world, and we have no detection mechanisms for locating a counterpart in a putatively supernatural world. The empirical detection of supernatural phenomena—the very concept of which is contradictory—would mean that those phenomena are actually natural, if perhaps anomalous (see Forrest 2000, 17).

Invoking scientific concepts of his day, Hume fashioned a demarcation criterion that incorporates the epistemic humility now recognized as integral to science. Explanations of natural phenomena can progress no further than "elasticity, gravity, cohesion of parts, communication of motion by impulse"—that is, other natural phenomena—that "are probably the ultimate causes and principles which we shall ever discover in nature" (Hume [1772] 2000, 27).[6] The mind's natural, spontaneous credulity can be trained and tempered by "Academic or Sceptical philosophy"—the *mitigated* (moderate) epistemological skepticism that Hume famously embraced—which requires "confining to very narrow bounds the enquiries of the understanding, and . . . renouncing all speculations which lie not within the limits of common life and practice" (Hume [1772] 2000, 35, 120). Hume thus reveals a line of demarcation between science ("natural philosophy") and religion, and by extension between science and religious pseudoscience.

Crossing the Epistemic Divide

Hume's clarification of the limits of cognition lays the epistemological groundwork in the *Enquiry* for his criticisms of miracle claims and other supernaturalist ideas. Summarizing Hume's position, Force points out that "to believe in a miracle requires evidence that is impossible to obtain because it runs counter to our unalterable experience to the contrary" (Force 1984, 530; see also Forrest 2000, 16). In constructing an ontology, the mind's only cognitive baseline is the limited data of sense experience, a baseline that Hume recognizes as all too easy for the mind to cross. Presaging modern creationism, he explains how even "philosophers" (who in his day included both natural scientists and speculative philosophers) get swept up into religious pseudoscience: "They acknowledge mind and intelligence to be, not only the ultimate and original cause of all things, but the immediate and sole cause of every event . . . in nature. They pretend . . . that the true and direct principle of every effect is not any power or force in nature, but a volition of the Supreme Being" (Hume [1772] 2000, 56). Ignoring the cognitive boundary of sense experience, "we are got into fairy land, long ere we have reached the last steps of our theory" (Hume [1772] 2000, 57). If the minds of philosophers could so easily accommodate the supernatural, ordinary laypersons' susceptibility to religious pseudoscience is all the more understandable.

Hume also captures the essence of religious pseudoscience: its scientific sterility. The cognitive inaccessibility of the supernatural means that "while we argue from the course of nature, and infer a particular intelligent cause . . . [of] the universe, we embrace a principle, which is both uncertain and useless" (Hume [1772] 2000, 107). Religious philosophers, enthralled with "the order, beauty, and wise arrangement of the universe," are thus moved by their own incredulity to conclude that "such a glorious display of intelligence" cannot be the product of chance (Hume [1772] 2000, 102). Moreover, Hume's recognition of the propensity for supernatural belief includes the paradoxical fact that people believe supernatural narratives not *despite* but *because* they contradict experience.

In "Of Miracles," Hume recognizes that, concerning ordinary experience, "we may observe in human nature a principle"—a cognitive rule—according to which "objects, of which we have no experience, resemble those, of which we have; that what we have found to be most usual is always most probable" (Hume [1772] 2000, 88). Yet humans love stories of the *extra*ordinary; if the stories are interesting enough, we simply create a new rule to facilitate belief. The mind rejects evidentially sound but boring explanations in order to accommodate implausible but entertaining or emotionally "agreeable" narratives; consequently, "when any thing is affirmed utterly absurd and miraculous, [the mind] rather the more readily admits of such a fact, *upon account of that very circumstance*, which ought to destroy all its authority" (Hume [1772] 2000, 88; emphasis added). And when "the spirit of religion join[s] itself to the love of wonder, there is an end of common sense" (Hume [1772] 2000, 89).

When this natural propensity to supernaturalism is reinforced by what Hume recognizes as the "great force of custom and education, which mould[s] the human mind from its infancy" (Hume [1772] 2000, 66), religious pseudoscience such as creationism can acquire a tenacious foothold not only in individual minds but virtually an entire country.

From Hume to Cognitive Science

Although Hume doubted that "religious fears and prejudices" would loosen their grip on the human mind and that philosophers would "abandon such airy sciences" as metaphysics, he was optimistic that "the industry, good fortune, or improved sagacity of succeeding generations may reach discoveries unknown to former ages" (Hume [1772] 2000, 9). He also knew that philosophers needed some help, similar to Alvin I. Goldman's acknowledgment three centuries later that "epistemology needs help from science, especially the science of the mind" (Goldman 2002, 39).

Hume's optimism seems to have been rewarded. In *Minds and Gods*, comparative religion scholar Todd Tremlin (2006) surveys the cognitive science of religion, whose practitioners seek what Hume sought in *NHR*: to learn "what those principles are, which give rise to the original [religious] belief, and what those accidents and causes are, which direct its operation" (Hume [1757] 2008, 134). Tremlin discusses research by scientists such as anthropologist Pascal Boyer, whose findings are strikingly consistent with Hume's insights, as are those of anthropologist Scott Atran, whose work I include here. Although cognitive scientists have uncovered empirical data of which Hume never dreamed, they address many of the same questions as Hume, adopting his empirical approach toward the mind and religion. Their research helps to explain not only religion's natural origins but also the phenomenon of religious pseudoscience, suggesting that the mind's transition from experience of the natural world to supernatural belief is an intuitive glide rather than an irrational leap.

Empirical data suggest, ironically, that the mind's cognitive glide into supernaturalism was facilitated by evolution. Tremlin notes that cognitive science has produced "rapid growth in our knowledge of the brain" (Tremlin 2006, 7). Since the human brain is a product of evolution, so is cognition. Generated from the matrix of the brain's evolutionary history, the mind's higher capabilities include consciousness of self and other thinking beings, an ability that Daniel Dennett calls "the intentional stance" (Dennett 1987). We respond to other humans as "intentional systems" whose actions "can be predicted by . . . attributing beliefs, desires, and rational acumen" to them (Dennett 1987, 49). Human intentionality is so sensitively attuned that we attribute intentionality even to inanimate objects, imposing on the world what Stewart Guthrie calls an "anthropomorphic model": "A runner in the park wishes to know . . . whether a sound behind him is running human footsteps or blowing leaves" (Guthrie 1980, 187). Justin Barrett calls the attribution of agency where it does not exist a "hyperactive agent-detection device," which (citing Guthrie) may have had the positive adaptive value of alerting humans to potential danger (J. Barrett 2000, 31).

These evolved capabilities in turn enable the mind to create the idea of *supernatural* agency, the "god" concept: "Understanding the origin and persistence of supernatural beings requires first understanding the evolved human mind. . . . [I]deas about gods and religion are not 'special' kinds of thoughts; they are produced by the same brain structures and functions that produce all other kinds of thoughts" (Tremlin 2006, 6–7). Moreover, Atran contends that rather than "bypass our own [evolutionary] hard-wiring" in order to generate religious beliefs, we actually exploit it; in a kind of cognitive exaptation (see below), we "conceptually parasitize" our "commonsense understanding" of the world in order "to transcend it" (Atran 2006, 311). This echoes Hume's argument that humans can generate the idea of "*an infinitely intelligent, wise, and good Being* . . . from . . . the operations of our own mind" (Hume [1772] 2000, 14). Hume thus prefigured Tremlin's contention that "understanding the way god concepts are constructed reveals that they closely resemble other kinds of ideas that people entertain. Though those who believe in them treat gods as unique beings, 'gods,' as concepts go, are not unique at all" (Tremlin 2006, 87). If supernatural concepts are unique in any sense, they are just uniquely human.

The mind's ability to generate supernatural ideas is not the only idea in cognitive science that Hume's work foreshadowed. He seems to have recognized what Tremlin calls the mind's "mental architecture" (Tremlin 2006, 64), even if, in Hume's simplistic empiricism, that architecture is minimal. He recognized mental processes corresponding to what Tremlin calls "mental tools that are present at birth and mature in the first years of life" (Tremlin 2006, 75). Of course, Hume had no conception of the brain as an "operating system prepared by evolution that contains all the instructions for human computation" (Tremlin 2006, 66). However, although he rejected Cartesian innate ideas, he need not be understood as viewing the mind as a Lockean blank slate given his recognition of its spontaneous ordering of impressions and association of ideas. In fact, arguing that Locke's understanding of "innate" was unclear, Hume offers a clarification similar to Tremlin's idea of the "crucial innate skills" (Tremlin 2006, 65) that enable newborns to immediately process sensory information. He proposes that "innate" may simply mean "natural": "If innate be equivalent to natural, then all the perceptions and ideas of the mind must be . . . innate or natural, in whatever sense we take the latter word" (Hume [1772] 2000, 16).

This is roughly consistent with Tremlin's position that, although religious ideas are not innate—therefore not inevitable—the mechanisms, or tools, for creating them are: "Religious ideas, like all other kinds of ideas, owe their existence to a raft of specialized tools used in the brain's mental workshop" (Tremlin 2006, 74). Hume likewise understands that religious ideas are natural but not inevitable products of cognitive processes. Religion as "belief of invisible, intelligent power" stems not from "an original instinct or primary impression of nature" but is a "secondary" product of the mind that "may . . . be altogether prevented" (Hume [1757] 2008, 134).

Moreover, Hume recognized that the mind's generation and ordering of basic ideas begins early—an infant learns the painful lesson of touching a candle flame before it is capable of "any process of argument or ratiocination" (Hume [1772] 2000, 33). Tremlin likewise notes that infants develop a prerational apprehension of the world that differentiates into "intuitive biology, intuitive physics, and intuitive psychology" (Tremlin 2006, 66), or what Boyer calls "intuitive ontology" (Boyer 2000, 196). From this intuitive ontology, the mind generates supernatural concepts, which "*activate a set of ontological categories* that . . . are present . . . from an early state of cognitive development" (Boyer 2000, 196). Echoing Hume's association of ideas, Boyer says that "objects in the environment are identified as belonging [not only] to kind-concepts ('telephone,' 'giraffe') but also to ontological categories (PERSON, ARTEFACT [*sic*], ANIMAL, etc.)" (Boyer 2000, 196). Because imaginary entities "are intuitively associated with particular ontological categories," the "concept of 'spirit' activates the category PERSON"; when one prays to "a particular statue of the Virgin," the category of "ARTEFACT" is activated (Boyer 2000, 197).

So supernatural concepts grow easily—one might say naturally—out of the mind's intuitive ontology. According to Boyer, they are generated from a few basic cognitive "templates," which are not themselves concepts but "*procedures* for the use of information provided by intuitive ontology" (Boyer 2000, 198). But what explains the mind's propensity for generating supernatural concepts at all?

Atran's explanation of the origin of supernatural belief (drawing from Boyer) elucidates the origin of religious pseudoscience and its tenacity once it establishes a foothold in the mind. He argues that "religion, in general, and awareness of the supernatural, in particular" are a "by-product of several cognitive and emotional mechanisms that evolved under natural selection for mundane adaptive tasks" (Atran 2006, 302). Religion is not "an evolutionary adaptation per se" but a cultural by-product of evolution, which "sets cognitive, emotional, and material conditions for ordinary human interactions" (Atran 2006, 304).

Atran thus views religion as an exaptation (Atran, pers. comm.), a biological concept that Wesley Wildman discusses within the context of evolutionary psychology: "An *exaptation* is a feature of an organism that originated not as an adaptation but as a side effect of an adaptation that proved (often much later) to have a secondary adaptive function" (Wildman 2006, 253–54). For Atran, evolution created the mind's capacity to generate supernatural concepts, which have enabled humans "to solve inescapable, existential problems that have no apparent worldly solution, such as the inevitability of death" (Atran 2006, 302), making religion psychologically and culturally adaptive. Drawing support from Justin Barrett's studies of "theological correctness" (see J. Barrett 1999), Atran notes that although some religions such as Buddhism and Taoism "doctrinally eschew personifying the supernatural," religions "invariably center on concepts of supernatural agents," with the "common folk . . . routinely entertain[ing] belief in . . . gods and spirits" (Atran 2006, 304–5). If Atran is correct, then, unhappily for creationists, the mind's capacity to conceive of an Intelligent Designer is the product of evolution.[7]

Even more, if cognition is the product of evolution, so are the epistemic boundaries of cognition. Indeed, according to Tremlin, cognitive scientists have a term for these boundaries that helps to explain God's being conceived anthropomorphically as a designer:

> Cognitive scientists refer to these restrictions and their ramifications as "cognitive constraint." The conceptual range of the human mind is constrained by its own processing methods and by the patterns and tools it uses to interpret and organize the world. . . . [C]hildren asked to draw "aliens" from other planets will yield predictable category similarities. Psychologist Thomas Ward describes the same result . . . examining the products of adults asked to create imaginary, otherworldly animals. Our mind's natural design makes it rather difficult for us to "think outside of the box." (Tremlin 2006, 91–92)

These examples illustrate Boyer's observation that the supernatural concept of "spirit" activates the intuitive category of "PERSON." With the addition of Tremlin's points that "gods are first and foremost intentional agents, *beings with minds*" and that "the human mind is prone to suspect agency given the slightest excuse" (Tremlin 2006, 102), we get the concept of an Intelligent Designer—and evolution built the cognitive box.

Evolution also helps to explain why the idea of a supernatural designer is so widely (though not universally) intuitive. (Concern with "the [ultimate] origin of things *in general*" is not universal [Boyer 2001, 13].) Tremlin notes, citing Boyer, that "the notion of superhuman entities and agency is the only substantive universal found in religious ideas" (Tremlin 2006, 144; see Boyer 2001, 18–19). Hume, of course, intended his analysis of cognition to apply universally. The survival of only one human species thus implies the universality of the still-relevant aspects of Hume's work. As Tremlin observes, "the genus *Homo* is a set of one," which means that "these structures and functions [of human minds] are *species typical*" (Tremlin 2006, 38). So human mental architecture is planet-wide, varying only according to specific but superficial influences of geography, culture, and the like.

Although other parallels exist between Hume's work and the cognitive science of religion, perhaps the most striking one helps to explain why people cling to pseudoscience such as creationism when overwhelming evidence should rationally compel them to abandon it. As mentioned earlier, Hume's critique of miracles includes the paradoxical observation that the mind finds miracle claims persuasive *because* they contradict experience. Empirical data presented by both Atran and Boyer show that in humans, the very counterintuitiveness of supernatural belief is among its most universally *persuasive* features. Attention-grabbing counterfactuality is actually an epistemic incentive when the belief involves the supernatural, an incentive that appears to be absent in the case of beliefs about the natural world. According to Barrett, counterintuitive concepts of "mundane" objects such as "invisible sofas" rarely acquire religious significance, such status usually being accorded only to counterintuitive concepts of intentional agents (either human or nonhuman) (J. Barrett 2000, 30–31). This perhaps explains the tendency to find counterintuitive religious concepts more persuasive than evidence-based, scientific ("mundane") concepts such as evolution.

Atran defines the "counterintuitiveness" of religious belief as the violation of "universal expectations about the world's everyday structure, including such basic categories of 'intuitive ontology' . . . as PERSON, ANIMAL, PLANT and SUBSTANCE" (Atran 2006, 308). Yet, another important feature of counterintuitive beliefs explains why the mind accommodates them so easily: "Religious beliefs are generally inconsistent with fact-based knowledge, *though not randomly*. . . . [O]nly if the resultant impossible worlds *remain bridged to the everyday world* can information be stored, evoked, and transmitted" (Atran 2006, 308; emphasis added). This supports nicely Hume's contention that supernatural concepts are abstracted from ideas more closely tethered to the natural world. Moreover, the fact that supernatural narratives contradict experience makes even skeptics eager to transmit them because they enjoy hearing and repeating the stories: "The passion of *surprize* and *wonder*, arising from miracles . . . gives a sensible [perceivable] tendency towards the belief of those events. . . . [E]ven those who cannot . . . believe those miraculous events . . . yet love to partake of the satisfaction at secondhand . . . and . . . delight in exciting the admiration of others" (Hume [1772] 2000, 88–89).

Atran recounts data showing that counterintuitiveness is a mnemonic device that facilitates transmission of supernatural ideas—as long as they do not violate in too many different ways the intuitive ontology from which they spring (see also Boyer 2000, 199–201). He also notes that a kind of natural selection is integral to the survival of supernatural belief: "A small proportion of minimally counterintuitive beliefs give the story a mnemonic advantage over stories with no counterintuitive beliefs or with far too many. . . . Such beliefs grab attention, activate intuition, and mobilize inference in ways that greatly facilitate their mnemonic retention, social transmission, cultural selection, and historical survival" (Atran 2006, 311). If such findings continue to hold up empirically, Hume will have been right on the money in *NHR* in assessing counterintuitive stories as "natural" enough to be plausible:

> If we examine, without prejudice, the ancient heathen mythology, . . . we shall not discover in it any such monstrous absurdity, as we may at first be apt to apprehend. Where is the difficulty in conceiving, that the same powers or principles, whatever they were, which formed this visible world, men and animals, produced also a species of intelligent creatures, of more refined substance and greater authority than the rest? . . . [T]he whole mythological system is so natural, that, in the vast variety of planets . . . in this universe, it seems more than probable, that, somewhere or other, it is really carried into execution. (Hume [1757] 2008, 165)

From a cognitive standpoint, the supernatural is entirely natural.

Where Do We Go from Here?

Cognitive science has progressed far beyond Hume, enriched by scientific data showing that the mind has been shaped by evolutionary processes that cannot privilege humans with cognitive access to anything beyond the natural world. If we had such access, there should be at least as much consensus in theology as in science, buttressed by a body of cumulative theological knowledge to which ecclesiastical and scriptural authority are utterly irrelevant. The clearest evidence that humans lack cognitive access to anything beyond the natural world is the multiplicity of supernatural beliefs that contradict not only empirical fact but other beliefs by adherents of the same faith. In Hume's day, Christian disputes over transubstantiation—"so absurd, that it eludes the force of all argument" (Hume [1757] 2008, 167)—were a prominent example.

Today, creationists squabble about Earth's age and the length of a biblical "day" (Forrest and Gross 2007, 291).

Hume's insights concerning the epistemic boundary between the natural and the supernatural, supplemented by cognitive science, raise questions around which discussion of the distinction between science and religious pseudoscience can be structured:

1. Are there uniquely supernatural concepts that cannot, even in principle, be generated by known human cognitive faculties, including the ability to imaginatively transcend experience?
2. If so, by what method can they be investigated, including by investigators who reject the supernatural, and how does this method work?
3. If such a method exists, why have creationists and other supernaturalists not demonstrated it?
4. If humans have cognitive access to the supernatural, why do believers still require authoritative scriptures written by prescientific people, upon which creationists (including ID proponents) depend?

Although addressing these questions should be unnecessary given creationism's scientific sterility, the persistence of creationism mandates their being pressed at every appropriate venue and level. (For pedagogical and constitutional reasons, public school science classes below the university level are inappropriate.)

However, when these questions are addressed, the approach must incorporate another insight that Hume shares with Boyer and Atran: religious belief and pseudoscience are not marks of stupidity. Cognitive science indicates that both are products of evolved intelligence, reflecting a natural, imaginative curiosity about what lies beyond the horizon of experience and an ability to envision alternative possibilities. Despite his religious skepticism, Hume spoke respectfully of the design argument's intuitive—and to some extent rational—appeal: "A purpose, an intention, a design is evident in every thing; and when our comprehension is so far enlarged as to contemplate the first rise [origin] of this visible system, we must adopt, with the strongest conviction, the idea of some intelligent cause or author" (Hume [1757] 2008, 183). (He means that the argument is intuitively, not logically, compelling.) Hume also understood religion's power over brilliant minds such as Newton's, along with the influence of historical context, as in the case of the scholarly King James I.

If [James I] wrote concerning witches and apparitions; who, in that age, did not admit the reality of these fictitious beings? If he has composed a commentary on the [Book of] Revelations, and proved the pope to be Antichrist; may not a similar reproach be extended . . . even to Newton, at a time when learning was much more advanced than during the reign of James? From the grossness [primitiveness] of its superstitions, we may infer the ignorance of an age; but never should pronounce concerning the folly of an individual, from his admitting popular errors, consecrated by the appearance of religion. (Hume 1782, 196–97)

The intuitive appeal of an "invisible, intelligent power" is strong, and Hume's epistemological insights, supplemented by cognitive science's illumination of religion's deep intuitive and emotive roots, warrant moral if not epistemic respect for sincere belief. Ancient religious answers to questions prompted by natural curiosity about puzzling phenomena predate by many millennia the methods of critically scrutinizing those answers. Moreover, for believers, supernatural belief not only anchors social relationships and moral norms but also secures eternal, if not necessarily temporal, well-being, a fact Tremlin emphasizes: "Believable supernatural concepts . . . trigger important social effects and create strong emotional states" (Tremlin 2006, 140). So, while supernatural beliefs may have no genuine ontological object, they are not wildly irrational. Although some professional *purveyors* of creationist pseudoscience are themselves ethical reprobates (Hume criticized charlatans who exploit religious belief), rank-and-file believers who accept creationism are neither evil nor stupid; they are just wrong (even if often maddeningly so). Yet the idea of an Intelligent Designer who loves its creatures is much closer to the intuitive matrix of that idea than the principles of natural selection are, and Atran observes that "science is not well suited to deal with people's existential anxieties" (Atran 2006, 317). Religion's social and emotional power, combined with its intuitive plausibility, thus explains the tenacity of religious pseudoscience even among people with enough scientific literacy (and sometimes expertise) to "know better." So where does this leave us?

Hume realized that pitting "profane reason against sacred mystery" can be like trying to "stop the ocean with a bullrush" (Hume [1757] 2008, 166). Tremlin cautions that "beliefs—religious or otherwise—may only rarely meet the demands of thought that the intelligentsia deem reasonable and rational" (Tremlin 2006, 140). Dembski's "God-speaks-and-things-happen" creationism is the same magical thinking as the belief in miracles that Hume critiqued three centuries ago. However, the public policy implications of creationism's widespread acceptance are greater today given the broader access to primary and secondary education.

Hume offered no easy solution to the problem he so incisively addressed; indeed, there is no easy solution. Counteracting creationism's intuitive plausibility is difficult. Most children's religious views are established well before elementary school, and popular culture (including the ubiquity of creationism on the Internet) almost always gives religious pseudoscience a head start. Although Hume's insights and cognitive science can identify the boundary between science and religious pseudoscience, they also highlight the obstacles to overcoming creationism. There is no surefire way to *counteract* religious pseudoscience that rivals the intuitive ease of *accepting* it.

The outlook is not totally bleak. Others have offered useful strategies to improve K-12 science education, although that is a long-term goal at best. However, under the most optimistic scenario, only a small segment of university students and a proportionately smaller, probably infinitesimal, segment of the public will ever hear about Hume and cognitive science. Successfully reaching beyond academia requires an audience literate enough to grasp the concepts involved and an army of willing public intellectuals (who are effective communicators).

However, another point emerges from this analysis of the insights of Hume and cognitive science: creationism cannot be counteracted by merely debunking religion. Rather, believers must see a viable religious alternative that is intellectually honest about science while serving other needs religion addresses. This requires more engaged public outreach by scientifically literate theologians. Yet even that is no guarantee that average believers will be receptive. Boyer points out the ever-present tension between the beliefs of ordinary believers and what religious "guilds" (institutions) and theologians declare doctrinally acceptable: "People are never quite as 'theologically correct' as the guild would like them to be" (Boyer 2001, 283). Indeed, a study by Justin Barrett of "both believers and non-believers" in such disparate cultures as Delhi, India, and Ithaca, New York, showed "theological correctness" to be a quite real phenomenon (Boyer 2001, 87–89; see also J. Barrett 1999). Theologians can thus expect resistance from the laity.

Consequently, those of us who work against religious pseudoscience face a predicament that we may simply be forced to wait out while simultaneously trying to resolve it in favor of real science. On one hand, we must avoid the approach Philip Kitcher rightly criticizes: "If you start with the thought that the predominance of religion in human societies is to be explained by a cognitive deficiency, you will tend to see your campaign for the eradication of myths in terms of a return to intellectual health" (Kitcher 2011, 10). This approach is morally condescending and strategically wrong. On the other hand, creationism not only wastes valuable time and talent but also foments political agitation that threatens children's education and the country's scientific productivity. Teachers, scientists, clergy, and academics must persistently, but respectfully, contribute their respective expertise to help their students and fellow citizens see through it.

Confronting religious pseudoscience requires respecting believers enough to be truthful with them about science rather than, as Kitcher puts it, patronizing them by a "polite respect for odd superstitions about mysterious beings and their incomprehensible workings" (Kitcher 2011, 1). We will just have to work much harder, and for quite a long time.

Acknowledgments

I wish to thank my reviewers for their very helpful comments.

NOTES

1. By "supernatural," I mean "belonging to a higher realm or system than that of nature; transcending the powers or the ordinary course of nature" (*Oxford English Dictionary*). My discussion applies to the supernaturalism of the Abrahamic religions, especially Christianity, from which creationism emerged, not to metaphysical views such as pantheism. So my distinction between "science and religion" refers to science and supernaturalist religion. I use "religion" and "supernaturalism" interchangeably.

2. Todd Tremlin defines what is "broadly called 'cognitive science'" as encompassing "neurology, psychology, biology, archaeology, paleontology, anthropology, linguistics, philosophy, and other fields" (Tremlin 2006, 7).

3. Hume's reference to "experiments" reflects the word's eighteenth-century ambiguity. Samuel Johnson's *Dictionary of the English Language* (1785) defined the verb form as "To search out by trial" and "To know by experience." Above, Hume intends the first meaning, referring to the need for empirical research on the mind.

4. Hume's reference to "science" reflected its contemporary definition as "knowledge," which included all specialized areas of learning.

5. I am concerned with Hume's discussion of epistemological problems with supernaturalism, not his critique of specific arguments for theism.

6. Recent (unsuccessful) scientific attempts to test the therapeutic efficacy of intercessory prayer support Hume's point. They have merely reaffirmed the impossibility of detecting, thus measuring and controlling, the requisite supernatural phenomena (see S. Barrett 2003; see also Shermer 2006).

7. Hume does not present the mind's ability to generate the god concept as equivalent to a denial of its existence, which is a separate issue. Theistic evolutionists can interpret evolution as having produced the ability to understand the concept of the God who guided evolution (see Haught 2009).

REFERENCES

Atran, Scott. 2006. "Religion's Innate Origins and Evolutionary Background." In *Culture and Cognition*. Vol. 2 of *The Innate Mind*, edited by Peter Carruthers, Stephen Laurence, and Stephen Stich, 302–17. New York: Oxford University Press.

Bader, Christopher, Kevin Dougherty, Paul Froese, Byron Johnson, F. Carson Mencken, Jerry Z. Park, and Rodney Stark. 2006. *American Piety in the 21st Century: New Insights to the Depth and Complexity of Religion in the US*. Waco: Baylor Institute for Studies of Religion. http://www.baylor.edu/content/services/document.php/33304.pdf.

Barrett, Justin L. 1999. "Theological Correctness: Cognitive Constraint and the Study of Religion." *Method and Theory in the Study of Religion* 11:325–39.

———. 2000. "Exploring the Natural Foundations of Religion." *Trends in Cognitive Sciences* 4 (1): 29–34. doi:10.1016/S1364–6613(99)01419–9.

Barrett, Stephen. 2003. "Implausible Research: How Much Is Enough?" *Wien Klin Wochenschr* 115 (7–8): 218–19. doi:10.1007/BF03040319.

Beauchamp, Tom L. 2000. Introduction to *An Enquiry Concerning Human Understanding*, by David Hume, xi–cvii. Oxford: Clarendon Press.

Behe, Michael J. 1996. *Darwin's Black Box: The Biochemical Challenge to Evolution*. New York: Simon and Schuster.

Boudry, Maarten, and Johan Braeckman. 2011. "Immunizing Strategies and Epistemic Defense Mechanisms. *Philosophia* 39 (1): 145–61. doi:10.1007/s11406–010–9254–9.

Boyer, Pascal. 2000. "Functional Origins of Religious Concepts: Ontological and Strategic Selection in Evolved Minds." *Journal of the Royal Anthropological Institute* 6 (2): 195–214.

———. 2001. *Religion Explained: The Evolutionary Origins of Religious Thought*. New York: Basic Books.

Bunge, Mario. 1984. "What Is Pseudoscience?" *Skeptical Inquirer* 9:36–46.

Copley, Stephen, and Andrew Edgar. 2008. Introduction to *Selected Essays*, by David Hume, vii–xxii. New York: Oxford University Press.

Dembski, William A. 1999. *Intelligent Design: The Bridge Between Science and Theology*. Downers Grove, IL: InterVarsity Press.

Dennett, Daniel C. 1987. *The Intentional Stance*. Cambridge, MA: MIT Press.

Discovery Institute. 1998. "The Wedge Strategy." Accessed October 19, 2012. http://www .antievolution.org/features/wedge.html.

Force, James E. 1984. "Hume and the Relation of Science to Religion Among Certain Members of the Royal Society." *Journal of the History of Ideas* 45 (4): 517–36.

Forrest, Barbara. 2000. "Methodological Naturalism and Philosophical Naturalism: Clarifying the Connection." *Philo* 3 (2): 7–29.

———. 2010. Review of *Living with Darwin: Evolution, Design, and the Future of Faith*, by Philip Kitcher. *Journal of Value Inquiry* 44 (3): 425–32. doi:10.1007/s10790–010–9222–4.

Forrest, Barbara, and Paul R. Gross. 2007. *Creationism's Trojan Horse: The Wedge of Intelligent Design*. New York: Oxford University Press.

Goldman, Alvin I. 2002. *Pathways to Knowledge: Private and Public*. New York: Oxford University Press.

Guthrie, Stewart. 1980. "A Cognitive Theory of Religion." *Current Anthropology* 21 (2): 181–94.

Haught, John F. 2009. "Theology, Evolution, and the Human Mind: How Much Can Biology Explain?" *Zygon* 44 (4): 921–31.

Hume, David. (1739) 1968. *A Treatise of Human Nature*. Edited by L. A. Selby-Bigge. Oxford: Clarendon Press.

———. (1757) 2008. *The Natural History of Religion*. In *Principal Writings on Religion Including Dialogues Concerning Natural Religion and the Natural History of Religion*, edited by J. C. A. Gaskin, 134–93. New York: Oxford University Press.

———. (1772) 2000. *An Enquiry Concerning Human Understanding: A Critical Edition*. Edited by Tom L. Beauchamp. New York: Oxford University Press.

———. 1782. *The History of England, from the Invasion of Julius Caesar to the Revolution in 1688*. Vol. 6. London: T. Caddell. Google Books.

Kitcher, Philip. 2007. *Living with Darwin: Evolution, Design, and the Future of Faith*. New York: Oxford University Press.

———. 2011. "Militant Modern Atheism." *Journal of Applied Philosophy* 28 (1): 1–13. doi:10.1111/j.1468–5930.2010.00500.x.

Pennock, Robert T. 2011. "Can't Philosophers Tell the Difference Between Science and Religion? Demarcation Revisited." *Synthese* 178 (2): 177–206. doi:10.1007/s11229–009–9547–3.

Pew Forum on Religion and Public Life. 2009. *Many Americans Mix Multiple Faiths*. December 9. http://pewforum.org/Other-Beliefs-and-Practices/Many-Americans-Mix-Multiple-Faiths.aspx#1.

Shermer, Michael. 2006. "Prayer and Healing." *Skeptic*, April 5. http://www.skeptic.com/eskeptic/06–04–05/.

Tremlin, Todd. 2006. *Minds and Gods: The Cognitive Foundations of Religion*. New York: Oxford University Press.

Wildman, Wesley J. 2006. "The Significance of the Evolution of Religious Belief and Behavior for Religious Studies and Theology." In *Evolution, Genes, and the Religious Brain*. Vol. 1 of *Where God and Science Meet: How Brain and Evolutionary Studies Alter Our Understanding of Religion*, edited by Patrick McNamara, 227–72. Westport, CT: Praeger.

PART V

True Believers and Their Tactics

15

Argumentation and Pseudoscience

The Case for an Ethics of Argumentation

JEAN PAUL VAN BENDEGEM

Logicians, including myself, share the belief that there exist arguments that are valid in some kind of necessary way, be it on the basis of the form of the argument or on the basis of its content. Instances of the first sort are cherished by (formal) logicians because in such cases it is rather easy to establish whether the argument has this or that particular form, and hence equally easy to decide whether validity is guaranteed or not. The second case is unfortunately all too often messier, as it involves meanings and therefore semantical considerations.

From this belief a second belief follows, viz. that there exists an ideal kind of argumentation, debate, discussion, or any similar kind of verbal interaction. If there are two parties, the first party presents its thesis and presents the arguments in its favor. The arguments are evaluated, and if they stand up to a thorough scrutiny, they will be accepted by the other party, who will thereby accept the thesis itself. In the "truly" ideal case, even if the other party, before the start of the discussion, actually believed the negation of the thesis defended, he or she will nevertheless reject one's own thesis and accept the thesis of the first party.

Obviously, ideals differ from reality, but the question is how large the difference is (or can be) and why. As someone who has participated *in real life* as a debater and a lecturer, I have heard (and unfortunately continue to hear) many silly and few sound arguments. This huge difference between theory and practice creates a rather strong tension, and, in general terms, that tension is what I want to discuss here. More specifically, if we take into account all the real-life aspects of a debate, a discussion, or an argumentation, what does it mean to defend a thesis, a position, or a claim in an *efficient* way?

In section two, I am more explicit, though rather brief, about the above mentioned ideal reasoner or debater. Then I sketch the picture that comes closer to real-life situations. In section four, I outline what this new look entails for argumentation, discussion, and debate. Next, I present some concrete cases, and in the final section, I raise the ethical issues posed by all this.

The Standard View of the Ideal Reasoner

Astrologer: Astrology must be true because it has predicted all important events that happened in this century.

Skeptic: Has it predicted the assassination of president x in country y?

Astrologer: Well, I have to be honest here, it has not.

Skeptic: And you agree that this is an important event?

Astrologer: Sure, no discussion about that!

Skeptic: So, I can assume that you agree that this is a very good argument against astrology according to your own strategy?

Astrologer: I don't like saying this, but I have to agree, yes, it is a good argument against astrology. So, yes, I will have to revise my opinion on this matter.

I cannot imagine that anyone would be prepared to believe that the above discussion is a report of a real-life event. Nevertheless, when we talk about the ideal reasoner, debater, and so forth, what we have in mind is something or somebody who comes pretty close to the astrologer in the above example.

A more specific image of the ideal reasoner is to be found in the fields of logic or of mathematics.[1] The logical and mathematical proof is the ideal type of argument logicians and mathematicians deal with, and it involves (at least) the following elements:

(I1) The two parties involved in the discussion agree on a number of logical inference rules. Standard logic and mathematics rely on a set of rules such as, for example, *modus ponens* (from A and "If A, then B," to infer that B, where A and B are arbitrary sentences) or dilemma (from "A or B," "If A, then C," and "If B, then C," to infer that C, where A, B, and C are arbitrary sentences).

(I2) The meaning of the sentences is unambiguous. Whatever one party states is understood by the other party as it is intended in a unique way. Meanings are usually also literal. No metaphors, no analogies are allowed for, as they do not help the reasoning process.

(I3) Apart from the sentences stated, nothing else enters into the game. No hidden premises, no hidden suppositions, no hidden hypotheses, and, in addition, no hidden agendas, no hidden "devious" intentions. One does not want to confuse the opponent, to divert attention, to "create smoke," to cheat, abuse, mislead or downright lie.

(I4) Both parties agree on how the discussion has to proceed. Preferably, there is a unique format, explicitly codified so that inspection for all parties concerned is guaranteed.

(I5) Finally, and most important, both parties agree on winning and losing or, more generally, on how to evaluate the final stage of the debate.

Obviously, in this type of discussion, no fallacies can occur, except perhaps caused by not being attentive, but these errors do not count as deviations from the basic "game." It is interesting to mention that within the field of formal logic, this is an approach to logical reasoning framed in terms of dialogues. What I have written above is basically a translation in ordinary language of the (standard) formal description of such dialogues. As a concrete example, consider this schematic dialogue (where again A, B, and C stand for arbitrary sentences):

First party: I claim that "If A, then B," then "If A and C, then B."

Second party: I'm not sure about this. So, for argument's sake, I will accept that "If A, then B" is the case. Can you show me that "If A and C, then B" must be the case?

First party: Yes, most certainly.

Second party: OK, I will accept "A and C"; now show me that B has to follow.

First party: Since you have accepted A and C, you also accept A, right?

Second party: I do.

First party: But you have already accepted "If A, then B," remember? So you have to accept B. There you are!

Second party: OK, you win.

I will refer to (I1) through (I5) as the ideal logician's attitude (ILA). Let me make two more remarks about this:

1. It seems clear that hardly any real-life dialogues come close to the ideal situation. Nevertheless, this does not exclude the possibility that any real-life discussion can be rewritten in this or a similar format. The comparison is often made with mathematics. Although the formal demands we put to mathematics today were not applicable, say, a thousand years ago, so the argument goes, there has been no intrinsic difficulty in reformulating ancient proofs in terms of present standards. There is, however, a serious drawback. Some of the reformulated proofs contain reasoning errors or are incomplete and thus cease to be proofs at all. What holds for mathematics surely holds for everyday argumentation and reasoning. What seems convincing in real life very often does not after formal reconstruction. So even though one might argue that such a formal description is always present *in potentia (leaving aside all the practical details of setting up the translation), the outcome will be rather disappointing: most reasonings will be uniformly rejected.* But it is precisely these rejections we want to study and understand.[2]

2. As said above, the role of the two parties is quite clear. How the parties have to handle attacks and defenses is explicitly stipulated, and therefore there is no discussion about the *procedure* to follow. Hence, questions such as "What is the right move in this part or at that stage of the discussion?" need not be posed.

Surely ILA has some nice features but it is obvious at the same time that the distance between ILA and real life is huge. To judge that distance, the next section outlines the real-life situation.

The Revised Picture of the Human Reasoner

The procedure I follow is to have another look at the five elements (I1) to (I5) that characterize the ILA and see what changes are necessary to achieve a more realistic picture. To criticize (I1), we do not have to leave the field of formal logic. Actually, within the discipline itself, there is a consensus on the following claims:

(R1) To talk of *the* logic (in whatever sense) is rather nonsensical. It is easy to check in whatever (decent) survey of formal logic is available to see that a multitude of logical systems exist "on the market."[3] Different logics are generated by the simple fact that a particular logical axiom or rule is accepted or not. Some logics accept—for example, the principle of the excluded middle ("either A or not-A") or the rule of double negation (from not-not-A to infer A)—but other systems do not. Hardly any logical principle, axiom, or rule has not been questioned (even within the domain of mathematics). This does not imply an arbitrariness—not all logics are equal, so to speak—but it does imply that motivated choices will have to be made. That being said, I do realize that in daily practice a set of rules such as *modus ponens* (from A and "If A, then B" to infer B) is hardly ever questioned (if even mentioned explicitly at all). But even so, this does not and cannot rule out situations where part of the discussion between two parties will involve a debate about the logical rules that will be accepted and applied in the course of the debate. A simple example: does a statement of the form "There exists an x with this or that property" imply a full description or identification of x? Some logics, including classical logic, say no, but other logics, such as intuitionistic logic, say yes. Once we enter into a discussion about the rules themselves, the next question then becomes: according to what rules is this argumentation supposed to proceed? I return to this problem later.

(R2) I believe that about (I2) even less needs to be said. After the contributions of such philosophers as Willard Van Orman Quine, Hilary Putnam, Ludwig Wittgenstein, and so many others, it is quite clear that the idea that all words and sentences have unique meanings and that unique literal meanings exist should be left behind us. Semantic holism (in its various senses) is the modern expression of this idea. Note that this does not exclude preferred meanings or gradations in meaning, where a relative degree of literalness can be attributed. It is always possible—for example, through the use of such a gradation—to select this or that particular interpretation as *the* interpretation. Nevertheless, semantic holism does imply that we have to take into account as an irreducible component of our evaluation of arguments and debates that (sometimes radically) different meanings of the same words and sentences are being used. Why else the need for principles of charity?[4] An important consequence is that metaphors, analogies, and similarities become means of expression in their own right. Even a quick glance at the study of metaphors makes clear that problems here are impressive. How does a metaphor function? What are essential and what are accidental elements? What is a good, what is a bad metaphor? There is a whole bunch of proposals, none universally accepted by the philosophical community.[5] In terms of the ILA, these linguistic devices should not be used or, if unavoidable, be reducible to logical-literal terms.[6] We therefore do better to study them and integrate them as proper vehicles for meaning.

(R3) Let us now examine the level of intentions of the parties involved. Even when both parties have the most honorable intentions imaginable, it still remains the case—as the extensive work on speech acts of such linguistic philosophers as John Searle, John Austin, Daniel Vanderveken, and others clearly show—that these intentions do play a part in determining how words function. Take a command expressed by a factual statement, such as "The dog wants to go out." A person with the ILA will reply that it is sufficient to formulate that statement as a command, but even then we have to assume that the hearer will recognize it as such. In general terms, even if all the rules involved were made explicit, the problem would still remain according to what rules the process of making the rules explicit should proceed. At some stage, necessarily, we must end up with a *practice* of speaking, arguing, debating, and so forth. Definite skills are involved in identifying an argument and responding to it in a way that is (in some sense or other) appropriate. If we take into account intentions and the complete *repertoire* of linguistic behavior, including lying and misleading, then it becomes clear that to respond to a fallacy we require more possibilities than the ILA will allow for, as the first example in section 5 shows.

(R4–R5) Finally, as to (I4) and (I5), we can be rather brief about the question of the existence of a unique procedure and the question about winning or losing. Rare, if nearly nonexistent, are the situations where one party gracefully acknowledges that the other party has made a point and the one party will duly revise its opinion. Usually what happens is that the point of whether a point has been made needs to be debated as well, again assuming one has agreement on the procedure to be followed.

To appreciate the distance between the ideal and the real situation, let us make a comparison with visual perception. The ideal picture would be the description of an eye that has nearly infinite precision both in the extremely small and large. Call it *God's eye*. We note that it does not correspond to the human eye at all, and so we draw an "adapted" picture. We take into account that there are upper and lower limits and, to a point, we actually defend that this is a good thing. So now we have a picture of a real eye, our eye. This, however, is not the end of the story: precisely because we are now looking at a real eye, we have to note that from time to time, it does absolutely silly things.

It gets tricked by all sorts of illusions, and we would do better to take these illusions into account as well. They are part and parcel of human vision. The analogy holds up to the point that our reasoning powers, our thinking abilities, our grasping of meanings also suffer from illusions, in this case, *cognitive* illusions.

These phenomena have been extensively studied in the second half of the twentieth century, starting with the pioneering work of Peter C. Wason, Jonathan St. B. T. Evans, Amos Tversky, Daniel Kahneman, and many others.[7] I do not present here an overview of this fundamentally important work, but list instead some well-known features (without any attempt at completeness):

- We have serious problems identifying and reasoning with logical connectives, especially implicational statements of the form "If A, then B" are not well understood. The best-known example is to assume that A follows from "If A, then B" and B, as Peter C. Wason's famous card experiment shows (see Wason and Johnson-Laird 1968).[8]
- We tend to favor positive, supporting information over negative, falsifying information. We also tend to stick to first impressions and overestimate the correctness of our memory and hence of the things we believe we know, especially in terms of everyday knowledge.
- We tend to be quite horribly inaccurate in the estimation of probabilities. Moreover, we do not tend to think in terms of populations (or samples), but rather to reason in terms of stereotypes or typical representatives.
- We tend to see connections and relations everywhere. This, no doubt, must be a very familiar feature to anyone involved with pseudoscience and the like. Coincidental events must necessarily be linked in one way or another. "There is no such thing as coincidence" is the often-expressed phrase that reflects this view.

In short, the distance with the ILA is indeed dramatically large. It is therefore a very wise decision to use and rely on the real, though admittedly not optimistic, picture of the human reasoner. It is that picture that brings us to a "better" (unfortunately so!) version of the dialogue at the beginning of this section:

Astrologer: Astrology must be true because it has predicted all important
    events that happened in this century.

Skeptic: Has it predicted the assassination of president x in country y?

Astrologer: Oh yes, it most certainly has. Look here, it was predicted that somebody "wearing the crown" in country z was going to have serious problems. Amazing, isn't it?

Skeptic: I am sorry! A crowned head in country z has nothing to do with the president of country y. So you cannot claim that you have predicted it.

Astrologer: Why are you being so literal!? Of course, the astrological prediction speaks about a crowned head, but then is a president not metaphorically a crowned head? And have you noticed that country z has very tight commercial relations with country y? What more do you need?

Skeptic: But don't you see that with such reasoning you will always be right?

Astrologer: I am quite sorry, but if the prediction had said something about a farmer in country w—and, as you know, country w has nothing to do with country y—then it could not have been a president in country y, could it now?

Skeptic (getting desperate): Why am I doing this? . . .

What Does It Mean to Debate with "Real" Human Reasoners?

In this section, the core idea of this contribution is formulated: what does the revised picture imply for the practice of debate, discussion, and argument? Under the assumption that these real-life elements are here to stay, as they are so deeply ingrained in our thinking processes, we might as well accept that situations where we have to deal with ideal reasoners are and will always be exceptional. Of course, we need not be unduly pessimistic. In many cases, discussions and debates will proceed according to mutually accepted rules and procedures, close enough to the ILA to have a guarantee that the outcome will be accepted by the parties involved. In such cases, the ideal and real perspectives come close enough. There is no need, so to speak, to stress the differences. We, as skeptics, may perhaps not like the outcome of such a discussion—you firmly believed that the other party was wrong but you did not find the best arguments to support your case and you lost the discussion—but we will still derive a justifiably good feeling that at least all has proceeded well. Phrased differently, what interests us here are the situations where the divergence between ideal and real perspectives is obvious, and the most prominent and interesting case is that of argumentative fallacies and the difficulties posed by how to deal with them. It seems to me that there are two basic attitudes in response to that problem, which I will label the meta-level response (MLR) and the object-level response (OLR), corresponding respectively to the ILA and to the real attitude:

- The MLR corresponds to the procedure in a discussion that, whenever one party commits a fallacy (of whatever form), it is the "duty" of the other party to identify the fallacy and indicate to the one party that a fallacy has been committed and that therefore the argument should be withdrawn. I label this procedure meta-level because it steps out of the ongoing discussion to make a remark *about* an argument scheme. As I argued before, from the ILA, this is the only option available. Needless to say, in real life this is often a poor strategy.
- The OLR takes as a starting point that elements such as intentions, stylistic considerations such as metaphors and analogies (in short, rhetorical considerations), up to and including misleading behavior and deception, are to be considered integral ingredients of the debate. It then follows that responses to a fallacy can at least make use of these elements as well (or, at least, they could be investigated as to their efficiency).

How precisely are we to imagine an OLR in a real-life setting? The next section presents two case studies to clarify the matter.

Two Illustrations of Real Debating Techniques

The first case investigates what the possibilities are to attack or to counter fallacies with "fallacies" of our own, and the second case deals with misleading and cheating. If the first method is reasonably acceptable, the second one has been deliberately chosen to raise a deeper ethical issue, the topic of the final section of this chapter.

Attacking Fallacies with Fallacies

As a first example, consider the argument *ad verecundiam*. One of the basic formats of this fallacy is the following (where P and Q represent the two parties in the debate):

P: You must accept statement A because authority X has said so.

One possible way, according to the OLR, to counter this argument is to attack it with the same fallacy, thus to present an argument *ad verecundiam* as well. Apart from the basic form

Q: You must reject statement A because authority Y has done so.

more sophisticated reactions are possible. All too often, the authority Y is not accepted by P as an authority.[9] But suppose you know (a) that P accepts both X and Z as authorities and (b) that Z rejects A; then it is far more interesting to confront P with that fact, independent of whether you yourself accept Z as an authority. First, you obtain P's commitment to the fact that Z is accepted as an authority, and then the rejection of A by Z is presented. This should weaken P's argument.

An intriguing subtlety is at work here, and it is interesting to spell it out in some more detail. The initial fallacy is the use of authority by P. Q's response is another instance of the same fallacy since Q also invokes an authority. The subsequent fallacy consists of, first, accepting the initial fallacy—that it is a good and acceptable thing to argue relying on authorities—and second, reasoning logically about authorities. In this particular case, the logical argument looks like this:

Given:
  (a) If someone accepts someone as an authority, then if the authority says something, that something is accepted by the first.
  (b) P accepts X and Z as authorities.
  (c) X says that A is the case.
  (d) Z says that not-A is the case.
Then it follows:
  (e) From (a), (b), and (c), P accepts A.
  (f) From (a), (b), and (d), P accepts not-A.
  (g) Hence P is contradicting him- or herself.

The crucial point and the subtlety of the matter is that this piece of logical reasoning on its own may be perfectly valid (but see footnote 8 for an important caveat), even according to ILA's standards, but we do remain within the framework of fallacious reasoning.

In debates with astrologers, parapsychologists or the "paranormally-gifted," it is possible to bring one's opponent to reject all his or her colleagues as incompetent and not qualified because they all make claims that he or she does not accept or reject. So, in the end one and only one authority is left: the astrologer, parapsychologist, or 'paranormally-gifted' him- or herself. This is a variation on the authority theme, because we are now talking about self-authority. The fallacious argument runs like this:

Given:
   (a) There are two authorities, X and Y.
   (b) X says A, and Y says not-A.
   (c) I am X.
Then it follows:
   (d) I am right, hence A; and Y is wrong.

X has to confirm him- or herself not merely as an authority, but as the *sole* authority; hence X's argument reduces to a form of "Because I say so." This introduces a form of "self-promotion," which is unlikely to convince much of anyone. In addition, it creates space for the argumentative counterattack that Y does not require such self-promotion but gladly leaves it to others to judge her authority. "You must believe me because I say so" versus "You need not believe me, instead ask around to see whether I am trustworthy." It seems reasonable to assume that the latter holds a stronger position than the former.

A second example is the argument *ad ignorantiam*. The basic format here is

P: A must be the case as there is no proof against A.

To counter this argument with an argument of the same type, it is interesting to look for a statement B such that (a) there is no proof against B, and (b) P is not likely to accept B (e.g., because B is purely nonsensical, even for P).

Q: B must be the case as there is no proof against B. Therefore, you must accept B.

If P goes along, the burden of proof now rests on P to show the distinction between A and B to explain why A should indeed be accepted and B rejected.

If, for example, someone claims that earth rays exist because they have not been shown not to exist, it is an interesting strategy to invent equally mysterious "negative" earth rays that compensate exactly for the earth rays, claiming that it is therefore quite understandable that no earth rays have ever been detected. It is now up to the other party to show what distinguishes "normal" earth rays from "negative" earth rays.

A third example concerns the use of a mistaken *analogy*. In this case, according to the OLR, rather than pointing out that the analogy is indeed mistaken, it might prove very helpful to continue the analogy to make clear within the analogy itself why it is mistaken. To a certain extent, you thereby accept the analogy, but the end result should be that the party who proposed the mistaken analogy in the first place will be faced with having to accept a conclusion he or she is not likely to accept.

No doubt the most famous example of mistaken analogy is the abuse of Galileo Galilei's case resulting in his conviction by the Holy Inquisition. The basic strategy consists of equating Galileo with the poor astrologer or parapsychologist and equating the Inquisition with the scientific establishment. Rather than pointing out that it is absolutely silly to equate a religious institution with a research community, an alternative strategy, that I have actually been using a number of times, is efficacious. I first point out that I am truly happy to be compared to a representative of the Inquisition. This expectedly raises either some kind of suspicion or some kind of confirmation ("Scientists are persecutors!"). But then I point out that (a) contrary to what so many believe, the Inquisition did ask for confirmation of the telescopic observations of Galileo and accepted the fact that these were correct;[10] (b) they really took their time before reaching a decision, so they were extremely careful, for a number of reasons, in their considerations; and (c) the final sentence was not signed by all members of the Inquisition, so one can hardly speak of a unanimous decision.[11] Now if I read (a) as a willingness to control, repeat, and check observations and accept them being correct if they turn out to be so; (b) as a sign of being extremely careful in drawing conclusions; and (c) as an acknowledgment of the existence of different opinions, then these seem characteristics of good scientific practice. So yes, it is an excellent idea to be compared to the Inquisition.

It is obvious that these strategies are not without danger. They do not guarantee success, and one takes the risk of committing oneself to claims—such as the existence of rays of some kind or other in the first example—one has to deny at a later stage. On the other hand, if OLR fits better with the real-life reasoner, then it is a reasonable assumption, to be investigated empirically, that the chance for success should be greater than in the ideal setting.

Whatever the outcome, it is clear that fallacies have an important part to play in the discussion and should not be rejected without further ado.

There is an additional remark to make. The three examples seem to suggest that a debate or a discussion consists of a sequence of arguments, one at the time, taken from a basic "catalog" as it were. Argument types, however, are not isolated units but can interact in various ways. The argument *ad verecundiam* and the mistaken analogy can be combined, for example, to produce the following argumentative strategy:

Given:
  (a) P uses a case of mistaken analogy (e.g., the Galileo case) to prove point A.
  (b) X also uses the mistaken analogy to prove the same point A.
  (c) However, P does not accept X as an authority.
  (d) To use the analogy implies a recognition of the authority of X.

Then it follows:
  (e) Since the premises are inconsistent, one of these (or a combination thereof) has to be rejected (save for (b), which is a factual statement):
    (e1) Drop (a), that is, drop the mistaken analogy.
    (e2) Drop (b), that is, accept that X does have some authority.
    (e3) Drop (d), that is, diminish the importance of the fact that X uses the analogy.

In all outcomes, P has to weaken his or her position, producing an argumentative advantage for the other party. This one example strongly suggests that there is a complex world of possibilities to be explored here.

Deceiving and Misleading as a Form of Argument

The second method leads us into trickier territory. Suppose that you know that P believes that phenomenon A could have been caused only by B. Either you can spend a lot of time trying to show that other causes besides B are at least possible and/or that B cannot occur (so it has to be one of the other possibilities). It seems to be far more efficient to proceed as follows:

  (a) Create circumstances that produce phenomenon A by different causes C.

    (b) Make sure that P does not know that that is how A has been produced.

    (c) Present P with A and ask what he or she thinks has happened.

In normal circumstances, P will answer that A has been caused by B. Then the whole setup can be revealed. This should produce a devastating effect.

Here are two examples that are real-life cases, so in principle their efficiency can be evaluated:

1. *The construction of fake crop circles.* The best way to demonstrate the fact that humans are perfectly capable of constructing even the most elaborate crop circles that can fool the community of croppies, of course, is to actually construct one. Better still is to construct a crop circle and not let anybody know it has been done by humans. This has actually happened a couple of times now.[12]

2. *Any magician effectively playing the part of a psychic.* I have had the occasion to witness a stage magician playing the part of psychic without the audience being aware of the fact. When at the end of the "séance" the real state of affairs was revealed, the range of emotions demonstrated was truly impressive. Ranging from sheer anger to stupefaction, it showed how far more effective this tactic proved to be than a traditional argumentation aiming to show that all of these tricks could have been done by a magician.

It is clear from the literature that skeptics have thought about such examples and have tried to find general "recipes" for inventing and trying out similar cases. Here are two such attempts:

1. Anthony Pratkanis (1995) indicates the necessary ingredients for the recipe to start a pseudoscience.[13] Why not do so? Start some type of pseudoscience with the intent, if successful, to reveal what was actually happening.

2. James Randi (1990) in his book on Nostradamus produces a list of characteristics and ingredients that prophecies are supposed to have.[14] Why not invent your own prophecies?

Let me end this section with a comment. Sociologists and psychologists have invented all kinds of techniques to find out what someone believes, without that person knowing what it is that the scientist is trying to find out.[15]

There is a curious form of deception at play there as well, so one cannot claim that the approach is unknown in scientific practice. However, by deceiving, misleading, cheating, committing fraud, we seem to have ventured into dangerous areas, and the question must be asked: is all of this ethically acceptable?

Do We Need an Ethics of Argumentation?

It is clear that speaking the truth (or, at least, what one believes to be the truth) is important. It is hard to imagine how the scientific enterprise could function if its members did not at least believe and share the idea that everyone has the intention to report things as truthfully (or faithfully) as possible. On the other hand, if a lie, in whatever circumstances, can save a life, then on ethical grounds we should tell that lie. So we have a continuum, being always truthful or faithful on one end, regularly lying on the other, with a vast gray area somewhere in between.

The question whether we need an ethics of argumentation now comes down to the question where to situate this gray zone. In other words, should we postpone the use of deceit and other means of trickery until no other option is available, or are we willing to allow for such methods, starting almost right from the truthful end of the spectrum? To invoke "ethics" here might sound rather heavy-handed, but it does pertain to the situation. To use misleading techniques can have far-reaching consequences, up to and including bodily harm, such as in the case of "alternative" medicine.

One might remark that in scientific practice, scientists are fully aware of the ethical side of their work, and institutions such as the National Academy of Sciences publish on a regular basis explicit rules concerning "good" scientific conduct.[16] However, this does not necessarily cover the ethical issues related to argumentation as such. One might additionally remark that within the field of philosophy the ethics of argumentation is indeed being discussed. The most prominent figure is Jürgen Habermas (1991), who made discussion and argumentation the cornerstones of his philosophical system. However, such approaches remain on a highly theoretical level and to some extent close to the ILA, whereas the case I am defending here is almost the exact opposite: we should consider seriously the idea of conceiving a practical *manual* explaining how one is to proceed in a debate with the best chances to win. This manual would acknowledge the enormous differences between, for example, a two-person discussion in front of an educated audience and a participation in a panel on a television channel. Of course, some of the theoretical issues raised by philosophers such as Habermas should find their way into this practice as well. One particular example is the implicit power relations that have to be present for a lie to be efficient. If P tells a lie to Q, then Q is not supposed to have access to all the relevant information of P; otherwise, Q would know that a lie has been told. So Q has less power than P. Such a power relation is in first order an epistemic relation, as we are talking about access to knowledge, but the question seems legitimate as to what grounds these epistemic relations. There is no reason to assume that these grounding relations do not (in particular circumstances) extend beyond the epistemic realm and thereby introduce social, political, or economic power relations. In other words, our manual will also be sensitive to such issues, where relevant.

I conclude thus with an appeal for anyone interested to produce such a manual, as I firmly believe that we need it. In the meantime, I will end this chapter with a philosophical comment, which at the same time summarizes the overall argumentative strategy of this contribution. One might have the feeling that the advantage of the ILA at least is that these ethical issues about the possible (ab)uses of deception, lying, and misleading need not be discussed, and hence that although we know that the real world is a messy place, we should prefer the ILA to the revised picture because of its simplicity. It is true, of course, that one does indeed avoid these ethical issues, but I want to claim that the ILA suffers from a fundamental shortcoming that might even create its own ethical problems to deal with. First, observe that surely the most striking feature of the ILA is its uniformity, one of the elements that contributes to its simplicity. More specifically, this uniformity relates to the user of the ILA, in the sense that it is user independent. The ILA will identify an argument as correct or incorrect, independent of who uses the argument. Users, so to speak, become (epistemically) interchangeable. In practice, this is not the case, and the distance seems unbridgeable, as the interchangeability is an essential feature of the ILA. So, in a first-order analysis, either the ILA has to be adapted to the real situation or the real situation has to fit the ILA. The first alternative is precisely the revised picture we have argued for, thereby replacing the ILA. The second alternative, however, suggests some kind of "enforcement" to guarantee that the ILA is in some sense "obeyed." If such a scenario were to be the case, it seems clear that the ILA has to deal with ethical issues of its own, just as the revised picture has to do. It leads to the suggestive hypothesis that ethics on the one hand and logic, reasoning, argumentation, and debate on the other hand are more intertwined than we assume.

NOTES

1. I wish to emphasise that logic and/or mathematics is just one possibility. The same conclusions could be reached, I believe, by starting out with, for example, H. P. Grice's *conversational maxims* (see Grice 1975): that is, (i) *quantity*: be as informative as the current purposes of the exchange require; (ii) *quality*: do not say what you believe is false or what you lack adequate evidence for; (iii) *relation*: be relevant; (iv) *manner*: be perspicuous. Either one assumes that these maxims always apply and that any situation in which they do not needs to be considered as a deviation, or one accepts that they do not always apply and that therefore some necessary changes need to be made.

2. I have to emphasize most strongly that what I write here is my view of the matter, and that is not necessarily shared by logicians involved with dialogue logic. For some the idea of such a logic is not about reconstructing real dialogues, but about analyzing certain (usually formal) features of the dialogue process. Then again, for others the ideal case does reveal some aspects of real-life debates without any pretence at capturing all the details. To get some feeling of the complexities involved, see the excellent book by Barth and Krabbe (1982) with the revealing title *From Axiom to Dialogue*.

3. It is sufficient to have a look at, for example, the volumes of the *Handbook of Philosophical Logic*, edited by Gabbay and Guenthner (1983–89, 2001–11). In the first edition, there were actually four volumes, but in the second, entirely revised and extended edition, the projected number of volumes is between 24 and 26!

4. An example of such a principle is found in the work of Donald Davidson: "We make maximum sense of the words and thoughts of others when we interpret in a way that optimizes agreement" (2001, 197).

5. See, for example, Ortony (1993) for an overview.

6. An important distinction must be made: logicians do try to formalize metaphors to understand how they work, but that does not imply a reduction to a literal kind of language.

7. A good overview is found in Pohl (2004).

8. An important remark must be made at this point. Given my criticism of the ILA and my support of logical pluralism, I must make explicit a hidden assumption. We are taking for granted here that the rule mentioned does indeed not belong to the corpus of rules accepted in daily reasoning. In other words, there could be cases where a form of ordinary reasoning is judged correct by one logical system and rejected by another.

9. Typical situation: it does not help much in discussion with an astrologist to call in a Nobel laureate for physics in the specialized field of astronomy to try to convince the astrologist that he or she is wrong about a particular point.

10. Though, obviously, their interpretation was not.

11. The statements (a), (b), and (c) are based on the historical work that has been done by many authors, such as Stillman Drake, Pietro Redondi, Richard Westfall, and William Wallace, to name but a few. An excellent summary is Finocchiaro (1989).

12. See, e.g., Nickell and Fischer (1992).

13. The characteristics (in shorthand) are the following: (1) create a phantom, (2) set a rationalization trap, (3) manufacture source credibility and sincerity, (4) establish a granfalloon, (5) use self-generated persuasion, (6) construct vivid appeals, (7) use pre-persuasion, (8) frequently use heuristics and commonplaces, and (9) attack opponents through innuendo and character assassination.

14. The characteristics (equally in shorthand) are the following: (1) make lots of predic-

tions; (2) be very vague and ambiguous; (3) use a lot of symbolism; (4) cover the situation both ways and select the winner as the "real" intent of your statement; (5) credit God with your success and blame yourself for any incorrect interpretations; (6) no matter how often you're wrong, plow ahead; (7) predict catastrophes; and (8) when predicting after the fact but representing that the prophecy preceded the event, be wrong just enough. The full details are in Randi (1990, chap. 3, "The Secret of Success").

15. There is a deep connection here with the problem of (scientific) objectivity, but I will not go into this difficult matter here.

16. See National Academy of Sciences (2009).

REFERENCES

Barth, E. M., and E. C. W. Krabbe. 1982. *From Axiom to Dialogue: A Philosophical Study of Logics and Argumentation*. Berlin: Walter de Gruyter.

Davidson, Donald. 2001. *Inquiries into Truth and Interpretation*. 2nd ed. Oxford: Oxford University Press.

Finocchiaro, Maurice A. 1989. *The Galileo Affair: A Documentary History*. Berkeley: University of California Press.

Gabbay, D., and F. Guenthner, eds. 1983–89. *Handbook of Philosophical Logic*. 4 vols. Dordrecht: Reidel.

———, eds. 2001–11. *Handbook of Philosophical Logic*. 16 vols. Springer: New York.

Grice, H. P. 1975. "Logic and Conversation." In *Syntax and Semantics*, edited by P. Cole and J. Morgan, 3:41–58. New York: Academic Press.

Habermas, Jürgen. 1991. *Erläuterungen zur Diskursethik*. Frankfurt am Main: Suhrkamp. Translated by Ciaran P. Cronin as *Justification and Application* (Cambridge, MA: MIT Press, 1993).

National Academy of Sciences. 2009. *On Being a Scientist: A Guide to Responsible Conduct in Research*. Washington, DC: National Academies Press.

Nickell, Joe, and John F. Fischer. 1992. "The Crop-Circle Phenomenon: An Investigative Report." *Skeptical Inquirer* 16 (2): 136–49.

Ortony, Andrew, ed. 1993. *Metaphor and Thought*. 2nd ed. Cambridge: Cambridge University Press.

Pohl, Rüdiger F., ed. 2004. *Cognitive Illusions*. New York: Psychology Press.

Pratkanis, Anthony R. 1995. "How to Sell a Pseudoscience." *Skeptical Inquirer* 19 (4): 19–25.

Randi, James. 1990. *The Mask of Nostradamus*. New York: Charles Scribner's Sons.

Wason, Peter C., and P. N. Johnson-Laird, eds. 1968. *Thinking and Reasoning*. Harmondsworth: Penguin.

16

Why Alternative Medicine Can Be Scientifically Evaluated

Countering the Evasions of Pseudoscience

JESPER JERKERT

Potential medical treatments must be tested on real persons with real symptoms. The so-called clinical trial has been considered the standard procedure for assessing medical treatments. In a clinical trial, patients with specified symptoms are given either of two or more predetermined treatments (one of which is usually taken as a baseline against which the others are judged). Health endpoints in the groups are then compared using statistical methods.

As we shall see, some have called into question the validity of clinical trials for certain unconventional treatments. This situation occurs regularly with supporters of unconventional and pseudoscientific practices, who criticize the way their beliefs are investigated scientifically or the verdicts reached by science. There is a literature that discusses the possibilities of testing alternative medicine scientifically or notes (and sometimes solves) where problems arise in such research (Anthony 2006; Miller et al. 2004; Jonas 2005; Margolin, Avants and Kleber 1998). But I want to discuss an even more basic question than those normally addressed in this literature: what treatments can be scientifically investigated at all?

This chapter aims to contribute to a better understanding of what conditions medical treatments must fulfill to be eligible for scientific investigation. In particular, the text is a rejoinder to the claims put forward by adherents of alternative medicine that their treatments are inaccessible to scientific scrutiny.

Clinical Trial Criticisms

Much is written about the proper performance and interpretation of clinical trials (Jadad and Enkin 2007). This literature often takes a quite practical stand and is not necessarily informed from a philosophical point of view. Typically, distinctions are not made between methodological features that are *necessary* and those that are merely *recommended*, or between features that are good in their own right and those that are good because they correlate to something else that is good in its own right. For example, researchers are advised to sample participants who are similar with respect to background variables such as sex and age (e.g., Stommel and Wills 2004, 332–33). Is this necessary, or is it just recommended? If it is necessary, does that mean that treatments intended for heterogeneous populations cannot be assessed properly in a single clinical trial?

Or consider the advice that treatments should be standardized in the sense that all patients in the same group receive (approximately) the same treatment. What about treatments that in real-life situations (i.e., not in a clinical trial) would involve many individual adjustments? Can we test the efficacy of a treatment that heavily depends on doctor–patient interactions properly? How could it be done? Normally, the handbooks in the field do not tell, or even acknowledge any potential problem or challenge (Mason, Tovey, and Long 2002). But some adherents of alternative medicine express serious doubts as to whether clinical trials properly assess highly individualized or esoteric treatments (Carter 2003; Weatherley-Jones, Thompson, and Thomas 2004; Verhoef et al. 2005). This seeming skepticism among practitioners and supporters of alternative medicine may be quite cautiously formulated, such as this quotation suggesting that clinical trials do not normally take all important outcomes into account:

> RCTs [randomized clinical trials] usually omit the measurement of important elements of "what works" in alternative medicine, which often acts in a different way to biomedical drugs. By presenting ethnographic evidence, I wish to show how evidence, when seen from the perspectives of the users and practitioners of alternative medicine, hinges on a very different notion of therapeutic efficacy. (Barry 2006, 2647)

If this criticism is justified—which depends on what the "very different notion of therapeutic efficacy" exactly amounts to—it may be countered by including additional outcome measures in the trial design, provided that they are quantifiable. But the argument critical of RCTs may be given an even more skeptical form by claiming that alternative medicine practitioners do not need to care much about the outcomes of clinical trials, for there are other ways of knowing. Here is a quotation to this effect:

> The individual patient's perception of improvement may constitute direct evidence of benefit based on primary experience. To prefer indirect evidence, such as that obtained from clinical trials, over primary experience represents an epistemic choice, not a scientific necessity. CAM [complementary and alternative medicine] and CAM practitioners, therefore, can continue to emphasize individual outcomes without inconsistency even when the therapies they utilize have failed to demonstrate efficacy in controlled clinical trials. (Tonelli and Callahan 2001, 1216)

If the attitude shown by Tonelli and Callahan is combined with the view that "primary experiences" cannot be measured so as to act as endpoints in a clinical trial, it indeed may be true that alternative medicine is inaccessible to scientific investigation. But as it stands, the argument is hardly convincing. For example, pain should surely count as "primary experience," and pain is often measured by asking patients to rate their current level of pain on a scale, a procedure that has been used extensively in clinical trials.

The criticism against clinical trials, of course, may also take an entirely rhetorical and insubstantial form, such as this quotation from a homeopath:

> Hailed as the "gold standard" testing methodology, the RCT has in effect become a mindlessly worshipped golden calf, in front of which other forms of evidence—and homeopathy/complementary and alternative medicines (CAMs)—are supposed to kowtow and pay homage, if not disappear from consideration altogether. (Milgrom 2009, 205)

These three quotations form part of a spectrum of suspicion toward RCTs.[1] It seems appropriate to try to assess whether this suspicion is warranted or not.

Objective

A first step of assessment is to make clear what kinds of treatments are eligible for a clinical trial. The text aims to delineate and discuss two methodological desiderata in clinical trials, which impose restrictions on the treatments being tested. These conditions, roughly put, are (1) the proper distinction of the two treatment groups, and (2) the elimination of confounding variables or variations. After introducing a suitable notational apparatus, I address three interrelated topics. First, I formulate and defend a distinguishing criterion that is sufficient to separate the treatments involved in a clinical trial. Second, I formulate a principle of elimination of confounding variables, which contains a reference to the principle of distinguishing between treatment groups. Third, I discuss and counter a few misunderstandings in the light of my proposed principles.

I must emphasize that the principles I discuss in this chapter are related foremost to *treatments*. My intention is not to formulate methodological principles covering each and every aspect of the design and performance of a clinical trial. For example, I say nothing about the reliability of information available in a trial or about how to handle dropouts or other issues related to the behavior of the trial participants.

The term "treatment" is taken in a broad sense. It not only refers to the drugs administered and the doctor–patient interaction, but it may also refer to the preparation of the drugs or to characteristics and doings of the doctor, including activities for diagnosis. Also, "no treatment" counts as a treatment in what follows. My framework is inclusive in the sense that it presupposes that treatment methods ought to be tested according to the same evidential standards, irrespective of whether they originate from within or outside of biomedical research.

A Model Situation

To introduce a notational apparatus, I take the following situation as the point of departure. Normally, we have at least two treatments that we wish to compare. For the sake of argument, let us assume that we have exactly two treatments, denoted A and B, and we wish to determine which is the best for a given condition (disease) or collection of symptoms. The condition or collection of symptoms we call D.[2] It takes no more than a moment of reflection to be convinced that the treatments must then be tested on a (large) number of people, divided into two groups. In other words, it will not suffice to test the treatments on but one or two patients if the result is supposed to be generalizable to a larger population. Individuals may differ from one another in so many respects as to preclude certain inferences from single individuals to a larger population. Hence, we need two groups, called G_A and G_B, for patients receiving treatments A and B, respectively. At the completion of the trial, we compare the outcomes at group level in one or several predetermined endpoint measures using some statistical analysis.

The Distinguishing Criterion

Since G_A and G_B are compared in the trial, distinguishing them ought to be important. Obviously, G_A and G_B are different from one another in the sense that the treatments A and B are different. So one way of distinguishing G_A from G_B would be to specify what A and B exactly amount to. However, it may not be necessary to specify A and B in detail. It is sufficient that we can specify a difference (referring to one or several features, i.e., a unidimensional or multidimensional criterion) using available information with the aid of which we can place each participating patient in G_A or G_B. I therefore propose the following distinguishing criterion (DC):

> For each patient involved in the trial, one must be able to tell, with the aid of a uni- or multidimensional criterion formulated before the commencement of the trial, whether treatment A or B was given, using any available information recorded before or during the trial.

To rule out an unwanted situation in which A and B are both given to the same patient, I would like to add a principle that excludes multiple treatments (EMT) for a single patient:

> No participant in the trial receives both treatment A and treatment B.

DC and EMT taken together are simply referred to as DC+EMT. As an example of the uses of DC+EMT, consider a situation where patients in G_A are given alternative (e.g., "anthroposophic") treatment and patients in G_B are given nonalternative (i.e., "ordinary") treatment for headache. Let us also assume that we know nothing about the doctors involved in the trial, so that we cannot use any of their characteristics as part of DC. Instead, all we know about A and B is contained in annotations left by the doctors involved, so these annotations constitute the "available information" referred to in DC. We then do not want to face a situation in which the treatment given by the alternative doctor is identical to the treatment given by the ordinary physician, according to our available information. For example, a treatment in the form of a short conversation followed by the prescription of sleeping pills, and nothing else, must not occur in both groups G_A and G_B. If it does, we will be unable to distinguish A from B.

If we are unable to distinguish A from B, the result of the trial is useless. Therefore, a distinguishing criterion is necessary. However, I do not claim that the criterion I have proposed, DC+EMT, is a necessity as it stands. But I do claim that DC+EMT is a sufficient condition for the separation of the treatments.[3]

Why must the criterion in DC be formulated *before* the trial starts? The answer is that once the outcomes are known, it may be possible to formulate a (more or less far-fetched and complicated) distinguishing criterion that divides the trial participants into whatever groups and hence yields whatever result is desired. This would amount to an ad hoc handling of data.

The Relevance Issue: Eliminating Confounding Variables

DC+EMT is not a very complicated principle (though it has some interesting implications for the possibility of testing alternative medicine, as we shall see). It only states that as long as we can tell that either A or B was given to each of the trial participants, there is nothing to complain about in terms of the distinction of treatment groups. But such a distinction is clearly not enough for a trial to be interesting or relevant. For example, let A be a homeopathic drug combined with a kind reception by the doctor, whereas B is a sugar pill along with a cold reception. Perhaps we will find that the average G_A health (in some measurable form) increased significantly more than the average G_B health. This result cannot be invoked as evidence for homeopathy since a fully plausible explanation why the G_A health increased more than the G_B health is the difference in reception. Therefore, we need to supplement DC+EMT with a principle that eliminates confounding variables.

In the course of a medical treatment, many factors might affect the patient: the chemical composition of any drugs administered, the frequency with which they are taken, the general health of the patient, the patient's expectations, the environment of treatment, and so on. Of course, we do not want any of these factors to destroy the possibility of making valid and useful inferences. I therefore suggest the following principle of elimination of confounding variables (ECV):

> There must be no variable present in the trial such that (i) there is a systematic discrepancy between G_A and G_B in this variable, (ii) the health endpoint records in G_A and G_B have been substantially affected by the variable, and (iii) the variable is not part of the criterion in DC.

The need for clauses (i) and (ii) should be fairly obvious. Note that clause (ii) does not mention the existence of a *difference* between G_A and G_B in the endpoint measures; it merely refers to the endpoint records, whether they are different from one another or not. This reflects the fact that an unwanted confounding variable may not only result in an endpoint difference, but may as well make it vanish.

Why is (iii) needed? Because we want to compare groups that have received treatments differing with respect to at least one predetermined variable. (Remember, according to DC we need to specify in advance the criterion for distinguishing A from B.) We cannot then claim that such a variable is a confounder and should be eliminated.

The way I have formulated ECV may not make it obvious how to apply it practically. One faces questions like: How does one know whether a difference is systematic? How does one know whether a variable has *substantially* affected the endpoint records in G_A and G_B? These are important but tricky questions. Space does not allow them to be analyzed here. Statistical methods as well as mechanistic reasoning could be helpful in sorting them out (Bland 2000; La Caze 2011).

DC was supplemented with a principle ruling out A and B being given to the same patient. I wish to make a similar move for ECV by adding an auxiliary principle, viz. a principle of consistency under scrutiny (CS):

> Any treatment effect must be such that it does not regularly disappear when included in a trial.

This principle prohibits the performance of a clinical trial to be taken as a confounding factor in itself. ECV and CS taken together are referred to as ECV+CS.

Treatments Inaccessible to Scrutiny?

Are there treatments that cannot be evaluated fairly in a clinical trial manner? There are. The problem most often discussed is that of a relevant comparison treatment against which the treatment of interest is to be judged. If the comparison treatment is not relevant, then at least one confounding variable is present, violating ECV. Psychotherapy treatments are much discussed in this connection. Acupuncture research is another case in point, for it is difficult to design a "sham" acupuncture that will control for all nonspecific effects. The most recent research using nonpenetrating sham acupuncture indicates that the evidential support for acupuncture is vanishing for many conditions (Ernst 2009).

There are also other types of problems for which cases of impossibility of evaluation can be made. For example, consider a treatment whose effect is claimed to be substantial under normal circumstances but negligible when part of a clinical trial. If this is true, the treatment effect violates CS, and any conducted trials would misrepresent the true potential of the treatment.

Another example could be a treatment claimed to have so many possible (positive) effects that no aggregated statistics can be collected. Although real-life examples may not be so common, some supporters of alternative medicine share one line of thought that could easily be extended in this direction. Take this quotation:

> Statistically based inferences about the likelihood of outcomes for typical cases are of little use in the treatment of individual cases. The homeopath follows patient symptoms over a length of time and the analysis of patterns of change requires holistic logic and practice. That is, in individual cases it is not possible to isolate symptoms and causes from the whole person. (Jagtenberg et al. 2006, 327)

Hundreds of health-related endpoints could be monitored and measured in clinical trials: mortality, number of sick days, amount of some specified substance in the blood, muscle strength, self-assessed pain, or what have you. But normally, only one or a few of them are used in any particular clinical trial. What if, according to the supporters of the treatment, the "holistic logic" requires us to monitor, say, 500 endpoint measures simultaneously to fairly represent the effects of the treatment, with no promises being made that any single endpoint will be affected? Such a scenario would make it practically impossible to assess the overall efficacy of the treatment. But as soon as a *particular* effect is singled out and is claimed to result (in some statistically defined meaning) from being treated, the treatment will again be possible to evaluate scientifically.

In the last two examples, a critic would naturally question the credibility of an effect that mysteriously disappears under scrutiny and the soundness of "holistic logic." Hence, if the supporters' power of interpretation is challenged, it becomes harder to find examples of treatments immune to scientific testing (apart from cases involving difficulties of finding relevant comparison treatments, as discussed above in relation to acupuncture).

In sum, there are treatments inaccessible to scrutiny through clinical trials, but the set of treatments unfit for testing is smaller than some proponents of alternative medicine would have us believe. The vast majority of treatments used in alternative medicine are offered with promises of health improvements that are measurable in ways compatible with clinical trials.

Countering Three Misunderstandings

Having presented the principles regulating which treatments are eligible for clinical trials, I now counter three misunderstandings concerning proper group comparisons in the light of DC+EMT and ECV+CS. I believe that these misunderstandings are quite common among practitioners and pseudo-scientists discussing the methodology of group comparisons. Of course, this list of misunderstandings is not exhaustive.

The first misunderstanding is that treatments A and B need to be completely distinct; in other words, no treatment feature can appear in both A and B. For example, if it is part of both A and B to give each patient a glass of water to drink, then the treatments are not completely distinct. Forcing A and B to be completely distinct has no support from DC+EMT (nor from ECV+CS). Many treatment features could appear in both treatments without ruining the possibility of performing a clinical trial. Normally, one would even recommend a fairly large overlap between A and B since one or only a few differences between A and B facilitate causal reasoning. If A is a standard treatment and B is the same standard treatment plus an added feature, a marked difference in the group outcomes would invite us to believe that the added feature may have caused the difference.

The second misunderstanding is that the treatments under investigation must be somehow known and explicable within ordinary science. This view allows no room for evaluating "spiritual" or otherwise unorthodox treatments, for they cannot be compared meaningfully to orthodox treatments. Statements to this effect are often advanced by those opponents to ordinary medicine who have a religious outlook. Here is an example:

> Only phenomena assumed to have existence according to orthodox Western science are regarded as suitable for biomedical research. Therefore, biomedical research methodology is in effect "closed" to the investigation of putative mechanisms of action resting on phenomena that potentially affect health or illness but are not reducible to categories of existence described in contemporary Western science. (Lake 2006, 68)

No support for this view may be gained from DC+EMT or from ECV+CS, nor from what has been said about the trial endpoint measurements (unless the treatment effect is annihilated by the very act of testing it, in which case it is unfit for a clinical trial according to CS). It is thus fully possible to put treatments with clearly "spiritual" elements to test. For instance, suppose that all patients in a trial suffer from legionellosis and that those enrolled in G_A meet an anthroposophic physician who tries to view the patients' etheric bodies using clairvoyance, but gives no other treatment except prescribing antibiotics. For the patients in G_B an ordinary physician prescribes antibiotics. The anthroposophic treatment (remember that we take "treatment" in a broad sense, so that diagnosis may be included) can be evaluated in a clinical trial: we will simply have to decide in advance whether information on any etheric body investigations will be collected and will be available for the application of DC or not. There is nothing in DC+EMT or ECV+CS that would eliminate the possibility of such a trial. On the contrary, it seems obvious that information on etheric body investigations should definitely be available to DC since such an investigation would surely be made only in the anthroposophic group.

Another example where it is obvious that alleged spiritual features should be taken as distinguishing could be a trial in which homeopathic drugs are tested against placebo. As far as orthodox science is concerned, the only difference is the method of preparation of the drug. Chemically, most homeopathic drugs are identical to placebo since repeated dilutions have effectively eliminated any original active ingredient. However, the preparation of homeopathic drugs is different from the preparation of nonhomeopathic placebos.

The homeopathic dilution process involves ritual shaking in each step. Such shaking will not have occurred for the preparation of the nominal placebo drug. The preparation differences may be used in a criterion for distinguishing between a homeopathy group and a placebo group (provided that information on the shakings can be made available). This example shows that it is not necessary that the distinction between treatments A and B can be made by looking at the actual physician–patient interaction or the chemical composition of any drugs involved. According to DC, it is sufficient that the distinction can be made through consulting available information, in whatever form. A sufficient piece of information might be "the drug given to patient p was prepared according to homeopathic regulations." As long as the necessary information is available, it does not matter that the information ought to be irrelevant from the point of view of established science. Hence, even medical claims with absurd features could be amenable to scientific investigation.

The third misunderstanding is that all patients within the same treatment group must be treated identically. Consider this quotation from a 2009 parliamentary bill, signed by two members of parliament from the Centre Party of Sweden:

> Since clinical trials with high scientific acceptance are not possible to apply to homeopathic drugs, as opposed to conventional drugs (due to the way in which homeopathic drugs are administered and selected for the individual patient), it is almost unthinkable . . . to grant access to them at the Swedish market. (Wallin and Kornevik Jakobsson 2009; my translation)

Members of parliament Gunnel Wallin and Maria Kornevik Jakobsson seem to believe that all patients must be treated in a predetermined, standardized manner for science to be able to pass judgment. As soon as a physician makes individual adjustments, the possibility of a scientific evaluation is destroyed, according to this view.

But this position is unfounded. The individual selection and administration of drugs do not by themselves block the performance of clinical trials, at least not according to DC + EMT or ECV + CS. Actual clinical trials of individualized treatments have been reported in the scientific literature (e.g., White et al. 2003), and there is even a meta-analysis published (Linde and Melchart 1998). It may be noted that clinical trials involving individualized treatments for animals have been reported as well (Hektoen et al. 2004).

Deviations from Statistical Ideality

An ideal trial is a trial with the greatest probability of finding a real outcome difference between G_A and G_B. In other words, the ideal trial is a trial with maximized statistical power. Many departures from ideality may be imagined, all of which decrease the statistical power of the trial. For this reason, not too large or too many deviations from ideality should be accepted since they undermine reliable and useful inferences. It may be helpful to list the four main departures from ideality in a situation where treatments A and B and the symptoms D are considered as given:

1. Increased variation within treatments A and B.
2. Increased overlap between A and B, that is, more common treatment features. This may be viewed as a special case of number 1 since sufficiently large variations within A and B will eventually lead to treatment features being common to both A and B.
3. Increased variation in the patients' background characteristics.
4. Increased variation within D, the symptoms that are considered as fitting for the treatment in question.

All of these variations in principle are allowed in a trial, but the greater and more numerous they are, the more difficult to demonstrate a true treatment effect. In this chapter, I am concerned mainly with trials in which numbers 1 and 2 are present.

What about Blinding, Randomization, and Other Clinical Trial Features?

Discussions concerning the performance of clinical trials frequently throw around certain concepts. In particular, four features of clinical trials are mentioned more often than others: *control*, *(single) blinding*, *double blinding*, and *randomization*. (Indeed, they are often found in the titles of articles reporting the outcomes of clinical trials.) They are not explicitly mentioned in my proposed conditions DC+EMT and ECV+CS. However, the ECV condition is all about enhancing control, so that anyone adhering to ECV is essentially controlling the trial in the ordinary sense of the word as applied to clinical trials. As for blinding and randomization, I believe they should be viewed as more or less practical ways of satisfying DC+EMT, ECV+CS, and possibly other implicit methodological conditions. In other words, I would maintain that DC + EMT and ECV + CS are the more fundamental principles here, and that blinding and randomization are no sine qua non but are recommendable practices to the extent that they contribute to satisfying DC + EMT and ECV + CS.

For example, it is obvious that blinding in the sense of making sure that participating patients do not know their group attribution in many cases comes down to eliminating a possible confounder and hence contributing to satisfying ECV. But even if blinding is a good methodological feature fitting the ECV desideratum well, is it not true that aspirations for *double* blinding are often challenging in alternative medicine, not least for individualized treatments? Ursula Flatters, senior physician at the anthroposophical Vidar Clinic (near Södertälje, Sweden), believes that "the anthroposophical treatment cannot be investigated in double blind trials since the therapy cannot be properly given without the doctor having met the patient" (Flatters 2002, 40; my translation).

In response, I offer two points. First, the doctor normally does meet the patient in ordinary medical practice as well (not only in anthroposophical medicine), and since clinical trials are performed for many nonanthroposophic treatments, this fact does not seem to constitute an insurmountable difficulty. Second, even if double blinding would not be possible in a particular trial, that does not mean that a clinical trial satisfying DC + EMT and ECV + CS cannot be performed. There may well be trials without double blinding that still adhere to DC + EMT and ECV + CS.

An obvious example would be a trial in which there is no doctor–patient interaction at all. Let *A* amount to taking a specified drug, which can be administered wholly by the trial participants themselves (though the exact identity of the drug may be concealed to the participants, hence maintaining single blindness). *B* amounts to taking another specified drug in the same way. Although there may be physicians and other people involved in the realization of the trial, there is little need to keep them ignorant of who is belonging to which group. Hence double blinding seems not to be necessary. Any counterargument to the effect that double blinding would still be a good thing would have to explain why this is so, if not precisely for maintaining DC + EMT or ECV + CS. It is true, for example, that double blinding might make fraud attempts more difficult to realize, but if this is the reason for adopting double blinding, then it is not a principle of double blinding that is advocated, but a principle of no frauds. I doubt, however, that a principle of no frauds would be helpful in practice. A principle of no frauds could probably be invoked to discard any trial design.

Another example where double blinding is not necessary could be the following. Patients are given individualized treatments for common cold, where G_A patients get individual treatments from a doctor prescribing conventional drugs and G_B patients get individual treatments from the same doctor prescribing nonconventional drugs. We may have selected the doctor in such a careful manner—and we may even monitor the doctor's behavior for added certainty—so that we are convinced that there is no difference in the doctor–patient interaction between G_A and G_B, although the doctor would know whether the patient at hand belongs to G_A or G_B. If so, single blinding may be judged sufficient and the lack of double blinding (i.e., blinding of the doctor) is no violation of DC+EMT or ECV+CS.

I would also like to briefly mention the concept of *bias*, which is much discussed in relation to clinical trials and their performance. Bias is a wide concept. According to Jadad and Enkin (2007, 29ff), there are numerous types, such as selection bias, ascertainment bias (bias due to lack of appropriate blinding), intervention choice bias, and many more. If this broad notion of bias is accepted, then there are biases unrelated to DC+EMT and ECV+CS. For example, among the biases discussed by Jadad and Enkin we find publication bias and even fraud bias, both unrelated to DC+EMT and ECV+CS. But if only biases related to the realization of the trial and to the nature of the treatments (in a wide sense) are considered, avoiding biases seems to me to be identical to satisfying DC+EMT and ECV+CS.

Conclusion

I have identified and discussed two criteria imposing restrictions on the treatments (in a broad sense) eligible for clinical trials. I have called these criteria DC and ECV, or DC+EMT and ECV+CS when supplemented by extra conditions. Neither DC+EMT nor ECV+CS eliminates the possibility of investigating alternative medical treatments, whether individualized or not, through clinical trials. The argument from individual adjustments is definitely untenable. There is no reason whatsoever to suspect that alternative medicine in general cannot be evaluated through well-designed clinical trials. Naturally, there may be difficulties in satisfying the criteria for each and every alternative medical treatment, especially if the treatment supporters are allowed to influence the trial design. But the overall picture is one of inclusiveness. Alternative medicine can be evaluated with science, and that is increasingly what has happened during recent decades. Insofar as adherents of alternative medicine are unhappy with the scientific verdict, their excuses need to be far more elaborate than sweeping statements on the purported impossibility of scrutinizing unconventional treatments with objective methods.

NOTES

1. I am not suggesting that the three quotations represent all existing positions critical of clinical trials. Since the disbelief may come in many versions and flavors, focusing on different treatments and highlighting various features of those treatments, assessing all criticisms would be a formidable task. It is beyond the scope of the present chapter.

2. It might be noted that some adherents of alternative medicine deny the very existence of diseases known to scientific medicine. Nonetheless, we need to define at least a collection of symptoms for which a certain treatment is recommended in order to perform a clinical trial at all.

3. In fact, DC can be made even weaker by allowing a criterion that does not with certainty decide whether a given patient belongs to G_A or G_B, but only increases the probability of assigning it to either group above chance level.

REFERENCES

Anthony, Honor M. 2006. "Some Methodological Problems in the Assessment of Complementary Therapy." *Statistics in Medicine* 6 (7): 761–71.

Barry, Christine A. 2006. "The Role of Evidence in Alternative Medicine: Contrasting Biomedical and Anthropological Approaches." *Social Science and Medicine* 62:2646–57.

Bland, Martin. 2000. *An Introduction to Medical Statistics.* 3rd ed. Oxford: Oxford University Press.

Carter, Bernie. 2003. "Methodological Issues and Complementary Therapies: Researching Intangibles?" *Complementary Therapies in Nursing and Midwifery* 9:133–39.

Ernst, Edzard. 2009. "Acupuncture: What Does the Most Reliable Evidence Tell Us?" *Journal of Pain and Symptom Management* 37 (4): 709–14.

Flatters, Ursula. 2002. 'Svar från FALK' [Reply from FALK (Swedish Association of Anthroposophic Medicine)]. *Folkvett* (2): 35–41.

Hektoen, Lisbeth, Stig A. Ødegaard, Torleiv Løken, and Stig Larsen. 2004. "Comparison of Homeopathy, Placebo and Antibiotic Treatment of Clinical Mastitis in Dairy Cows—Methodological Issues and Results from a Randomized-Clinical Trial." *Journal of Veterinary Medicine A* 51:439–46.

Jadad, Alejandro R., and Murray W. Enkin. 2007. *Randomized Controlled Trials: Questions, Answers and Musings.* 2nd ed. Malden, MA: Blackwell.

Jagtenberg, Tom, Sue Evans, Airdre Grant, Ian Howden, Monique Lewis, and Judy Singer. 2006. "Evidence-Based Medicine and Naturopathy." *Journal of Alternative and Complementary Medicine* 12 (3): 323–28.

Jonas, Wayne B. 2005. "Building an Evidence House: Challenges and Solutions to Research in Complementary and Alternative Medicine." *Forschende Komplementärmedizin* 12:159–67.

La Caze, Adam. 2011. "The Role of Basic Science in Evidence-Based Medicine." *Biology and Philosophy* 26 (1): 81–98.

Lake, James H. 2006. *Textbook of Integrative Mental Health Care.* New York: Thieme.

Linde, Klaus, and Dieter Melchart. 1998. "Randomized Controlled Trials of Individualized Homeopathy: A State-of-the-Art Review." *Journal of Alternative and Complementary Medicine* 4 (4): 371–88.

Margolin, Arthur, S. Kelly Avants, and Herbert D. Kleber. 1998. "Investigating Alternative Medicine Therapies in Randomized Controlled Trials." *Journal of the American Medical Association* 280 (18): 1626–28.

Mason, Sue, Philip Tovey, and Andrew F. Long. 2002. "Evaluating Complementary Medicine: Methodological Challenges of Randomised Controlled Trials." *British Medical Journal* 325:832–34.

Milgrom, Lionel R. 2009. "Gold Standards, Golden Calves, and Random Reproducibility: Why Homeopaths at Last Have Something to Smile About." *Journal of Alternative and Complementary Medicine* 15 (3): 205–7.

Miller, Franklin G., Ezekiel J. Emanuel, Donald L. Rosenstein, and Stephen E. Straus. 2004. "Ethical Issues Concerning Research in Complementary and Alternative Medicine." *Journal of the American Medical Association* 291 (5): 599–604.

Stommel, Manfred, and Celia E. Wills. 2004. *Clinical Research: Concepts and Principles for Advanced Practice Nurses.* Philadelphia: Lippincott, Williams and Wilkins.

Tonelli, Mark R., and Timothy C. Callahan. 2001. "Why Alternative Medicine Cannot Be Evidence-Based." *Academic Medicine* 76:1213–20.

Verhoef, Marja J., George Lewith, Cheryl Ritenbaugh, Heather Boon, Susan Fleishman, and Anne Leis. 2005. "Complementary and Alternative Medicine Whole Systems Research: Beyond Identification of Inadequacies of the RCT." *Complementary Therapies in Medicine* 13 (3): 206–12.

Wallin, Gunnel, and Maria Kornevik Jakobsson. 2009. 'Konkurrensneutral marknad vad gäller läkemedel' [Nonbiased Market for Medical Drugs]. Parliamentary bill 2009/10S0328, Swedish Parliament. http://www.riksdagen.se/sv/Dokument-Lagar/Forslag/Motioner/Konkurrensneutral-marknad-vad-_GX02So328/.

Weatherley-Jones, Elaine, Elizabeth A. Thompson, and Kate J. Thomas. 2004. "The Placebo-Controlled Trial as a Test of Complementary and Alternative Medicine: Observations from Research Experience of Individualised Homeopathic Treatment." *Homeopathy* 93:186–89.

White, Adrian, P. Slade, C. Hunt, A. Hart, and Edzard Ernst. 2003. "Individualised Homeopathy as an Adjunct in the Treatment of Childhood Asthma: A Randomised Placebo Controlled Trial." *Thorax* 58:317–21.

17

Pseudoscience

The Case of Freud's Sexual Etiology of the Neuroses

FRANK CIOFFI

> Psychoanalysis is the paradigmatic pseudoscience of our epoch ... with its facile
> explanation of adult behavior by reference to unobservable and arbitrarily pos-
> ited childhood fantasy.
>
> —Frederick Crews (1995, 9)

The question of whether a theory is pseudoscientific arises whenever its proponents believe that it is as credible as fully accepted scientific theories, but
its critics believe that the claim is unfounded. Is untestability the main criterion by which we are to judge whether such a claim is blatantly unfounded?
Although an enterprise as distinctive as psychoanalysis warrants the introduction of considerations that do not arise in connection with other candidates
for the status of pseudoscience, untestability is a concept common in condemnations of both psychoanalysis and other candidates for pseudoscience
(Hines 1988, 107–40; Cioffi 1995). Some commentators deny that psychoanalysis is untestable and therefore conclude that it is not pseudoscientific.
Adolf Grünbaum (1984) and Edward Erwin (2002, 428–32) are among those
who take this view.

Grünbaum argues that to infer pseudoscientific status from the "malfeasances" of the advocates of a theory is to overlook that it is the status of the

"theory-in-itself" that is at issue. He therefore attempts to demonstrate that the "theory-in-itself" is "testable" by producing an example which he thinks is indisputably testable (Grünbaum 1984, 38–40). Grünbaum's dealings with the untestability indictment invite an adaptation of the classical polemical jibe, "Freud has been charged with having murdered a man and his dog. Well, here is the dog. Does he look dead to you?" Grünbaum's living dog is the peripheral claim that paranoia is repressed homosexuality. Notturno and McHugh deny its centrality as well as its testability (1986, 250–51). In any case, Grünbaum avoids the most common candidate for untestability, Sigmund Freud's sexual etiology of the neuroses (Cioffi 1998, 240–64).

Erwin's invocation of the testability of psychoanalytic theory as a defense against the charge of pseudoscience encounters an objection distinct from the unrepresentativeness of his examples. This is the inconclusiveness of the criterion of testability itself. Erwin's criterion would result in a view of pseudoscience which would permit an apologist to argue that although psychoanalytic discourse might abound in spurious claims of validation, it is nevertheless not a pseudoscience since it is testable.

But if a critic believes that he can show that more penetrating questions have been raised by seven-year-olds in a catechism class than were managed by several generations of Freudians with respect to the sacred texts of the sexual etiology, what more can be asked of him? He may be guilty of hyperbole, but the considerations he raises of uncritical reading and mechanical reiteration are the relevant ones. He need not focus solely on the issue of untestability. The undeserved status that a theory enjoys and the unwarranted claims made on its behalf—rather than its relation to testability—are what incite critics to speak of it as pseudoscientific. It is these claims which provoke the epistemically derogatory appraisals that must be rebutted.

Grünbaum also denies that Freud's behavior in dealing with falsification reports is any different from that of a physical scientist (Grünbaum 1984, 113). Freud was moreover hospitable to adverse evidence (though he need not have been, any more than a physicist). This would seem to raise characterological rather than strictly logical issues. Those unpersuaded by Grünbaum as to Freud's hospitality to adverse evidence see Grünbaum as succumbing to the hagiographic conception of Freud as the indefatigable pursuer of truth and thus as giving Freud the benefit of the doubt whenever the question of the adequacy of his response to criticism comes up. Frederick Crews sees things differently and speaks of Freud's "pseudo-hospitality to objections" (Dufresne 2007, 80).

What is the bearing of hagiography on the epistemic status of Freud's theories? Simply this: it is on Freud's authority that the credibility of his theories is mainly based.

The strongest grounds for considering a theory pseudoscientific are that its advocates regularly and gratuitously imply that the theory has survived all attempts at falsification. I hope to clarify the notion of pseudoscience by examining what is repeatedly held to be Freudian theory's most defining characteristic, its "shibboleth," as Freud put it, namely, the privileged etiological role of sexuality.

Dr. Freud and Dr. Jazz

Hello Central—Give me Dr. Jazz.
He's got just what I need; I know he has.

—Jelly Roll Morton

At one stage, Freud's conception of sexual gratification was indistinguishable from that provided by Dr. Jazz, Jelly Roll Morton's pimp. The problem this sets us is by what stages and with what justification Freud's conception became less like that of Dr. Jazz, and why this progressive attenuation so often went unrecognized so that, though Freud's extension of sexuality beyond the gross—to encompass devotion to ideals—was hailed in some quarters as a natural extension of the everyday sense, in other quarters it continued to be celebrated as an attack on prudery and restraint.

Although the importunacy of sexuality is emphasized by both Freud and Jelly Roll, there is an important distinction between them. Though both Freud's patients and Jelly Roll needed the sexual relief provided by "Dr. Jazz," Jelly Roll knew this, but Freud's patients, the doctor tells us, need not have. "On being told that their complaint results from 'insufficient satisfaction,' patients regularly reply that that is impossible for precisely now all sexual need has become extinguished in them" (1895, *SE* 3:107). How in such cases did Freud establish the fact of sexual need since the patients themselves assured him that they were unaware of any such need? Freud could say they had repressed their awareness. But he might also say, as he in fact came to, that he had not maintained that it was need of sex in the Jelly-Roll sense that was pathogenic, but the need for love.

Freud's Invocation of "Love" as a Candidate for Falsification Evasion

Was Freud's post hoc elucidation of sexuality as "love" an example of falsification evasion? When objections were raised against the generality of Freud's sexual etiology of the neuroses or counterexamples cited, Freudians explained that the objections were based on a misconstrual of what Freud had meant by the terms "sexuality," "erotic," and "libido," and so his critics mistakenly excluded the influence of sex in the wider sense of "love." But Freud's original paradigm of sexual consummation was orgasm, the failure of which to achieve after sexual excitation was pathogenic. Is Freud implying that we are to seek for the analogy to orgasm phenomena as apparently remote from sexual desire as "love for parents and children, friendship, love for humanity in general, devotion to concrete objects and abstract ideas" (1921a, *SE* 18:90)? It would seem so.

A common objection to Freud's response to apparent falsification of his sexuality etiology is that he was guilty of a wanton extension of the term "sex." But there is a stronger objection: his equivocations. For example, in January 1909, Freud writes to Carl Jung that once the Americans "discover the sexual core of our psychological theories they will drop us" because of "their prudery" (Freud 1974, 196). This provokes some critics to accuse him of bad faith when he later invokes his enlarged conception of sexuality—what pastors call "love"—to deal with nonsexual counterexamples. How could what pastors call "love" offend the prudery of the Americans? (Freud wrote to an adherent whom he thought was backsliding on the issue of sexuality, "you know that our word 'erotic' includes what in your profession [Protestant pastor] is called 'love'" [Jones 1955, 2:489].)

In 1914 Charles Burr also went beyond the charge of counterintuitive conceptual innovation to the charge of equivocation: "When anyone now accuses the disciples of the newer psychology of laying greater stress on sexual matters as a cause of mental trouble than they deserve, the word 'libido' is claimed to be used symbolically. But on reading the interpretation of the dreams reported in books and papers one finds 'libido' is used in its common, ordinary, everyday meaning." And he goes on to accuse the Freudians of disingenuousness (1914, 241). T. A. Ross made the same objection. Though the Freudians now tell us that "by sex they mean . . . all that spiritual affection which may or may not be accompanied by physical passion . . . this is simply untrue. By sex they mean just what everyone else means" (1926, 169). Burr and Ross are denying the veracity of the Freudian post hoc elucidations of the term "sexual," and this goes beyond untestability and is a more characterological ("disingenuous") than conceptual objection. The objection is not that sex has a definite meaning and Freud wantonly suspends it. It is that at certain points Freud himself gives sexuality a force, which he then denies whenever it is politic to do so. When he is making claims as to the novelty of his discoveries or insisting on the boldness and distinctiveness of his views as compared to Jung's and Alfred Adler's, he gives sexuality a meaning continuous with its normal force. But when coping with skeptical objections or anticipated prudery, he invokes Plato and St. Paul on charity (1921a, *SE* 18:91).

For evidence of the opportunistic character of Freud's enlarged conception of sexuality, we need only look at the 1912 exchange between Morton Prince and several Freudians to see how remote from what Freudians then considered eligible for pathogenic status was Freud's wider conception. Ernest Jones, Smith Ely Jellife, and W. A. White, in their exchange with Prince, see Prince's invocation of his patient's guilt over her relation with her dead mother as a counterexample to Freud's sexual etiology and so reject it, insisting on carnal sexuality as the only eligible pathogen (Prince 1913–14, 334–35).

In his 1914 "Narcissism" paper, Freud treats the patient's claim that she fell ill because no one could love her as a counterexample to his libido theory and so rejects it, insisting that the real source of her difficulties was her "aversion to sexuality" (1914, *SE* 14:99).

The following charge, as Freud himself phrases it in his *Introductory Lectures* of 1916, is much more pertinent than the accusation of untestability: "the term sexual has suffered an unwarranted expansion of meaning at the hands of psychoanalysis, in order that its assertions regarding the sexual origins of the neuroses and the sexual significance of symptoms may be maintained" (1916–17, *SE* 16:304). The charge which Freudians had to meet was not merely that of attenuating the content of the term "libido" until it was empty, but of a disingenuous alternation in the scope of the term, now employing it in its narrower, and then in its unconscionably wider sense, as it suited them.

The Sexualization of the Self-Preservative Impulse as a Candidate for Falsification Evasion

As a result of his war experience, one of Freud's earliest English partisans, W. H. Rivers, wrote:

We now have abundant evidence that those forms of paralysis and contracture, phobia and obsession, which are regarded by Freud and his disciples as preeminently the result of suppressed sexual tendencies, occur freely in persons whose sexual life seems to be wholly normal and commonplace, who seem to have been unusually free from those sexual repressions which are so frequent in modern civilization. (Rivers 1924, app. 1)

Several other war psychiatrists also felt that the war neuroses refuted Freud's sexual etiology. Freud tried to meet this objection by suggesting that the apparent counterexamples in which conflicts over self-preservation rather than sexuality were the source of the war neuroses were really illustrations of the correctness of his theory, since self-preservation was at bottom a sexual impulse. In his *Autobiographical Sketch*, Freud wrote:

The war neuroses, they said, had proved that sexual factors were unnecessary to the aetiology of neurotic disorders. But their triumph was frivolous and premature. . . . [P]sycho-analysis had long before arrived at the concept of narcissism and of narcissistic neuroses, in which the subject's libido is attached to his own ego instead of to an object. (1925, *SE* 20:54–55)

The invocation of "narcissistic libido" (i.e., self-love) or, alternatively, the claim that self-preservation was itself libidinal, became the standard tactic for coping with the case of the war neuroses.

Forty years later Robert Waelder was announcing the exceptionless character of the sexual etiology:

It is not any kind of inner conflict, or any kind of impulse at variance with the rest of the personality, that can start the neurotic process. Careful examination of the pathogenic conflict shows that it is a conflict over a sexual impulse—the term "sexual" being used in a broader sense . . . equivalent with sensual excitement or gratification. (Waelder 1960, 39)

But had not the war neuroses shown it to be false that to produce a neurosis an inner conflict must involve "sexual excitement or gratification"? Waelder dealt with this objection by a quotation from Freud's preface to the book on the war neuroses:

Freud pointed out that since the formation of the theory of narcissistic libido, a sexual energy concentrated upon the ego, it could no longer be taken for granted that ego conflicts did not involve the libido; i.e., the wish to escape from the trenches could have been the motive of the traumatic neuroses, and the neurosis formation may yet have to do not just with self-preservative tendencies but with the narcissist libido. (Waelder 1960, 165)

Shall we say that, with the declaration that the self-preservative instincts were themselves libidinal, Freud met the objection that the war neuroses constitute refutations of the libido theory—or shall we invoke a category of William James ([1890] 1950, 1:163) and say rather that in their remarks on narcissistic libido the Freudians created "a bog of logical liquefaction into the midst of which all definite conclusions of any sort can be trusted ere long to sink and disappear"?

In his preface to the book on the war neuroses, Freud wrote that "narcissistic sexual hunger (libido) . . . [is] a mass of sexual energy that attaches itself to the ego and satisfies itself with this as otherwise it does only with an object" (1921b, 3). Satisfaction is usually found in objects by rubbing up against them, squeezing them, sucking them, licking them, penetrating them, and otherwise maximizing oral, genital, and tactile contact with them; it is difficult to do many of these things to an ego. "But so what?!" a Freudian apologist might reply. "Does not science develop its concepts in unheard of counterintuitive ways?" But there is an objection that this reply does not meet. This is the objection that the Freudians did not merely enlarge the concept of sexuality to encompass self-preservation; they equivocated about it.

In 1920, after Freud's announcement that self-preservation was libidinal, Ernest Jones—in the official journal of the movement—reproached Charles Burr, the author of a book he was reviewing, for his ignorance in taxing Freud for failing to distinguish the impulse of self-preservation from that of the sexual impulse: "'He evidently does not know that this division has always been made by Freud who never tires of insisting on it. As Freud's whole theory of the psychoneuroses is based on the conception of a conflict between the sexual and the non-sexual (ego) impulses this is rather a fundamental misrepresentation" (Jones 1920, 324). And yet both before and after this pronouncement, self-preservation was represented as sexual. It is episodes like these—and there are several—that take us beyond unfalsifiability and justify Burr's suggestion of bad faith.

Freud does not explicitly state what his critics' grounds for rejecting the libidinal character of the war neurotics' symptoms had been, but his invocation of the sexuality of the self-preservative instinct suggests that their objections were based on their failure to find the genital, oral, and anal ideation which Freud invariably found in the transference neuroses of peacetime and whose repression he held to be required for the formation of neurotic symptoms. It is therefore surprising that Freud never gives any illustration of what repressed self-preservational but nevertheless sexual ideas and impulses would be like were they to be uncovered. What repressed self-preservational but nevertheless sexual ideation might explain the tremors, paralyses, and contractures of the war neuroses, as Freud's patient Dora's fellatio fantasies explained the tickle in her throat? (Freud 1905a). Or as the Wolfman's repressed wish to be sodomized by his father explained his chronic constipation (Freud 1918)?

The Testability of Freud's Infantile Sexual Etiology

In *Three Essays on the Theory of Sexuality*, Freud wrote that "the years of childhood of those who are later neurotic need not differ from those who are later normal except in intensity and distinctness" (1905b, *SE* 7:176n2). A few years later, "the neuroses have no psychical content peculiar to them.... [N]eurotics fall ill of the same complexes against which we healthy people struggle as well. Whether that struggle ends in health or neurosis depends on quantitative considerations, on the relative strengths of the conflicting forces" (1910, *SE* 11:50). The introduction of "quantitative considerations" may not entirely preclude the testability of the infantile sexual etiology, but what it does preclude is the claim to have tested it since no means of measuring the quantitative factor are as yet known (Freud 1912, *SE* 12:236).

Freud's conception of what would constitute a falsification of his infantile etiology is so indeterminate that he himself seems unsure whether he has encountered any. For example, in 1928 Freud asserted of the castration complex that "psychoanalytic experience has put these matters in particular beyond doubt and has taught us to recognize in them the key to every neurosis" (1928, *SE* 21:184). And yet he had earlier testified that he knew of cases in which the castration complex "plays no pathogenic part or does not appear at all" (1914, *SE* 14:92–93). Freud must later have decided that what appeared at the time to be a falsification was not, on reflection, really one. So much for testability.

Erwin discusses attempts to find evidence for the influence of oral infantile history on adult character but shows no awareness of the way in which the theory that he finds unproblematically testable is qualified by Freud into unconstruability. In his final pronouncement on the subject of the influence of infantile oral experiences, Freud wrote that "the phylogenetic foundation has so much the upper hand in all this over personal accidental experience that it makes no difference whether a child has really sucked at the breast or has been brought up on the bottle and never enjoyed the tenderness of a mother's care. In both cases the child's development takes the same path" (1940, *SE* 23:188–89).

So much for the prospect of testing via epidemiology. What of particular reconstructions? Might they not fall foul of experience? Let us suppose that the analyst infers, from the patient's dreams or other analytic material, that his early history was one of oral deprivation but that historical investigation fails to confirm this. This is how Freud deals with such a contingency: "For however long a child is fed at its mother's breast he will always be left with a conviction after he is weaned that his feeding was too short and too little" (1940, *SE* 23:188–89).

This is not an isolated remark. It is Freud's habit to anticipate incongruities between the environmental circumstances of the patient's infantile past and the mental history to which the analyst had committed himself by invoking an autonomous source for such a history. For example, "In studying reactions to early traumata we often find to our surprise that they do not keep strictly to what the individual himself has experienced but deviate from this in a way that would accord much better with their being reactions to genetic events and in general can only be explained through the influence of such." (1939, *SE* 23:99).

This device of Freud's—the universalization of his pathogenic factors— has been generalized by his followers. A contribution to the *Psychoanalytic Review* tells us that "among the assumptions of psychoanalysis" is that "psychoanalytic principles of mental functioning are considered to be universal. Not merely the incest taboo, but all of the dynamism and contents of the unconscious are held to be alike for all men . . . all cases will have all things. Every defense as well as every wish. All anal and phallic needs exist in every normal, neurotic or psychotic individual" (Zippin 1967, 142).

What are the implications of Freud's infantile sexual etiology for prophylaxis? Grounds for thinking that Freud was at pains to avoid the risk of falsification are found in the evasiveness of his remarks on the prevention of neuroses through enlightened child rearing. If Freud could boast in the first edition of *The Interpretation of Dreams*, "I had discovered the infantile etiology of the neurosis and had thus saved my own children from falling ill" (1900, *SE* 5:469), why in the ensuing decades did he not state what the prophylactic implications of his etiological discovery were?

Why is Freud so bold in his etiological claims yet so timid in the expression of the prophylactic regimen that is entailed by them? As further evidence of Freud's suspiciously evasive attitude toward prophylaxis it is suggestive that, though from 1910 on many of his patients were normal men and women who were undergoing training analyses, he never communicated any information as to how the childhoods of such normal individuals differed systematically from those of his neurotic patients, although this might have afforded an empirical basis for a prophylactic regimen.

In one of his last works, *The Outline*, Freud wrote of the threat of castration that it was "the central experience of the years of childhood, the greatest problem of early life and the strongest source of later inadequacy" (1940, *SE* 23:191). Yet he held that a proscription on castration threats would be of no avail, for the threat of castration is not one that need have occurred in the early life of the patient; it is sufficient that the patient's ancestors would certainly have been subject to such threats. On Erwin's view (1996), we should refrain from concluding from Freudians' use of this tactic that we are dealing with pseudoscience because the theory to whose rescue it comes is nevertheless testable. However, the thesis about the privileged etiological role of sexuality is pseudoscientific not because it is untestable but because of the spurious claim to have tested it.

When Is the Modification of a Theory Opportunism Rather than Falsificationist Rectitude?

Grünbaum and others have argued that Freud's repeated modifications of his theory show that he was pursuing an empirical method. Frank Sulloway once supported Grünbaum: "Not only is psychoanalysis falsifiable, Grünbaum demonstrates, but it also underwent many conceptual changes in Freud's own lifetime that can only be understood as the product of new information contradictory to Freud's previous views" (1985, 24). (Sulloway no longer thinks this.) Theory modification in itself does not entail a falsificationist methodology. If a thesis such as Grünbaum's is to be sustained, the grounds for modification must be shown to have been empirical. You do not make the existence of God falsifiable by agreeing to abide by the toss of a coin. Failing a demonstration of modification by freshly observed facts, Freud's theory changes are susceptible to the charge of opportunism.

The most notorious cases that charge Freudians with opportunism concern their dealings with homosexuality and with penis envy. Was the modification of Freud's claims with respect to homosexuality and penis envy the outcome of investigation or of market research? Consider the revocation of the notion of penis envy as the key to female psychology. Revisionist Freudians argue that psychoanalysts have abandoned penis envy because of new evidence. But no new evidence in support of this withdrawal has been proffered. And it is hard to imagine a form it could take that has not always been available. Though it would be neatly symmetrical to say that Freudian ideology collided with feminist ideology and lost, it is more plausible that current sensitivities exposed the thesis of penis envy to the normal processes of critical appraisal from which other psychoanalytic theses were piously insulated. The pathological status of male homosexuality was revoked in a similar manner. As with penis envy, no new evidence in favor of the altered status was produced, but merely its offensiveness to homosexuals.

Contrast these political adjustments with respect to penis envy and homosexuality with the unmodified standing of masturbation. In 1896 Freud referred to masturbation as "this pernicious form of sexual satisfaction" (1896, *SE* 3:150). He continued to insist on its aberrant and harmful nature throughout his career. A paper by Alden Bunker (1930, 113) provides grounds for this harsh appraisal: "Since in masturbation external stimuli are meager and phantasy must be drawn on, it requires a great consumption of psychic energy and so may readily result in exhaustion of its supply." What Bunker might have been thinking of is the large amount of effort expended in masturbatory fantasizing since the masturbator has to do all the work, not just inventing the script but casting all the parts as well as enacting many of them. The masturbator thus stands to his fantasies as Orson Welles to his films, "produced, directed, written, cast and acted by 'yours truly.'" No wonder he is exhausted.

But Bunker overlooks an alternative, less pathogenic account. If masturbation really were a more frequent cause of neurasthenia than excessive intercourse, as Freud maintained, this may have been due not to its constituting "a pernicious form of sexual satisfaction," but because it was more likely to be excessively indulged in since, to vary Shaw's epigram, it is masturbation rather than marriage that provides the maximum of temptation with the maximum of opportunity. (Cheaper too.) And since unlike homosexuals and feminists, masturbators do not form a constituency likely to take to the streets to make its disapproval felt, their disparagement was not made the subject of belated apologies.

How Does the Probity or Prudence of Analysts Bear on the Credibility of Their Findings?

It is easy to demonstrate the frequency of characterological disputes about Freud in particular and analysts in general, but are these biographical excursions gratuitous, or do they bear on the issue of credibility? Why does it matter to other than his biographers whether Freud was disingenuous and/or mendacious? It matters because psychoanalysis is a testimonial science. We credit Freud, and analysts in general, not on the public grounds they give but on the private grounds they say they have. (On at least two occasions, Freud himself acknowledged the probative insufficiency of his reports of analyses, 1909a, *SE* 10:103; 1918, *SE* 17:13. See also Jones 1923, 414.) Their trustworthiness is thus of the essence.

Yet significant disagreement centers on this very issue. For example, in his contribution to the Hook NYU symposium, Gail Kennedy writes: "I do not see how anyone who has tried to read without bias in the extensive literature of psychoanalysis can fail to arrive at the conclusion that psychoanalysis is an attempt by responsible enquirers to establish a new branch of science" (1959, 272). Roy Grinker, on the other hand, refers to the "worn-out hackneyed reiterations and reformulations of Freudian literature and the stultifying stereotypes stated as positive facts" (1965, ix). Since the subjects of these appraisals are anonymous, the issue is not easily dealt with in this form.

However, it does often take a form more amenable to discussion, that of judgments as to the reliability and trustworthiness of Freud himself. Here are some remarks on this issue:

> Frederick J. Hacker: [Freud] mercilessly dragged [problems] to the surface and exposed them, with the fanaticism of the incorruptible truth-searcher. (1956, 107)
>
> Walter Kaufmann: Freud had extra-ordinarily high standards of honesty and I know of no man or woman more honest than Freud. (1980, 90)
>
> Erich Fromm: Freud, ever the sincere thinker, always offers us undistorted data. (1971, 91)

> Robert Holt: [Freud] did undoubtedly make up many constructions from
> theoretical whole cloth, later presenting them as what his patients told
> him. (1997, 404)

The inconsistency of these judgments compels us to resort to the historical record on which they were presumably based. Some grounds for thinking that Robert Holt is closer to the mark than the others are provided by Joseph Wortis and Abram Kardiner. This is Joseph Wortis on Freud's procedure during the analytic hour: "I would often give a whole series of associations to a dream symbol, . . . and he would wait until he found an association which would fit into his scheme of interpretation and pick it up like a detective at a line-up who waits until he sees his man" (Wortis 1940, 844–45). Abram Kardiner reports of his analysis, "Freud made a beeline for the Oedipus conflict" (qtd. in Bromberg 1982, 135).

However, there are those for whom facts like these have no pertinence to the question of pseudoscience, for they do not bear on the formal testability of the theory. Grünbaum argues that Freud's "intellectual honesty or methodological rectitude" is irrelevant to the scientific status of Freud's theories since these are nevertheless capable of investigation (Grünbaum 1983, pers. comm.). But why bother to investigate theories if one does not credit the intellectual honesty or methodological rectitude of their advocates? This question can be answered only by determining whether sufficiently good evidence suggests that the theories in question might be true. And this in turn depends on our estimate of the prudence, if not the probity, of the analysts who assure us, as Arlow puts it, that "psychoanalytic therapy is a meticulous painstaking investigation into human mental processes" (1959, 211).

Falsification Evasion as a Matter of Judgment

How is the charge of falsification evasion to be substantiated? How can we distinguish a complexity that does not admit of ready testing from self-protective obfuscation? The issue this raises is as much characterological as methodological. We may generate intuitive agreement as to instances where rejecting falsification was justified and instances where it was not, but this is not the same as having a rule. And even more doubtful is our possession of a rule that will tell us when noncapitulation is not merely unwarranted, but wanton and tendentious. In *Conjectures and Refutations*, Karl Popper wrote:

I do not believe that it is possible to give a completely satisfactory definition [of confirmation]. My reason is that a theory which has been tested with great ingenuity and with the sincere attempt to refute it will have a higher degree of confirmation than one which has been tested with laxity; and I do not think that we can completely formalize what we mean by an ingenious and sincere test. (2002, 288)

This issue—the manner of falsification evasion—is raised by Notturno and McHugh: "It is not so much that Freud denies falsification that is bothersome; it is the way he does so" (1986, 250–51). Stephen Jay Gould, in criticizing Lombroso's theory of criminality, spoke of Lombroso's "fudging and finagling" (1978, 227). It is the suspicion of "fudging and finagling" rather than a formal charge of falsification evasion that is invited by Freud's defensive elucidations of what "sexual" means.

Falsification evasion may be a methodological category, but obfuscation is not. It is the discrepancy between what people believe they were told and what they are belatedly informed they were really told after they complain that they had been misled. This is the central issue in Freudian test-evasion disputes, rather than the strictly logical implications of what was written. It is the Freudians' alternation between their "read my lips" mode and the "you should have read the small print" mode that requires extenuation. The apposite question is not the deceptively determinate, "Did Freud lower the testability of his theory?" but the impressionistic, Jamesian, "Was Freud 'turning the area roundabout into a bog of logical liquefaction'? Was he 'fudging and finagling'?"

Spurious Confirmation Claims as Criteria of Pseudoscience

There is a stronger ground than falsification evasion for judging Freud's sexual etiology to be pseudoscientific. It is the spurious claim that the sexual etiology has survived repeated attempts to falsify it. Freud announced at five-year intervals—from 1905 to 1920—his failure to discover counterexamples to his sexual etiology. This raises issues that take us beyond falsification evasion to disreputable confirmation claims.

In a remark first made in the three essays of 1905 and updated in successive editions, Freud wrote:

[A]ll my experience shows that these neuroses [hysteria, obsessional neurosis, dementia praecox, paranoia] are based on sexual instinctive forces. By this I do not merely mean that the energy of the sexual instinct makes a contribution to the forces that maintain the pathological manifestations (the symptoms). I mean expressly to assert that that contribution is the most important and only constant source of energy of the neurosis.... The evidence for this assertion is derived from the ever-increasing number of psychoanalyses of hysterics and other neurotics which I have carried out during the last 25 years. (1905b, *SE* 7:163)

In the case history of Dora, he wrote, "I can only repeat over and over again—for I never find it otherwise—that sexuality is the key to the problem of the psychoneuroses and of the neuroses in general" (1905a, *SE* 7:115). One question this raises is whether Freud had a sufficiently determinate conception of what would constitute its being found "otherwise" to justify his claim that he had never encountered such. What would it be to encounter a case in which the energy of the sexual instinct was not the "only constant source of energy of the neurosis"?

One fairly determinate claim that Freud might be making was to have found sexual episodes in the infancy of his patients that distinguished them from nonneurotics. But it is a peculiarity of Freud's infant etiology that he is committed to no such differentiating episodes. What he is committed to is a "quantitative" difference in intensity that is only to be discerned post hoc.

What makes his confirmation claims spurious and not just mistaken is his awareness of the plethora of reports of the failure to confirm his etiology. While this need not have compelled him to acknowledge the falsity of his etiology, the most it entitled him to claim was that the matter was sub judice. Instead he arbitrarily disqualified those who reported falsifications on the grounds of their prejudice or incompetence, and stuck by his claim to invariable confirmation—claims, moreover, that were almost universally accepted. In 1939, *Time* magazine, which has some entitlement to representativeness, told its readers that Freud's discovery of the "astonishing fact" that the neuroses were due to "the sex experiences of early childhood" was "painstakingly confirmed in hundreds of cases" ("Medicine" 1939).

Is the Oedipus Complex a Discovery or a Myth?

Several psychoanalytic researchers have pronounced the Oedipus complex untestable, including Peter Madison (1961, 190–91) and Sybil Escalona (1952). Neither, however, concluded that it was pseudoscientific on that account. What may have prevented them is their assumption that, though erroneous or inadequate, the Oedipus complex served no other purpose than that of empirical explanation. That it might be declared a myth seems not to have struck them. Nor that there are those who while not going so far could nevertheless not even imagine its falsity.

At the NYU symposium on psychoanalysis and scientific method, a prominent analyst, Jacob Arlow, when asked to imagine coming across a child who did not lust after his mother and consequently hate his father, could not do so. The most he could imagine, on being asked what would persuade him the child did not have an Oedipus complex, was that the child was an idiot (Arlow 1959, 208; Hook 1959b, 217).

When no choice is left but to acknowledge that theses that for decades have enjoyed the status of institutionally accredited truths are in fact baseless, a common strategy is to insist that, reconstituted in some alternative nonliteral fashion—as myth or parable, say—they are defensible. For example, D. M. Thomas (1982) says of a Freudian etiology that had been shown to be false: "It is beautiful, which means it has a different, deeper kind of truth."

Jonathan Lear asks, "What after all, is Oedipus's complex? That he killed his father and married his mother misses the point" (1995, 24). Lear thinks that what Oedipus was avoiding was not his repressed parricidal and incestuous wishes but "the reality of the unconscious." Freud, on the other hand, invokes Denis Diderot's dialogue "Rameau's Nephew" to convey his conception of what the infant would do if he had the strength of a grown man: "strangle his father and lie with his mother" (1940, SE 23:192).

Is the relevant observation to make of pronouncements such as Lear's just that by retreating from its literal sense he has rendered untestable the privileged etiological role Freud conferred on familial sexuality? What we have in the transmutation of the Oedipus complex from a hypothesis as to grossly sexual incestuous fantasies to an edifying myth is a determination to hang onto a verbal formula while leaving it unclear what content is assigned to it. Invoking untestability does not capture the source of our suspicion that something is going on other than hypothesis propounding, something which the term "myth" vaguely aims at. (For some additional episodes of the retreat from literality, see Cioffi 2004, 371–73.)

Conclusion

When critics have described Freud as a pseudoscientist, what have they been trying to call attention to? A common view is that the theses advanced were untestable. An alternative is that spurious claims had been made that because the theory in question had survived attempts at falsification, it was entitled to the same degree of credence as those deemed scientific.

It is undesirable to adopt a conception of pseudoscience that would permit someone to argue that, though psychoanalysis might be "the very opposite of an authentic investigative instrument" (Crews, qtd. in Dufresne 2007, 77), it is nevertheless testable and therefore not pseudoscience. Grünbaum's insistence on the distinction between the "theory-in-itself'" and the malfeasances of those who uphold it is reminiscent of the diagnosis of Sir John Falstaff's urine: "the water itself was a good healthy water; but, for the party that owned it, he might have more diseases than he knew for" (Shakespeare, *2 Henry IV*, 1.2.2–4). Those who defended psychoanalysis against the charge of pseudoscience did not invoke its mere testability, but rather denied the spuriousness of the confirmation reports and defended the probity and judiciousness of those who advanced them. However, the thesis as to the privileged etiological role of sexuality was declared pseudoscientific not because it was untestable, but because of the spuriousness of the claim to have tested it.

Yet in spite of the objections repeatedly lodged against it, Freud's sexual etiology retained its privileged status in the eyes of admirers. Consider, for example, the warm reception afforded Richard Wollheim's celebratory *Sigmund Freud* (1971). The admirers' veneration had a potent source in what Philip Rieff called Freud's "authority for a new ease in enunciating the sexual fact" (Rieff 1961, 163).

The source of the unreasoning attachment to Freud in general is a puzzle, but the appeal of his dealing with sexuality is more readily explained. It has been suggested that over and above the welcome libertarian implications of the balefulness of sexual repression ("I stand for an incomparably freer sexual life" [Jones 1955, 2:465]), the vocabulary in which Freud dealt with sexual life and its anomalies and vicissitudes may have permitted a more detached, less troubled view of it (Bettelheim 1967, 217).

One poignant tribute to Freud runs, "I now acknowledge the shakiness of the scientific premises on which many of his theories have been erected. But ... Freud reconciled me to much that was baffling in others and, more important, in myself; and if, with a personality as anomalous as my own, I have none the less contrived to have an almost continuously happy life along admittedly austerely narrow lines, I owe it largely to him" (King 1975, 103). We can only guess as to how the reconciliation Francis King speaks of was effected, but it may be that a crucial role was played by the abstract distancing idiom with which the anomalies (of which King thought himself an example) are treated.

Freud may have been practicing (unconsciously) a species of exorcism and taking the sting out of our shameful erotic fantasizing through the invention of an abstract, distancing notation—deviant objects, aims, and sources replace the concrete, all-too-vivid, furtive, and shameful life of erotic reverie. The Freudian vocabulary enables those who assimilate it to contemplate their sexuality with less disquiet.

Consider the example of sexual masochism. Those who go to Freud for enlightenment on this topic will find a discussion of the fusion of Eros with the death instinct. But consider some standard sadomasochistic fare with titles such as "Miss Floggy's School for Naughty Boys." What have Miss Floggy and her ilk to do with the fusion of Eros with the death instinct? How did the fusion of Eros and the death instinct bring about Jean-Jacques Rousseau's craving to be spanked by Mlle Lambercier? Is it not significant that talk of the fusion of Eros and the death instinct raises no embarrassed smile whereas acknowledgment of a craving for the attentions of Miss Floggy surely would?

But what of the more general veneration for Freud in spite of the efforts of generations of skeptics? Freud's joke about the brandy drinker whose indulgence impaired his hearing is apposite here. On the advice of his doctor, he refrained from brandy and regained his hearing, but he nevertheless returned to drinking brandy. When his doctor remonstrated with him, he produced the understandable defense that nothing he heard while refraining from brandy was as good as the brandy. For many, Freud is their brandy.

REFERENCES

Arlow, Jacob. 1959. "Psychoanalysis as Scientific Method." In Hook 1959a, 201–11.

Bettelheim, Bruno. 1967. *The Empty Fortress*. New York: Free Press.

Bromberg, Walter. 1982. *Psychiatry between the Wars, 1918–1945: A Recollection*. Contributions in Medical History, No. 10. Westport, CT: Greenwood Press.

Bunker, Alden. 1930. "From Beard to Freud." *Medical Review of Reviews* 36:108–14.

Burr, Charles. 1914. "A Criticism of Psychoanalysis." *American Journal of Insanity*, October, 234–41.

Cioffi, Frank. 1995. "Pseudoscience" and "Psychoanalysis, Philosophical Problems of." In *The Oxford Companion to Philosophy*, edited by Ted Honderich. Oxford: Oxford University Press.

———. 1998. "'Exegetical Mythmaking' in Grünbaum's Indictment of Popper and Exoneration of Freud." In *Freud and the Question of Pseudoscience*, 240–64. Chicago: Open Court.

———. 2004. "The Evasiveness of Freudian Apologetic." In *Who Owns Psychoanalysis?*, edited by Ann Casement, 363–84. London: Karnac.

Crews, Frederick. 1995. *The Memory Wars: Freud's Legacy in Dispute*. New York: New York Review.

Dufresne, Todd. 2007. *Against Freud: Critics Talk Back*. Stanford, CA: Stanford University Press.

Erwin, Edwin. 1996. *A Final Accounting: Philosophical and Empirical Issues in Freudian Psychology*. Cambridge, MA: MIT Press.

———. 2002. "Pseudoscience, and Psychoanalysis." In *The Freud Encyclopaedia*, edited by Edward Erwin. New York: Routledge.

Escalona, Sybil. 1952. "Problems in Psychoanalytic Research." *International Journal of Psychoanalysis* 33:11–21.

Freud, Sigmund. 1895. "On the Grounds for Detaching a Particular Syndrome from Neurasthenia under the Description 'Anxiety Neurosis.'" *SE* 3:85–115.

———. 1896. "Heredity and the Etiology of the Neuroses." *SE* 3:143–56.

———. 1900. *The Interpretation of Dreams. SE* 5:1–621.

———. 1905a. "A Fragment of an Analysis of a Case of Hysteria." *SE* 7:7–122.

———. 1905b. *Three Essays on the Theory of Sexuality. SE* 7.

———. 1909a. *Two Case Histories. SE* 10.

———. 1909b. Sigmund Freud to Oskar Pfister. In *The Life and Work of Sigmund Freud*, vol. 2, by Ernest Jones. New York: Basic Books, 1955.

———. 1910. *Five Lectures on Psycho-analysis. SE* 11.

———. 1912. "Types of Onset of Neurosis." *SE* 12.

———. 1914. "On Narcissism." *SE* 14:67–102.

———. 1916–17. *Introductory Lectures on Psychoanalysis. SE* 16:9–496.

———. 1918. *From the History of an Infantile Neurosis. SE* 17:3–204.

———. 1921a. "Group Psychology." *SE* 18.

———. 1921b. Introduction to *Psycho-analysis and the War Neuroses*, by Sandor Ferenczi, Karl Abraham, Ernst Simmel, and Ernest Jones, 1–4. London: International Psychoanalytical Press.

———. 1925. *Autobiographical Sketch. SE* 20.

———. 1928. "Dostoievski and Parricide." *SE* 21.

———. 1939. *Moses and Monotheism. SE* 23.

———. 1940. *An Outline of Psychoanalysis. SE* 23:144–207.

———. 1953–74. *The Standard Edition of the Complete Psychological Works of Sigmund Freud (SE)*. Edited by James Strachey et al. 24 vols. London: Hogarth Press.

———. 1974. *The Freud-Jung Letters*. London: Hogarth Press.

Fromm, Erich. 1971. *The Crisis of Psychoanalysis*. Greenwich, CT: Fawcett.

Gould, Stephen Jay. 1978. *Ever Since Darwin*. London: Penguin.

Grinker, Roy. 1965. Foreword to *Sigmund the Unserene: A Tragedy in Three Acts*, by Percival Bailey. Springfield, IL: Charles C. Thomas.

Grünbaum, Adolf. 1984. *The Foundations of Psychoanalysis*. Berkeley: University of California Press.

Hacker, Frederick J. 1956. "The Living Image of Freud." *Bulletin of the Menninger Clinic* 20 (3).

Hines, Terence. 1988. *Pseudoscience and the Paranormal*. Buffalo, NY: Prometheus Books.

Holt, R. R. 1997. Review of *Freud Evaluated: The Completed Arc*, by Malcolm Macmillan and Frederick Crews. *Psychoanalytic Books* 8 (Winter): 397–410.

Hook, Sidney, ed. 1959a. *Psychoanalysis, Scientific Method, and Philosophy: A Symposium.* New York: New York University Press.

———. 1959b. "Science and Mythology in Psychoanalysis." In Hook 1959a, 212–24.

James, William. (1890) 1950. *Principles of Psychology.* 2 vols. New York: Dover.

Jones, Ernest. 1920. Review of *The Elements of Psychoanalysis*, by Paul Bousfield. *International Journal of Psychoanalysis* 1:324–28.

———. 1923. "Reflections on Some Criticisms of the Psychoanalytic Method of Treatment." In *Papers on Psychoanalysis.* 3rd ed. London: Baillière, Tindall, Cox.

———. 1953–57. *The Life and Work of Sigmund Freud.* 3 vols. New York: Basic Books.

Kaufmann, Walter A. 1980. *Discovering the Mind.* Vol. 3, *Freud versus Adler and Jung.* New York: McGraw-Hill.

Kennedy, Gail. 1959. "Psychoanalysis: Protoscience and Metapsychology." In Hook 1959a, 269–81.

King, Francis. 1975. "Francis King." In *Bookmarks: Writers on Their Reading,* edited by Frederic Raphael, 103–9. London: Quartet Books.

Lear, Jonathan. 1995. "A Counterblast in the War on Freud: The Shrink Is In." *New Republic,* December 25.

Madison, Peter. 1961. *Freud's Concept of Repression and Defense, Its Theoretical and Observational Language.* Minneapolis: University of Minnesota Press.

"Medicine: Intellectual Provocateur." 1939. *Time,* June 26.

Notturno, Mark, and Paul R. McHugh. 1986. "Is Freudian Psychoanalytic Theory Really Falsifiable?" *Behavioral and Brain Sciences* 9:250–52.

Popper, Karl. 2002. *Conjectures and Refutations: The Growth of Scientific Knowledge.* London: Routledge.

Prince, Morton. 1913–14. "The Psychopathology of a Case of Phobia." *Journal of Abnormal Psychology* 8 (December–January).

Rieff, Philip. 1961. *Freud: The Mind of the Moralist.* Garden City, NY: Doubleday.

———, ed. 1963. *Freud: Early Psychoanalytic Writings.* New York: Collier.

Rivers, W. H. R. 1924. *Instinct and the Unconscious.* Cambridge: Cambridge University Press.

Ross, T. A. 1926. *The Common Neuroses.* London: Edward Arnold.

Sulloway, Frank. 1985. "Grünbaum on Freud." *Free Inquiry,* Fall.

Thomas, D. M. 1982. Review of *The Wolfman: Conversations with Freud's Patient Sixty Years Later*, by Karin Obholzer. *Sunday Times,* November 14.

Waelder, Robert. 1960. *Basic Theory of Psychoanalysis.* New York: International Universities Press.

Wollheim, Richard. 1971. *Sigmund Freud.* New York: Viking Press.

Wortis, Joseph. 1940. "Fragments of a Freudian Analysis." *American Journal of Orthopsychiatry* 10 (4): 843–49.

Zippin, D. 1967. "Constructions in Analysis: Implications for Methodology." *Psychoanalytic Review* 54D: 141–49.

18

The Holocaust Denier's Playbook
and the Tobacco Smokescreen

Common Threads in the Thinking and Tactics

of Denialists and Pseudoscientists

DONALD PROTHERO

You are entitled to your own opinion, but you are not entitled to your own facts.
—Former US Senator Daniel Patrick Moynihan, 2003
To treat your facts with imagination is one thing, but to imagine your facts is
another.

—John Burroughs, American naturalist and essayist

Let's imagine a scenario:

- Scientific consensus about a certain theory ranges from 95 to 99 percent of
all the scientists who work in the relevant fields.
- This scientific topic threatens the viewpoints of certain groups in the
United States, so it is strongly opposed by them and people they influence.
- Their antiscientific viewpoint is extensively promoted by websites and
publications of right-wing think tanks.
- The opponents of this scientific consensus cannot find legitimate scientists
with expertise in the field who oppose the consensus of qualified scientists, so they beat the bushes for "scientists" (none of whom have relevant
training or research credentials) to compose a phony "list of scientists who
disagree with Topic X."

- Deniers of the scientific consensus resort to taking quotations out of context to make legitimate scientists sound like they question the consensus.
- Deniers of the scientific consensus often look for small disagreements among scholars within the field to argue that the entire field does not support the latter's major conclusions.
- Deniers often pick on small errors by individuals to argue that the entire field is false.
- Deniers of the scientific consensus often take small examples or side issues that do not seem to support the consensus and use these to argue that the consensus is false.
- Deniers of the scientific consensus spend most of their energies disputing accepted scientific evidence, rather than doing original research themselves.
- By loudly proclaiming their "alternate theories" and getting their paid hacks to question the scientific consensus in the media, they manage to get the American public confused and doubtful, so less than half of US citizens accept what 99 percent of scientists with relevant expertise consider to be true.
- By contrast, most modern industrialized nations (Canada, nearly all of Europe, China, Japan, Singapore, and many others) have fewer problems with the scientific consensus and treat it as a matter of fact in both their education and in their economic and political decisions.
- The deniers are part of the right-wing Fox-news media bubble and repeat the same lies and discredited arguments to themselves over and over.
- Powerful Republican politicians have used the controversy over this issue to try to force changes in the teaching of this topic in schools.

Reading through this list, most people would immediately assume that it describes the creationists and their attempts to target the scientific consensus on evolution. Indeed, the list does describe creationists or "evolution denialists"—but it also describes the actions of the climate change denialists (who deny global climate change is real and human caused). In fact, the membership lists of creationists and climate-change deniers overlap, with both causes being promoted equally by right-wing political candidates, news media (especially Fox News), and religious organizations such as the Discovery Institute and many others.

Let's make an important distinction here: these denialists are not just "skeptics" about climate change or any other scientific idea that they do not like. A skeptic is someone who does not believe things just because someone proclaims them, but rather tests them against evidence. Sooner or later, however, if the evidence is solid, then the skeptic must acknowledge that the claim is real. Denialists, by contrast, are ideologically committed to attacking a viewpoint they do not agree with, and no amount of evidence will change their minds. As astronomer Phil Plait posted in a recent blog entitled "I'm Skeptical of Denialism" (2009):

> I have used the phrase "global warming denialists" in the past and gotten some people upset. A lot of them complain because they say the word denial puts them in the same bin as Holocaust deniers.
>
> That's too bad. But the thing is, they do have something in common: a denial of evidence and of scientific consensus.
>
> Moon hoax believers put themselves in this basket as well; they call themselves skeptics, but they are far from it. Skepticism is a method that includes the demanding of evidence and critical analysis of it. That's not what Moon hoax believers do; they make stuff up, they don't look at all the evidence, they ignore evidence that goes against their claims. So they are not Moon landing skeptics, they are Moon landing deniers. They may start off as skeptics, but real skeptics understand the overwhelming evidence supporting the reality of the Moon landings. If, after examining that evidence, you still think Apollo was faked, then congratulations. You're a denier.

Belief versus Reality

Reality is that which, when you stop believing in it, doesn't go away.
— Philip K. Dick, science fiction writer

Facts do not cease to exist because they are ignored.
— Aldous Huxley

Reality must take precedence over public relations, for Nature cannot be fooled.
— Richard Feynman

Climate denialism and creationism have a lot in common with many other kinds of denialism. In each case, a well-entrenched belief system comes in conflict with scientific or historic reality, and the believers in this system decide to ignore or attack the facts that they do not want to deal with. As Shermer and Grobman (2000) detail in their book *Denying History*, Holocaust deniers are a prototypical example of this. Despite the fact that there are still hundreds of survivors who were victims and witnesses of the Holocaust (sadly, fewer and fewer of them still survive as time marches on) and many detailed accounts written by the Nazis themselves, the deniers keep on pushing their propaganda to a younger generation that has no memory of the Holocaust and does not hear about it in school. When you dig deep enough, the Holocaust deniers are nearly all hardcore anti-Semites and neo-Nazis who want to see the return of the Third Reich, but for public appearances they try to put on a façade of legitimate scholarship (Shermer and Grobman, 2000). Most people regard the Holocaust deniers as a minor nuisance, but to the Jewish community they represent the problem of anti-Semitism through cultures and time. In Germany and in several other European countries, it is illegal to deny that the Holocaust happened, and prominent deniers (such as David Irving) have been convicted and gone to prison. Yet in the Muslim world, Holocaust denial is very common as a tool to incite Muslims against Israel. Just in the past few years, we have heard numerous Muslim leaders (such as President Mahmoud Ahmadinejad of Iran) make statements of Holocaust denial with full approval of his government and many other Muslims.

As Shermer (1997, 212–16) and Shermer and Grobman (2000, 99–120) point out, the Holocaust deniers use a standard "playbook" or strategy to fight their battle with historical facts. Creationists have used the same playbook for more than fifty years, and now it is the standard method used by any denialist group that wants to push its views against the widely accepted consensus.

Here are some of the main tactics used:

- Quote mining from an authority by taking the quote out of context to mean the opposite of what the author intended. Creationists are notorious for quote-mining works by real scientists, using these quoted words dishonestly without revealing that the context of the quote. Whenever you spot someone quote-mining out of context, it is a sure indicator that they either do not understand what they have read, or they do understand but are deliberately trying to mislead people who do not have the time or inclination to go back and check what the quote really says or means. Climate denialists are guilty of the same tactic, for example, pulling small pieces of prose out of the stolen e-mails from the Climate Research Unit of the

University of East Anglia and using the quote to mean the exact opposite of what the author intended (as is clear from reading the entire text).

- Attacking the weak points of their opponents' position while being careful not to state their own position unless forced to do so. This puts the opponent on the defensive. While the denialists exploit small inconsistencies in their opponent's position, they lead the audience into the "either/or fallacy"—if the consensus position is not 100 percent solid, then the audience might think that denialists could be right.

- Mistaking honest debate within the scholarly and scientific communities as evidence that scientists cannot get their stories straight or that there is no consensus. In reality, any good scientific or scholarly community is always debating the details of important ideas (such as evolution), but that does not mean that they do not agree that evolution occurred or that the Holocaust really happened.

- Focusing on what is not known and ignoring, diminishing, or disregarding what is known. Holocaust deniers and creationists alike will target the more problematic, unsolved questions in a research field (like some details of Holocaust history or evolutionary biology), and claim that because scholars have not resolved it, their solution is better. But scholarship is a constantly changing, open-ended process, and the problems that are not yet solved provide all the more reason to do further research, not to abandon a well-established body of scientific knowledge altogether.

If this list looks very similar to the one given at the beginning of the chapter, this is no accident. Starting with this core list of tactics, the denialists have built up a long menu of techniques and strategies they can use to fight their ideological battles, no matter how clearly their ideas conflict with accepted historical or scientific consensus.

The Tobacco Smokescreen Strategy

Doubt is our product since it is the best means of competing with the "body of fact" that exists in the minds of the general public. It is also the means of establishing a controversy.

—Tobacco company memo (Oreskes and Conway 2010)

So far, we have discussed groups with ideologies or belief systems that they sincerely hold for religious or political reasons, ideologies that lead to denial of any reality that conflicts with their worldview. But there is a second category of science deniers as well: people who recognize reality but, for political or economic reasons, do all they can to obscure that reality. The most famous such example is the case of the tobacco companies, but the same considerations apply to energy companies cynically funding right-wing global warming denialists (Oreskes and Conway 2010), and many other examples.

The tobacco companies pioneered the prototype of this "smokescreen strategy" more than sixty years ago. The reality of smoking and its addictive and deadly nature had been suspected since the 1930s and conclusively established since the 1950s, both by independent researchers and even by scientists funded by tobacco interests. In a perfect world, the discovery that smoking was dangerous would have led to all sorts of immediate efforts to restrict or ban it, as the Food and Drug Administration does right away when a medication is found to be even remotely risky. But this did not happen. Smoking remained a common habit for at least fifty years after its dangerous nature was discovered in the 1930s and confirmed over and over, and its reduced usage in the United States has occurred only during the past decade or so. In most other countries of the world, smoking is actually on the rise as people become wealthier and can afford cigarettes, even though most governments around the world warn their people of the dangers of smoking. Why didn't the conclusive evidence of the dangers of smoking immediately curtail its use?

Part of the answer to this curious dilemma is well known: *tobacco companies actively fought scientific research and spread lies and disinformation to protect themselves* (see Oreskes and Conway 2010; Michaels 2008; McGarity and Wagner 2010). Back in the 1930s, German scientists had studied the link between smoking and cancer, and the Nazi government was one of the first to push an active antismoking campaign. Unfortunately, this research was considered tainted by its Nazi origins and had to be rediscovered in the 1940s and 1950s. By 1953, the studies of the link were conclusive. It was publicized in the *New York Times*, *Life* magazine, and even *Reader's Digest*, one of the most widely read publications in the world at that time; its article was entitled "Cancer by the Carton." As Oreskes and Conway (2010, 14–16) document, the tobacco companies were thrown into a panic, and on December 15, 1953, they met with Hill and Knowlton, one of the biggest public relations firms in the country, to do whatever it took to prevent science from changing people's deadly habits. They began a PR strategy that is the blueprint for nearly every effort since then by powerful vested interests to deny or corrupt science they do not want to hear:

"No proof."   It's a great PR tactic to claim that science has not "proved" smoking causes cancer with 100 percent certainty or that anthropogenic climate change has not been "proven" (or whatever inconvenient scientific reality people try to deny). But this mistakes the fundamental nature of science. *Nothing in real science is 100 percent proven.* Science is always about tentative hypotheses that are tested and retested, and only after a considerable number of studies are conducted and a large majority of scientists agree to a common conclusion do scientists regard something as *very likely,* or *very well established,* or *well corroborated*—but never "proved." That word is appropriate only in everyday conversation and in the world of mathematics, where definite proofs are possible because of the discipline's reliability on deductive logic. The empirical world is too complex and messy to allow "absolute proof." If something has a 99 percent probability of occurring, or of being true, then such a level of confidence is so overwhelming that it would be foolish to ignore it. We can tell a person about to jump off a building that the odds are 99 percent that he will be seriously injured or killed, and this should be sufficient level of confidence for a nonsuicidal person to avoid jumping. Sure, there is a possibility that someone will suddenly put a safety net or giant airbag below the jumper after he leaps, but such miracles are extremely improbable. Likewise, the link between cancer and smoking is virtually certain, as is the scientific consensus that global climate change is real and anthropogenic, and that evolution has occurred.

"Other causes."   Another tactic by denialists is to claim that because other causes may contribute to the problem, we should not try to blame or regulate the cause under discussion. In some cases, a phenomenon does indeed have multiple complex causes (such as the cancers not related to smoking or the multiple causes of autism-spectrum disorders), and it is hard to isolate one in particular. But in the case of smoking-related cancers, the evidence is overwhelming, and it is dishonest to deny that smoking is the main cause of these forms of cancer. Nevertheless, denialists of all stripes have long tried this form of misdirection to get people to look away from the ugly reality in center stage. Suppose it were true that there are additional causes? In cases like smoking and cancer, doesn't it make sense to eliminate this primary, well-established cause (smoking) and then see whether the other potential causes are really important?

"Both sides of the question."   Purveyors of pseudoscience who are trying to protect their favored idea or product always appeal to journalistic "fairness" and argue that we need to listen to "both sides" of a controversial question. Most journalists, not knowing the facts of the case, play along. In instances where the arguments for each side are inconclusive or equally balanced, this is appropriate. But it is not in the case of topics where the scientific evidence is overwhelming and conclusive. Journalists do not give "equal time" or "present both sides" of the question as to whether the earth is flat, or whether it is the center of the universe,, even though the "flat earthers" and modern geocentrists sincerely believe they are right and also demand equal time. No journalist runs a story in which the Holocaust deniers get equal time for spouting their anti-Semitic bile. Likewise, the "junk science" presented by medical quacks, anti-vaxxers, creationists, and global warming deniers does not deserve the same credibility because the overwhelming majority of the scientific community has found these ideas wanting and has rejected them.

"Alternate research."   One of the main strategies used by tobacco companies and their PR firms to fight the scientific evidence was to pay for their own research, hoping that these scientists might find something that helped their cause. It is well documented (based on many different sources, especially the famous study by Glantz et al. 1996) that the tobacco companies paid a lot of money for such research, and certainly did fund studies that found "other causes" of cancer. They then publicized these studies to the hilt, assuming that if "other causes" were mentioned, the strong link between smoking and cancer would be ignored by the public. However, not everything went as planned. By the 1960s, most scientists funded by the tobacco industry had bad news for their patrons: the research clearly showed the cancer-smoking link was real. So what did the tobacco companies do? Did they publicize these results as well, as any honest scientific organization would do? No, they actively suppressed and buried the inconvenient truth. The scientists could not prevent this since the tobacco companies had paid for the research and had made them sign agreements about its release and publication. Meanwhile, the companies' PR machinery and huge advertising budgets continued to crank out denials of the link well into the 1990s, more than thirty years after their own scientists (and outside researchers) had clearly demonstrated that the companies were lying. The same strategy can be seen when energy companies fund research (Oreskes and Conway 2010) that promotes their ends (especially research that might question the reality of anthropogenic global warming) or by creationists who fund strange studies by fringe "scientists" (such as the odd research into "created kinds") that support their cause.

"Dissenting experts."   As Oreskes and Conway (2010) document through-out their book, one of the key strategies of tobacco companies and other organizations trying to deny an inconvenient scientific reality is to look for anyone with credentials who will serve as a "front person" for their cause and give them scientific credibility. These "experts" often turn out to be scientists with no relevant training in the field in question, yet because of their past (ir-relevant) scientific laurels, they are taken seriously by the press and public. The shocking thing that Oreskes and Conway (2010) document is that just a few individuals (Fred Seitz, Fred Singer, William Nierenberg, Robert Jas-trow, and a few more) were at the front of every single one of these attempts to deny scientific reality. Most gained their reputation as nuclear physicists, building the hydrogen bomb. They retained the cold warrior mentality that anything threatening capitalism and free enterprise is bad—even when the scientific case for it is overwhelming. Thus, we find nuclear scientists at the head of panels and commissions (often secretly and lavishly funded by special interest money) defending tobacco companies, energy companies, chemi-cal companies, and the like against the evidence for smoking-related cancer, secondhand smoke, anthropogenic global warming, the ozone hole, acid rain, and the "nuclear winter" scenario. Never mind that a background in nuclear physics gives you absolutely no qualifications to evaluate studies in medicine or climate science. These few men have done more to harm the US public and stunt the dissemination of scientific research than any Soviet threat could ever have accomplished.

"Doubt is our product."   When all else fails, denialists use a smokescreen and try to confuse the public. In the 1953 PR report that Hill and Knowlton prepared for the tobacco companies, the firm made it clear that its primary strategy was to muddy the waters of public opinion and confuse people so that "scientific doubts must remain" (Oreskes and Conway 2010, 6). For decades afterward, the tobacco companies pursued this strategy, spending huge amounts of money to trumpet "scientific" studies that cast doubt on the smoking-cancer link (not re-vealing that they themselves had paid for these studies, so that impartiality was compromised). Even in the late 1970s and early 1980s, the tobacco company denials were uncompromising, and the research of their paid hacks continued to be publicized. But the lawsuits from victims of smoking kept mounting, one after another. Although none had managed to win yet, the evidence kept piling up. In 1964, the Surgeon General first made the link official on behalf of the US government, and as the years went on, the Surgeon General's warnings got more and more scary. Finally, the tobacco companies had to be prosecuted under the RICO act (Racketeer Influenced and Corrupt Organizations Act, originally passed to give prosecutors and congressional committees the power to break up organized crime and racketeering). During these hearings in 1999, the tobacco executives repeatedly lied under oath before Congress, denying their product was harmful and that they knew about this fact many years earlier. Their own internal company documents showed that their statements were lies and that their companies had long suppressed research they themselves had funded that demonstrated the tobacco-cancer link. The Supreme Court upheld the verdict against them in 2006. The most revealing document of all was an internal memo that came to light during the investigation, revealing the tobacco executives' full knowledge of what they were doing. It read: "Doubt is our product since it is the best means of competing with the 'body of fact' that exists in the minds of the general public. It is also the means of establishing a controversy." As Oreskes and Conway (2010) document, we have the energy companies funding research to create doubt about anthropogenic climate change and many other organizations that are not interested in scientific truth but only in protecting the status quo—and their bottom lines.

So how do we avoid fooling ourselves? How do we avoid getting caught up in "weird beliefs" and find out what is real? How do we get past the "smoke-screen" of powerful interests that create doubt about a scientific issue to stall any policy action? Many people have their own ideas about this, from religious beliefs to political dogmas, but the one strategy that has worked time and again is the *scientific method*.

Science and Reality

Science is nothing but developed perception, interpreted intent, common sense rounded out and minutely articulated.

—George Santayana, philosopher

The good thing about science is that it's true whether or not you believe in it.
—Astrophysicist Neil DeGrasse Tyson,
on *Real Time with Bill Maher*, February 4, 2011

So what is science, and why do we consider it so useful and important? Despite the Hollywood stereotypes, science is not about white lab coats and bubbling beakers or sparking apparatuses. Science is a way of looking at the world using a specific toolbox—the *scientific method*. There are many definitions of it, but the simplest is a method by which we generate explanations about how the natural world works (*hypotheses*), and then try to test or shoot down those ideas using evidence of the real world (*testability* or *falsifiability*). Scientific hypotheses must always be tentative and subject to revision, or they are no longer scientific—they are dogma. Strictly speaking, science is not about final truth, or about certainty, but about constructing the best models of the world that our data allow and always being willing to change those models when the data demand it.

Since Karl Popper's time, a number of philosophers of science have argued that his definition of science as only falsifiable hypotheses was too narrow and excluded some fields that most would regard as legitimate science. Pigliucci (2010) proposed a broader definition of science that encompasses scientific topics that might not fit the strict criterion of falsifiability. All science is characterized by three elements:

1. *Naturalism*: we can examine only phenomena that happen in the natural world because we cannot test supernatural claims scientifically. We might want to say that "God did it" explains something about the world, but there is no way to test that vague notion.

2. *Empiricism*: science studies only things that can be observed by our senses, things that are accessible not only to ourselves but also to any other observer. Science does not deal with internal feelings, mystic experiences, or anything else that is in the mind of one person and no one else can experience.

3. *Theory*: science works with a set of well-established theories about the universe that have survived many tests. Gravitation is a "theory," as much as evolution is a theory. This is very different from the popular use of the word "theory" as a "wild speculation," as in "theories of why JFK was assassinated." From well-established, highly successful, and explanatory theories like gravity, evolution, or plate tectonics, scientists then make predictions as to what nature should be like if the theory is true, and go out and test those predictions.

In this way, science is very different from dogmatic belief systems like religion and Marxism, which take certain absolute statements to be true and then try to twist the world to fit their preconceptions. None of these other belief systems are open to be critically tested and to allow us to discover that they might be false because their core beliefs are sacrosanct and unchanging. By contrast, science is constantly changing not only the small details of what it has discovered, but occasionally even its core beliefs.

The scientific viewpoint on the natural world is in many ways a humble one: we do not have absolute truths but are trying to understand nature as best we can. As scientists, we must be ready to abandon any cherished hypothesis when the evidence demands it. As Thomas Henry Huxley put it, it's "the great tragedy of science—the slaying of a beautiful hypothesis by an ugly fact." As scientists, we must be careful when we use words like "truth" and "belief" because science is not about *believing* accepted truths, but about *accepting* extremely well-corroborated hypotheses concerning nature that approach "truth" in the everyday sense. In the vernacular, scientists are comfortable using words like "real" or "true" to describe phenomena that are so well established that it would be perverse not to admit they exist. We all agree that gravity is real, but we still do not understand how it works in detail. Despite this fact, objects fall through the sky no matter whether we fully understand why or not. Likewise, evolution happens all the time around us (Prothero 2007) whether we fully understand every detailed mechanism or not.

Why do we think science is a better descriptor of nature and the natural world than religion? For one thing, science tells us what is real, not what we want to hear. If science were just like any other form of human thought, it would play to our biases, conform to our worldview, and give us the answers we feel comfortable with. But more often than not, science gives us "inconvenient truths," answers we do not want to hear, and this suggests we are dealing with an external reality, not just our preconceptions about the world. Astronomy has told us we are not the center of the universe, but a tiny planet in a small solar system on an insignificant galaxy nowhere near the middle of the universe. Geology has shown us that human existence spans only a few million years at best, while the earth is 4.6 billion years old, and the universe is 13.7 billion years old. In 1859, Darwin showed that we are not specially created by a deity but are just another species among millions, so closely related to the other two species of chimpanzees that in molecular terms, we are a "third chimpanzee" (as Jared Diamond put it). More recently, climate and environmental science has shown that we are capable of destroying our own environment and making the planet uninhabitable not only for other species alive today (and quickly going extinct), but even for our grandchildren. None of these conclusions fit our preferred view of the world, suggesting that science finds out what is real, not what we want the world to be like. As the British comedian and actor Ricky Gervais (2009) put it,

> Science seeks the truth. And it does not discriminate. For better or worse it finds things out. Science is humble. It knows what it knows and it knows what it doesn't know. It bases its conclusions and beliefs on hard evidence—evidence that is constantly updated and upgraded. It doesn't get offended when new facts come along. It embraces the body of knowledge. It doesn't hold on to medieval practices because they are tradition. If it did, you wouldn't get a shot of penicillin, you'd pop a leech down your trousers and pray.

Baloney Detection

Skeptical scrutiny is the means, in both science and religion, by which deep thoughts can be winnowed from deep nonsense.

—Carl Sagan

As mature adults, we have learned not to be naïve about the world. By hard experience, we are all equipped with a certain degree of healthy skepticism. We know that politicians and salespeople are often dishonest, deceptive, and untruthful, and that most advertisements mislead, lie, exaggerate, or distort the truth; we tune most of these messages out. We are always cautious when buying something, worried that the vendor might cheat us. We all follow the maxim *caveat emptor*, "Let the buyer beware." Such a view may seem cynical, but we learn about human nature the hard way, and it is essential to our survival to be skeptical and not too trusting. The list below gives some of the keys we look for in assessing the credibility of an authority and detecting "baloney":

Quote mining.   As discussed already, quoting out of context is a sure sign that the claim is bogus, especially when the quote miner has blatantly misused the passage to mean the opposite of the author's intent.

Credential mongering.   Another red flag is when an author makes a claim and waves his credentials in front of us to intimidate us by his "authority." This is particularly common in creationist books that flaunt the author's PhD on the cover, but it occurs even in science when a fringe scientist-author wants to be taken seriously. When you see "PhD" on the cover of a book, it is often an indicator that the book cannot stand on the strength of its own arguments and evidence.

Irrelevant expertise.   A related problem is that the general public is often impressed when someone has a PhD and assume that it makes the degree holder somehow smarter and more expert than the average person. The primary thing that getting the PhD demonstrates is that the degree holder was able to struggle through five to seven years of graduate school, jump through all the hoops, and finish research and writing in a very narrow topic of her dissertation. As most of us with doctorates know, all this focus on a narrow topic actually makes you *less* broadly trained than you were before you started. More important, a PhD degree *qualifies you to critique only the topics in which you were trained*. When you hear a creationist with a PhD in hydraulic engineering or biochemistry attacking evolutionary biology, your baloney detector should be warning you. If someone does not have relevant training in biology or paleontology, they are no more qualified to critique evolutionary biology than they are qualified to write a symphony or build a skyscraper. Another example is provided by denialists who trot out "lists of scientists who disagree with" evolution, global climate change, or whatever. A thorough analysis of these lists show that very few of the "scientists" on it have any relevant training, whether it be in climate science research or evolutionary biology. Yet the denialists know that pointing to this phony "list of dissenting scientists" will impress the layperson who does not understand the difference.

Conflict of interest.   We find that medical studies that deny problems with a drug or deny the dangers of smoking were paid for directly by the companies that benefited from this research. Likewise, most of the "experts" who deny global warming have no relevant qualifications at all, or work in the oil, coal, or mining industries (like the mining geologist Ian Plimer, who has no training in climate science), or work for conservative "think tanks" heavily supported by industries that stand to gain from their advocacy. As Upton Sinclair put it, "it is difficult to get a man to understand something, when his salary depends upon his not understanding it!"

Misunderstanding the burden of proof.   When scientific consensus has been reached and a large amount of evidence supports a particular explanation of a phenomenon, then the burden of proof falls on the dissenter who seeks to overthrow the consensus. It is not enough for a climate denialist to point to one small piece of contrary data or a creationist to pick on one little inconsistency in the huge data set supporting evolution. Like a lawyer in a civil court case, they must show that the *preponderance of the evidence* supports their view before scientists and others will take their arguments seriously. Likewise, the Holocaust denier cannot just point to a few inconsistencies in the documentation of certain events in 1942 and claim that the Holocaust never occurred. If there is overwhelming evidence showing that events like the Holocaust and evolution occurred, we expect the dissenters to argue otherwise on the basis of even stronger evidence before we give them much credence. As Carl Sagan put it, "extraordinary claims require extraordinary evidence."

Whom Can We Trust?

For every expert there is an equal and opposite expert; but for every fact there is not necessarily and equal and opposite fact.

—Thomas Sowell, American economist

If you tell a lie big enough, and keep repeating it, people will eventually come to believe it.

—Nazi propaganda chief Josef Goebbels

This discussion about authority and experts raises a good point: whom can we trust on issues of science and pseudoscience? How do we tell who are the truly qualified, unbiased experts and who the paid hacks or phony con artists? How can we judge the back-and-forth arguments among creationists and scientists or between climate scientists and their critics?

The media, unfortunately, make the situation worse by their attempts at "fairness" and "equal time." They feel obligated to let both sides have an equal hearing, no matter what the merits of each case. Where the evidence is inconclusive, this might make sense, but in most scientific issues, the consensus has been reached; it is foolish to give pseudoscientists equal time to spout their nonsense. Global climate change denialists try to get an equal hearing every time global warming is mentioned, no matter how fallacious their arguments or data are. The anti-vaxxers get their celebrity spokespeople like Jenny McCarthy and Jim Carrey on TV all the time, just because they are celebrities, but the relatively unglamorous medical researchers who have the data on their side do not seem to have the same media impact. This is an inherent problem with the media, which are driven by a kind of "if it bleeds, it leads" mentality, featuring sensationalism over sound science. The scientific community does not have much chance to change this as long as the general population is willing to pay attention and give good ratings to shows that feature UFOs, Bigfoot, quack medicine, and creationist nonsense.

Goldman (2006) and Pigliucci (2010) provide an interesting set of criteria by which we can evaluate expertise. In some cases, it is sufficient for laypersons to listen to the arguments from both sides and judge for themselves. But many arguments are beyond the education and training of most people to evaluate, so then the next best criterion is the evidence of *agreement or consensus by experts in the field*. This is where the scientific method works best. In general, scientists are a very hard-boiled, skeptical group, criticizing each other and spotting the flaws in the work of other people. The entire process of peer review is a very rigorous and often nasty one whereby all scientific ideas must run a gauntlet of intense scrutiny.

Goldman's fourth criterion is to *look for bias or conflict of interest*, and this too is very revealing when it comes to evaluating the critics of mainstream science. Almost without exception, every one of the "scientific experts" who disagree with evolution reached their positions because of a prior religious conviction. I know of no anti-evolutionists who looked at the evidence fairly and came to doubt it, because they always start with their religious biases distorting their perception. The ones who claim they were atheists first and then rejected evolution did so because of a religious conversion experience, not because they found the evidence lacking, as I documented in Prothero (2007). Likewise, nearly all the global climate change critics come from right-wing think tanks. Their opinions were influenced by their libertarian or conservative antigovernment laissez-faire attitudes, not from a dispassionate study of the climate data. Those employed by the oil, coal, and mining industries clearly have a conflict of interest since their livelihoods depend on denying that their products are causing global climate change. The leading anti-vax doctor, Andrew Wakefield, did his research after he was retained by lawyers trying to win a class-action suit against the MMR vaccine; he was also trying to develop his own vaccine to replace it. By contrast, most research scientists who support evolutionary biology or climate science are paid (usually much less) by nonprofit universities and nonprofit government organizations, none of which tell their scientists what to believe or what conclusions they want them to reach.

Finally, Goldman (2006) says that if all else fails, look at the track record of the expert. A classic case is Kent Hovind, the creationist minister who calls himself "Dr. Dino," even though he has no training in paleontology; his phony "doctorate" is from a diploma mill that sends "degrees" to whomever pays the fee. Hovind has been repeatedly caught lying about evolution in his debates and books, but the pattern of deception and cheating runs even deeper: he is now serving a ten-year prison sentence in the ADX Florence Prison in Florence, Colorado, for tax evasion. Creationists like Duane Gish have long track records of distortions and lies and double-talk, all documented in books and in the blogosphere (see Prothero 2007). Creationists almost never publish in peer-reviewed journals or present at professional scientific meetings, but avoid the scrutiny of the real scientific community and publish only in their own sympathetic house journals and websites. Anti-vax doctors like Wakefield have track records of bad research that has been repudiated by the medical community. Due to his many transgressions, Wakefield has been banned from practicing medicine, and his work has been shown to be fraudulent. His paper on the alleged vaccine-autism link was retracted by the journal where it was first published.

With all the screaming and shouting and name calling in the media and blogosphere, it is often hard to tell who is telling the truth and who is just a shill for a powerful industry or political faction or religious group. But with a careful examination of who has actual expertise, it is possible to find out what science really shows us. As Thomas Henry Huxley advised, "sit down before fact as a little child, be prepared to give up every preconceived notion, follow humbly wherever and to whatever abysses nature leads, or you shall learn nothing."

REFERENCES

Gervais, Ricky. 2009. "Why I'm an Atheist." *Speakeasy* (blog). *Wall Street Journal*, December 19. http://blogs.wsj.com/speakeasy/2010/12/19/a-holiday-message-from-ricky-gervais-why-im-an-atheist/.

Glantz, S. A., J. Slade, L. A. Bero, P. Hanauer, and D. E. Barnes, eds. 1996. *The Cigarette Papers*. Berkeley: University of California Press.

Goldman, Alvin I. 2006. "Experts: Which Ones Should You Trust?" In *The Philosophy of Expertise*, edited by Evan Selinger and Robert P. Crease, 14–38. New York: Columbia University Press.

Michaels, David. 2008. *Doubt Is Their Product: How Industry's Assault on Science Threatens Your Health*. Oxford: Oxford University Press.

McGarity, Thomas O., and Wendy E. Wagner. 2010. *Bending Science: How Special Interests Corrupt Public Health.* Cambridge, MA: Harvard University Press.

Oreskes, Naomi, and Conway, Erik M. 2010. *Merchants of Doubt: How a Handful of Scientists Obscured the Truth on Issues from Tobacco Smoke to Global Warming.* New York: Bloomsbury Press.

Pigliucci, Massimo. 2010. *Nonsense on Stilts: How to Tell Science from Bunk.* Chicago: University of Chicago Press.

Plait, Phil. 2009. "I'm Skeptical of Denialism." *Bad Astronomy* (blog). *Discover*, June 9. http://blogs.discovermagazine.com/badastronomy/2009/06/09/im-skeptical-of-denialism/.

Prothero, Donald R. 2007. *Evolution: What the Fossils Say and Why It Matters.* New York: Columbia University Press.

Shermer, Michael. 1997. *Why People Believe Weird Things: Pseudoscience, Superstitions and Other Confusions of our Times.* New York: Holt.

Shermer, Michael, and Alex Grobman. 2000. *Denying History: Who Says the Holocaust Never Happened and Why Do They Say It?* Berkeley: University of California Press.

PART VI

The Cognitive Roots of Pseudoscience

19

Evolved to Be Irrational?

Evolutionary and Cognitive Foundations

of Pseudosciences

STEFAAN BLANCKE AND JOHAN DE SMEDT

People believe the weirdest things. Forty percent of the US population endorses the claim that Earth and all life on it was created by God six to ten thousand years ago (Newport 2010); three in four Americans accept some form of paranormal belief such as astrology or extrasensory perception (Moore 2005). Europeans are no less gullible: two Britons in five believe that houses can be haunted, and one in five thinks that aliens have visited our planet at some point in the past (Lyons 2005). Pseudomedical treatments such as homeopathy are widely practiced and, in some countries like Belgium, even refunded by health care. Horoscopes are consulted in numerous popular magazines and newspapers. In sum, there seems to be no end to the irrational propensities of the human mind.

In this chapter, we examine in four parts how an evolutionary and cognitive perspective might shed light on the pervasiveness and popularity of irrational beliefs that make up pseudosciences. First, we set up the general theoretical framework, explaining what an evolutionary and cognitive approach entails. Second, we explore how this framework adds to our understanding of why the human mind is so vulnerable to systematic reasoning errors. Third, we demonstrate how concrete pseudosciences tap into particular cognitive dispositions. And fourth, we explain why a number of irrational beliefs take on the form of pseudosciences. To conclude, we turn to the question raised in our title and briefly discuss how the evolution of the mind relates to human (ir)rationality.

The Evolved Mind

Understanding the human mind as a product of evolution was first proposed by Charles Darwin. In his seminal work *On the Origin of Species*, which actually makes little direct mention of human evolution, he professed that "psychology will be based on a new foundation, that of the necessary acquirement of each mental power and capacity by gradation" (Darwin 1859, 488). Twelve years later, in *The Descent of Man*, Darwin (1871) argued that humans share particular cognitive faculties with other animals, differing only in degree, which showed that the mind had indeed evolved. But for more than one hundred years, despite the enormous potential for explaining human thought and behavior, Darwin's radically new approach to the human mind was largely ignored, notwithstanding a few unsuccessful and premature attempts to Darwinize psychology, such as Freudian psychoanalysis. This situation changed during the second half of the previous century with the development of evolutionary psychology.

Evolutionary psychology emerged from several scientific traditions, synthesizing elements from research fields such as cognitive science, cognitive ethology, and sociobiology (Tooby and Cosmides 2005), as a consequence of the evidence that had been accumulating in those fields. It challenged the prevailing paradigm in the social sciences, identified by Tooby and Cosmides (1992) and others (e.g., Pinker 2002) as the Standard Social Science Model, which regards the human mind as a blank slate with a small number of general purpose learning mechanisms that are inscribed with any content culture provides (Pinker 2002; Tooby and Cosmides 1992).

Instead, this new evidence suggests that the human mind consists of a number of domain-specific, specialized mental inference systems that evolved in response to specific adaptive problems our ancestors had to solve during their evolutionary history. These were mainly problems dealing with survival, mating and sex, kinship and parenting, and group living (Buss 2008). One school of thought in evolutionary psychology (e.g., Pinker 1997) holds that cognitive evolution has not kept pace with cultural developments: the circumstances in which humans live have altered dramatically since the early Holocene (due to, for example, the invention of farming and the Industrial Revolution), but, according to evolutionary psychologists, our evolved mind is still mainly adapted to a hunter-gatherer way of life. Human evolution did not stop in the Pleistocene, as is evident, for example, in mutations in enzymes that allow for the digestion of starchy food and dairy products (e.g., Perry et al. 2007), but evolutionary psychologists (e.g., Tooby and Cosmides 1992) contend that the pace of cultural evolution over the last ten thousand years has outstripped that of organic evolution, so that human cognitive adaptations are still to a large extent fitted to a hunter-gatherer lifestyle. There has been some tentative genetic evidence for ongoing cognitive evolution over the past few thousand years (e.g., P. D. Evans et al. 2005; Mekel-Bobrov et al. 2005), but these findings have faced criticism (Currat et al. 2006; Yu et al. 2007). The structure of the human mind constrains and governs human thought and behavior in systematic ways. For example, people are more wary of spiders than of cars, even though the latter category forms a far bigger risk to one's health than the former in most human lives.

What is of interest here is that the mind is endowed with cognitive dispositions that are largely adaptive: they offer the ability to produce representations of particular aspects of the world that allow humans to respond quickly and aptly to specific situations. These predispositions are often pictured as "fast and frugal heuristics" (Gigerenzer et al. 1999) that result in *intuitive* ways of reasoning that are fast, automatic, and largely unconscious. To be sure, we believe that we control our thoughts, that an "I" does our thinking. This *reflective* way of thinking, which is mostly conscious and functions more slowly in comparison to intuitive reasoning, arises from the human capacity to represent representations. Because this meta-representational capacity does not deal with the outside world directly, it is regarded by some to be domain general (e.g., Sloman 1996), although according to Sperber (1996), it can be deemed a cognitive specialization that has evolved specifically to deal with representations. Humans do indeed seem to possess two distinct ways of processing information, intuitive and reflective, also called dual-process reasoning (J. S. B. T. Evans 2010). As we will see further on, this has important implications for our understanding of human rationality and thus for our present discussion of pseudosciences.

The Evolution of Cognitive Bias

Because the human mind has evolved to deal with adaptive problems in real-life situations, it focuses on specific cues in its environment that are relevant for solving these problems, rather than generating a perfectly accurate picture of the environment. Thus, we can expect human reasoning to exhibit trade-offs between speed and truth preservation, leading to fast but not always reliable heuristics. This prediction is borne out by ample studies under the banner of the "heuristics and biases" program, initiated by Tversky and Kahneman in the 1970s (for an overview, see Gilovich et al. 2002). Even in solving abstract reasoning tasks, people rely on their intuitive judgment (unless taught otherwise), which leaves them highly vulnerable to systematic errors. For instance, when evaluating probabilities, people tend to make judgments on the basis of representativeness (Tversky and Kahneman 1974). The effect of these heuristics is exemplified by the classical Linda problem (Tversky and Kahneman 1983). Participants are invited to read the following description: "Linda is 31 years old, single, outspoken, and very bright. She majored in philosophy. As a student, she was deeply concerned with issues of discrimination and social justice and participated in anti-nuclear demonstrations." Then, they are asked which of the two following options they think is most probable: (a) Linda is a bank teller, or (b) Linda is a bank teller and a feminist. Although a conjunction can never be more probable than either of its two constituents, around eighty-five percent of participants judge that the second option is more likely than the first, arguably because they consider the text to be more representative of a feminist than of a bank teller. This has been dubbed the conjunction fallacy. Fallacies like these have proven to be extremely robust and not easy to weed out (Tentori et al. 2004).

Gigerenzer and colleagues (1999) have argued that the appearance of fallacies like this does not reflect people's failure to think rationally, but rather results from researchers appraising people's reasoning skills by inappropriate standards. To return to the Linda problem, people are supposed to apply a content-free logical rule to arrive at the correct answer. The test, however, contains ambiguous terms like "probable" that trigger conversational heuristics that look for intended meaning and relevance, causing subjects to understand the word in nonmathematical terms such as "possible" or "conceivable." When asked for a frequency judgment ("How many?") instead of a probability judgment, as a result of which the ambiguity dissolves, people do infer the mathematical meaning, and the conjunction fallacy largely disappears (Hertwig and Gigerenzer 1999). According to Gigerenzer (2008), variations on experiments like this confirm that the mind should be regarded as a collection of specialized inference systems that have evolved in such a way that the human brain responds to the environment quickly, frugally, *and* efficiently.[1] Hence, according to dual-process theories of reasoning, a picture emerges of two forms of rationality. On the one hand, there is the slow and reflective mode of rationality that conforms to the norms and rules of logic and probability. On the other hand, we have an "ecological" or "bounded rationality" that conforms to the adaptive requirements set by the environments in which the human species has evolved (Hilton 2002). From this perspective, the appearance of irrationality does not result from flawed reasoning, but rather from evaluating the latter form of rationality by the standards of the former. However, when intuitive reasoning is applied to complex and abstract cognitive problems, irrational reasoning can still result (Haselton et al. 2005). The fast and frugal heuristics sometimes lead to error, as they continue to interfere with people's reflective inferences, in the form of well-attested kinds of irrationality (see above).

Keeping the above framework in mind, we argue that the tenacity and popularity of particular pseudosciences, even in the face of strong adverse evidence, are partly explained by the fact that pseudosciences tap into people's intuitive understanding, thereby exploiting the mental heuristics that have evolved to respond efficiently to particular environmental and social situations. Let us illustrate this point by taking a closer look at one of the most pervasive irrational belief systems of today, creationism.

Pseudoscience and Content Biases: Creationism as a Case Study

Here, we use the term "creationism" not in its common sense of young-earth creationism, but as a form of belief system that contends that there is evidence that God has purposively intervened in the natural world, creating or designing entities (species, adaptations) that could not have arisen through a naturalistic process. As such, creationism not only denotes young-earth creationism, but also includes old-earth and Intelligent Design (ID) creationism (Matzke 2010; Scott 2009). Note that each of these variants is presented as scientific by its adherents, or at least as scientific as evolutionary theory.

Although the various strands of creationism might differ in their theological specifics, our use of the term "creationism" depends on the idea that they share a minimal core of common assumptions. In the rest of this chapter, we argue that these core assumptions tie in closely with human intuitions concerning the origins and causal structure of the biological world. More specifically, creationism exploits or piggybacks on the human mind's essentialism, its preference for teleological explanations and its hyperactive tendency to detect agency. As we will see, each of these intuitions makes sense from an evolutionary perspective.

Psychological Essentialism

Essentialism is a hallmark of creationism. It is the view that entities, such as species, possess an immutable essence that guides their development and behavior. Essentialism can be described as a fast and frugal heuristic that instantly provides our mind with a rich inductive potential, not on the basis of apparent similarities but on the basis of an unobserved core that is believed to cause members of a given category to share particular behavioral and physical properties. As such, essentialism "allows one to exploit the causal structure of the world (of natural kinds, in particular), without necessarily knowing anything about the causes themselves" (Barrett 2001, 7). Historically, essentialism constitutes a major and recurrent theme in Western thought at least since Aristotle (Mayr 1982), a clear indication of its enduring appeal. Today, students' understanding of evolutionary theory is still hindered by essentialist inclinations (Shtulman and Schulz 2008): students with the highest essentialist tendencies have the least understanding of the mechanism of natural selection. Studies on essentialist reasoning in children indicate that this intuition develops early and in the absence of instruction, and that it is stable across cultures. Five-year-olds acknowledge that category membership remains unaffected by superficial changes. They consider a butterfly to belong to the same category as a caterpillar despite the dramatic developmental transformations the organism goes through (Gelman 2003). Also, essentialism is not restricted to Western culture: Yukatek Maya children reason as much about biological categories in terms of essences as children in the United States, a finding that suggests that essentialism is a universal feature of the human mind (Atran 2002). Moreover, young children often reason more in an essentialist fashion than adults, another indicator that this tendency is a stable part of human cognition (Gelman 2004). Although humans are capable of exploiting the causal structure of the world in other ways than through essentialism, it provides a quick and efficient heuristic to do so—for example, if one apple is edible, one can quickly generalize that all are edible; if one tiger is dangerous, one can infer that all are dangerous. Interestingly, humans are not the only species to use essential reasoning in this adaptive way: rhesus monkeys (*Macaca mulatta*) also infer that superficial changes to the exterior of a fruit do not alter its inside properties (Phillips et al. 2010).

Evans (2000a, 2001) found that young children until the age of ten have a preference for creationist accounts for the origin of species, and this is often accompanied with essentialist thinking. Creationists believe that God (or a "designer") has created the biological world, which is divided into distinct, nonoverlapping categories or kinds, the members of which share an unobserved essence that makes them belong to that particular category and which resists evolutionary change. For instance, in *Evolution? The Fossils Say NO!*, young-earth creationist Duane Gish (1978, 43) firmly asserts that "the human kind always remains human, and the dog kind never ceases to be a dog kind. The transformations proposed by the theory of evolution never take place." ID adherents are no different in this regard. Although some claim that they have no issue with common descent, they too state that natural selection is limited to microevolution, which has always been conceded by creationists as limited change *within* "kind." Toward naturalistic macroevolution ("the molecule-to-man theory," in the words of Gish), however, ID proponents are as skeptical as any other creationist. As one of the leading figures within the ID movement, biochemist Michael Behe (1996, 15), puts it, "the canyons separating everyday life forms have their counterparts in the canyons that separate biological systems on a microscopic scale. . . . Unbridgeable chasms occur even at the tiniest level."

Teleology

Intuitively, humans not only view the world in terms of essences, but they also assume that things in the world happen or exist for a purpose. This teleological tendency reveals itself from a young age. Four- and five-year-olds are more inclined to ascribe functions to biological wholes and natural objects than adults do. They assume that lions are "to go in the zoo" and that clouds are "for raining" (Kelemen 1999a). When asked "why rocks are so pointy," seven- to ten-year-olds prefer a teleological explanation ("so that animals wouldn't sit on them and smash them") over a purely physical explanation ("They were pointy because bits of stuff piled up on top of one another for a long time," see Kelemen 1999b). The teleological tendency wanes with age, which is probably due to the effects of science education. Scientifically untrained Romani adults were shown to be more prone to ascribe teleological explanations to nonbiological natural entities than their educated peers (Casler and Kelemen 2008). However, evidence suggests that education merely suppresses the teleological tendency, which continues to act as a mental default setting throughout the lifespan. Adults are more likely to endorse teleological explanations ("the sun makes light so that plants can photosynthesize") when questioned under time pressure (Kelemen and Rosset 2009). Also, Alzheimer's patients tend to revert to teleological thinking as a result of their condition (Lombrozo et al. 2007), indicating that the exposure to causal explanations affects only people's reflective but not their intuitive beliefs.

Understanding biological properties in teleofunctional terms, particularly in combination with our capacity to categorize, provides a rich and valuable source of information for making inferences about the environment. As such, the teleological stance can also be identified as a fast and frugal heuristic that may have added to our adaptive rationality. Some philosophers even argue that teleological reasoning forms an indispensable conceptual tool for acquiring a solid *scientific* understanding of the biological world (Ruse 2003). Nonetheless, teleological intuitions have also been shown to highly constrain students' understanding of evolutionary theory. Students tend to mistake natural selection for a goal-directed mechanism. Or they assume that evolution as a whole moves toward an end, which is commonly identified with the human species (see Bardapurkar 2008 for a review). Like essentialism, the teleological stance becomes an easy target for exploitation by irrational belief systems when it operates on unfamiliar terrain.

In creationist literature, the idea that things in this world exist because of a particular purpose is a strong and recurrent theme. In *Scientific Creationism*, under the subtitle *Purpose in Creation*, Henry M. Morris (1974a, 33–34) contends that "the creation model does include, quite explicitly, the concept of purpose" and that "the creationist seeks to ascertain purposes." Rhetorically, he asks his readers:

> Do both fish and men have eyes because man evolved from fish or because both fish and man needed to see, in order to fulfil their intended creative purpose? Can stars and galaxies be arranged in a logical hierarchy of order from one type to another because they represent different stages in an age-long evolutionary process, or because they were each specially created to serve distinct purposes, such purposes requiring different degrees of size and complexity?

The same notion of purposefulness also resonates throughout the entire ID literature. In fact, the basic claim of the movement is that complex biological systems can be compared with artifacts, implying that they too have been made to serve a particular purpose. Often, people's teleological intuitions are brought in as a justification for the design inference. As William Dembski (1999, 48), another important ID proponent, puts it:

Intelligent Design formalizes and makes precise something we do all the time. All of us are all the time engaged in a form of rational activity which, without being tendentious, can be described as "inferring design." Inferring design is a perfectly common and well-accepted human activity.

Naturally, being creationists, Morris and Dembski depict the alleged purposes in nature as resulting from the intentional actions of a supernatural agent. As such, creationism does not only hijack people's teleological intuitions, but it also taps into the strong inclination of the human mind to detect other agents and understand their behavior as motivated by intentions and desires. This makes creationism all the more cognitively appealing.

Detecting Agents and the Intentional Stance

The human mind is highly prone to detecting agency, and it often does so even in the absence of agents. Just think of the times you thought someone was near when it turned out to be only some piece of garment hung on a clothesline or a bush blown in the wind, or of the times you mistook a bag blown by the wind for a bird or a small animal. The opposite scenario, however, in which one mistakes an agent for a inanimate object, rarely occurs, even though it is in principle possible (e.g., mistaking a person for a mannequin or a bird for a lump of earth and some leaves). At least two good evolutionary reasons have been proposed as to why the mind is more likely to produce false positives than false negatives when it comes to detecting agency. First, we can expect that agency detection is hyperactive, based on game-theoretical considerations involving predator-prey interactions, in particular the costs of false positives and negatives and the potential payoffs (Godfrey-Smith 1991). For complex organisms that live in variable conditions and rely on signals in the environment that are not always transparent to make decisions, it is far less costly to assume that there is an agent when there is none than the other way around (Guthrie 1993)—this is the case not only for animals that need to avoid predators, but also for predators looking for potential prey, in which case the potential benefit outstrips the costs of a false positive. Because of the asymmetry between costs, natural selection favors organisms with an agency detection device that occasionally generates false positives rather than false negatives. Second, agency detection is not only related to predator-prey interactions, but is also highly relevant for the detection of the attention of conspecifics. Being watched may have consequences for one's reputation. Any reputational damage might entail a decrease in cooperation opportunities, thus limiting access to vital resources, which in turn affects reproductive success. This provides a plausible scenario for why the human mind is hypersensitive to cues of being watched by other agents. For example, a picture of two eyes suffices to induce people to put more money in a donation box (Bateson et al. 2006) or leave significantly less litter in a canteen (Ernest-Jones et al. 2011); stylized eyespots on a computer screen or an eye-like painting significantly increase generosity in a Dictator Game (Haley and Fessler 2005; Oda et al. 2011).

Evolutionary psychologists argue that the human mind has an evolved capacity to interpret the behavior of other agents as motivated by internal states, such as intentions and beliefs. Adopting the "intentional stance" (Dennett 1987) allows one to predict the behavior of complex organisms. To account for the origin of this capacity, two scenarios have been proposed—they are related to the scenarios set out above explaining human hypersensitivity to the presence of other agents. One is that the intentional stance has evolved to deal with complex social interactions. This Machiavellian intelligence hypothesis traces the evolution of human mind reading in the complex social interactions that most primates engage in. Given the large group sizes in humans compared to other primates, humans require more sophisticated mind-reading skills to successfully interact with group members (see, e.g., Byrne 1996; Humphrey 1976). The other suggests that this stance has evolved in relation to predator-prey interactions: the ability to remain undetected by predators or to find prey requires that one is able to accurately predict what other agents will do (Barrett 2005; Boyer and Barrett 2005).

For the purpose of this chapter, we need not decide between these hypotheses, which are also not mutually exclusive. The human mind does not have only the capacity to interpret the behavior of agents in term of their intentions; it also forms expectations as to what agents are capable of, in particular in relation to inanimate objects. Ten-month-old babies assume that only agents create order out of chaos (Newman et al. 2010), and ten- to twelve-month-olds expect an object's movement only to be caused by a human hand, not by an inanimate object (Saxe et al. 2005). These inferences add to the rich explanatory power that comes with human intuitive psychology, or theory of mind.

This intuitive psychology is easily triggered. Adults have been shown to overattribute intentions to purely natural events. Sentences like "she broke the vase" are by default interpreted as describing an intentional act, not something that happened by accident (Rosset 2008). However, it is unclear whether folk psychological intuitions are also invoked by and connected with the teleological intuitions discussed above. In the case of artifacts, there is an obvious link between the purpose of the artifact and the intention for making it, which results in the "design stance" (Dennett 1987). For instance, both children and adults privilege a creator's intent over later afforded usage when deciding which function to attribute to an artifact (Chaigneau et al. 2008; Kelemen 1999a). But concerning the natural world, the connection between the teleological and intentional stance is far less apparent. Both Evans (2000b) and Kelemen and DiYanni (2005) have established a link between these two stances in seven- to ten-year-old children from the United States and the United Kingdom respectively, independently of their being raised in a religious cultural environment. Based on these findings, Kelemen (2004) coined the term "intuitive theists," meaning that these children intuitively project an agent who is responsible for creating the world. However, the Dutch children who were probed by Samarapungavan and Wiers (1997) for their beliefs concerning the origins of species did not express such a creationist inclination. Furthermore, in the aforementioned studies with Alzheimer's patients (Lombrozo et al. 2007) and adults under time pressure (Kelemen and Rosset 2009), the teleological and intentional stance were not clearly correlated. Alzheimer's patients, despite their increased endorsement of teleological explanations, were not more likely to invoke God as an explanation compared to healthy control subjects. People who were more likely to endorse teleological explanations under time pressure were not more likely to believe in God. In sum, intuitive teleology cannot be equated with intuitive theism (De Cruz and De Smedt 2010). It seems that people's creationist intuitions are not as deeply ingrained as their teleological intuitions.

Even though theism is not intuitive in the sense of being an innate, untutored intuition, it is nevertheless easy to grasp and natural for minds like ours, which are hypersensitive to the actions of agents, readily infer intentionality, and consider only agents to be capable of creating movement and order (on the appeal of intentional explanations and agentive thinking, see Buekens, chapter 23, in this volume). The suggestion that the world is the result of a creative act by a hidden supernatural agent is something that makes intuitive sense. Indeed, creationists insist that the intentions of such an agent can be read off from both the order and the beauty in the universe and the functional complex systems found in nature. For instance, Morris (1974a, 33) writes:

The Creator was purposive, not capricious or indifferent, as He planned and then created the universe, with its particles and molecules, its laws and principles, its stars and galaxies, its plants and animals, and finally its human inhabitants.

He goes on:

The creationist explanation will be in terms of primeval planning by a personal Creator and His implementation of that plan by special creation of all the basic entities of the cosmos, each with such structures and such behavior as to accomplish most effectively the purpose for which it was created.

Hence, creationists compare the bacterial flagellum with an outboard rotary motor (Behe 1996) and conceptualize DNA as some kind of code, programmed by an Intelligent Designer (Davis et al. 1993; H. M. Morris 1974b). In biology textbooks, artifact metaphors are commonly used as explanatory tools to make sense of complex biological systems, which points to their strong intuitive appeal. However, because of this appeal, they can become an alluring piece of rhetorical equipment in the hands of creationists, who intend these metaphors to be taken quite literally (Pigliucci and Boudry 2011).

Discussion

Although we have limited our discussion of mental predispositions exploited by creationism to the essentialist, teleological, and intentional biases, other biases may be at play as well. For instance, the intuitions that humans are fundamentally different from other animals (De Cruz and De Smedt 2007) and that mind and body belong to two separate ontological domains (Bloom 2004; Slingerland and Chudek 2011) are other good candidates to explain widespread pseudoscientific thinking. Also, we have demonstrated only how creationism piggybacks on those inference systems, but we hold that the same reasoning goes for other pseudosciences. Essentialism, for instance, may contribute to explaining the persistence of homeopathy (Hood 2008)—even if a substance is diluted to the point that it is no longer chemically detectable, our intuitive essentialism can lead to the mistaken intuition that the essence of the product is still there (for other instances, see Talmont-Kaminski, chapter 20, in this volume). Note, however, that we do not intend to debunk the beliefs that make up pseudosciences simply by demonstrating that the latter tap into people's evolved intuitions. Doing so in a straightforward way would be committing the genetic fallacy. One could try to set up a debunking argument by claiming that our evolved inference systems are systematically off-track or unreliable, but this does not seem to be the case. After all, these cognitive predispositions at least produce ecologically rational solutions to recurrent problems the human mind has evolved to solve. Furthermore, scientific beliefs too rely on intuitive assumptions. For example, scientists share with young children (e.g., Saxe et al. 2005) the intuition that any contingent state of affairs has one or more causes to account for it. The search for (often nonobvious) causes is part of our intuitive understanding of the world that is continuous between scientific and everyday reasoning (De Cruz and De Smedt 2012). Hence, if dependence on evolved biases would count as a debunking argument, scientific beliefs would also be susceptible to debunking arguments, a conclusion we obviously do not want to draw. Rather, a cognitive and evolutionary approach to pseudosciences helps to explain why people steadfastly adhere to such belief systems, even in the face of strong defeating evidence.

Context Biases, or Why Pseudoscience?

Irrational (reflective) belief systems tend to mimic real sciences, sometimes down to the smallest detail. Biblical creationism has developed into scientific creationism or ID, osteopathy and the like are presented as alternative treatments on a par with modern medicine, and contemporary vitalistic theories use scientific terms like "energy" to leave a scientific impression. Obviously, these pseudosciences piggyback on the authority science has been endowed with in modern society. The question remains as to why it is so important for pseudosciences to seek that authority, and why they often succeed in attaining it. Again, an evolutionary and cognitive perspective can shed some light on these issues.

Humans are social rather than individual learners: they gain significantly more information through communication with conspecifics than by direct experience with the environment. Although the benefits of social learning, the extent of which is unique to humans, are huge (one has access to much more information at a much lower cost), such a capacity would not have evolved if humans did not have ways to protect themselves from being misinformed. Therefore, Mercier and Sperber (2011) have argued that humans are critical social learners, who exhibit epistemic vigilance with regard to socially transmitted information: they critically evaluate both the content and the source of the information received. As to the latter, both cues that signal competence and benevolence are important, although these are less easy to trace when one is confronted with information that is transmitted via cultural communication. As a result, the epistemic vigilance provided by the heuristics that track such cues might break down (Sperber et al. 2010). To deal with the resulting uncertainty and to restore protection against false beliefs, a predisposition might have evolved to trust epistemic authorities, that is, individuals (or, by extension, institutions) other people defer to as being competent and benevolent sources of information (Henrich and Gil-White 2001). Hence, people may put their epistemic trust in authorities, simply for the reason that the latter are commonly acknowledged as such. Why has science come to enjoy this epistemic authority? Undoubtedly, the tremendous instrumental efficacy of science, in the form of, for instance, medicine and communication technology, has been an important factor in its widespread public acceptance. However, it is important to point out that this trust is not universal and that in some communities people defer to religious authorities as a source of reliable information (Kitcher 2008). Religion is historically and socially well embedded in these communities, where it enjoys public support and is endorsed in education (denominational education, Sunday school). If people indeed place their epistemic trust in science, why is this trust not universal, and why are some pseudosciences like creationism widely endorsed? One reason is that creationists successfully present themselves as scientifically legitimate. Many of their proponents have PhDs and publish books and papers in scientific fields. Given that their claims enjoy the extra advantage of being in line with our evolved cognitive predispositions, such as essentialism, teleology, and the intentional stance—whereas real science often runs counter to these intuitions—they can successfully win converts among the general public.

Conclusion

Let us return to the question in the title. Are we evolved to be irrational? Given the ubiquity of pseudosciences, this seems a fair question to ask. However, from an evolutionary perspective, we should at least expect some rationality in ecologically relevant domains. The representations an evolved mind generates should allow an organism to at least respond aptly, and thus rationally, to environmental situations. The human mind is stacked with fast and frugal heuristics, the operations of which result in an adaptive, ecological rationality.

However, when these heuristics operate outside their proper domain in solving abstract and complex cognitive problems that require a reflective mode of thinking, their output becomes subjugated to the normative rationality of logic and probability theory. Hence, when their impact on reflective thinking remains unchecked, we are likely to endorse irrational beliefs. The tendency to endorse pseudosciences increases when they are given an air of scientific respectability, which allows them to take advantage of the epistemic authority that scientific theories enjoy. Therefore, to answer our title question, although we could not have evolved to be irrational, sometimes people are irrational because we have evolved.

Acknowledgments

The research for this chapter was supported by Ghent University (BOF08 /24J/041 and COM07/PWM/001). We would like to thank Johan Braeckman, Helen De Cruz, and the editors of this volume for their helpful remarks.

NOTE

1. However, this view is not widely shared in the psychology of reasoning. For example, Tentori et al. (2004) contend that Gigerenzer's frequency approach already provides participants with a part of the solution, prompting them to conceptualize the problem in terms of frequencies.

REFERENCES

Atran, Scott. 2002. "Modular and Cultural Factors in Biological Understanding: An Experimental Approach to the Cognitive Basis of Science." In *The Cognitive Basis of Science*, edited by Peter Carruthers, Stephen P. Stich and Michael Siegal, 41–72. Cambridge: Cambridge University Press.

Bardapurkar, A. 2008. "Do Students See The "Selection" In Organic Evolution? A Critical Review of the Causal Structure of Student Explanations." *Evolution. Education and outreach* 1 (3): 299–305.

Barrett, H. Clark. 2001. "On the Functional Origins of Essentialism." *Mind & Society* 2 (1): 1–30.

———. 2005. "Adaptations to Predators and Prey." In *The Handbook of Evolutionary Psychology*, edited by D. M. Buss, 200–23. Hoboken: John Wiley and Sons.

Bateson, M., D. Nettle, and G. Roberts. 2006. "Cues of Being Watched Enhance Cooperation in a Real-World Setting." *Biology Letters* 2 (3): 412–14.

Behe, Michael J. 1996. *Darwin's Black Box. The Biochemical Challenge to Evolution.* New York: Free Press.

Bloom, Paul. 2004. *Descartes' Baby: How Child Development Explains What Makes Us Human.* London: Arrow Books.

Boyer, Pascal, and H. Clarck Barrett. 2005. "Domain Specificity and Intuitive Ontology." In *The Handbook of Evolutionary Psychology,* edited by David M. Buss, 96–118. Hoboken: John Wiley and Sons.

Buss, David M. 2008. *Evolutionary Psychology. The New Science of the Mind.* 3rd ed. Boston: Pearson.

Byrne, Richard W. 1996. "Machiavellian Intelligence." *Evolutionary Anthropology: Issues, News, and Reviews* 5 (5): 172–80.

Casler, K., and D. Kelemen. 2008. "Developmental Continuity in Teleo-Functional Explanation: Reasoning About Nature among Romanian Romani Adults." *Journal of Cognition and Development* 9 (3): 340–62.

Chaigneau, Sergio E., Ramón D. Castillo, and Luis Martínez. 2008. "Creators' Intentions Bias Judgments of Function Independently from Causal Inferences." *Cognition* 109 (1): 123–32.

Currat, Mathias, Laurent Excoffier, Wayne Maddison, Sarah P. Otto, Nicolas Ray, Michael C. Whitlock, and Sam Yeaman. 2006. "Comment on 'Ongoing Adaptive Evolution of Aspm, a Brain Size Determinant in *Homo Sapiens*' and 'Microcephalin, a Gene Regulating Brain Size, Continues to Evolve Adaptively in Humans.'" *Science* 313 (5784).

Darwin, Charles. 1859. *On the Origin of Species by Means of Natural Selection: Or the Preservation of Favoured Races in the Struggle for Life.* London: John Murray.

———. 1871. *The Descent of Man, and Selection in Relation to Sex.* London: John Murray.

Davis, Percival, Dean H. Kenyon, and Charles B. Thaxton. 1993. *Of Pandas and People: The Central Question of Biological Origins.* Dallas: Haughton.

De Cruz, Helen, and Johan De Smedt. 2007. "The Role of Intuitive Ontologies in Scientific Understanding—The Case of Human Evolution." *Biology and Philosophy* 22 (3): 351–68.

———. 2010. "Paley's Ipod: The Cognitive Basis of the Design Argument within Natural Theology." *Zygon: Journal of Religion and Science* 45 (3): 665–84.

———. 2012. "Evolved Cognitive Biases and the Epistemic Status of Scientific Beliefs." *Philosophical Studies* 157:411–29.

Dembski, William A. 1999. *Intelligent Design. The Bridge between Science and Theology.* Downers Grove: InterVarsity Press.

Dennett, Daniel C. 1987. *The Intentional Stance.* Cambridge: MIT Press.

Ernest-Jones, Max, Daniel Nettle, and Melissa Bateson. 2011. "Effects of Eye Images on Everyday Cooperative Behavior: A Field Experiment." *Evolution and Human Behavior* 32 (3): 172–78.

Evans, E. Margaret. 2000a. "Beyond Scopes. Why Creationism Is Here to Stay." In *Imagining the Impossible: Magical, Scientific and Religious Thinking in Children,* edited by K. Rosengren, C. Johnson and P. Harris, 305–31. Cambridge: Cambridge University Press.

———. 2000b. "The Emergence of Beliefs About the Origins of Species in School-Age Children." *Merrill-Palmer Quarterly-Journal of Developmental Psychology* 46 (2): 221–54.

———. 2001. "Cognitive and Contextual Factors in the Emergence of Diverse Belief Systems: Creation Versus Evolution." *Cognitive Psychology* 42 (3): 217–66.

Evans, Jonathan St. B. T. 2010. *Thinking Twice: Two Minds in One Brain.* Oxford: Oxford University Press.

Evans, P. D., S. L. Gilbert, N. Mekel-Bobrov, E. J. Vallender, J. R. Anderson, L. M. Vaez-Azizi,

S. A. Tishkoff, R. R. Hudson, and B. T. Lahn. 2005. "Microcephalin, a Gene Regulating Brain Size, Continues to Evolve Adaptively in Humans." *Science* 309 (5741): 1717–20.

Gelman, Susan A. 2003. *The Essential Child: Origins of Essentialism in Everyday Thought.* Oxford: Oxford University Press.

———. 2004. "Psychological Essentialism in Children." *Trends in Cognitive Sciences* 8 (9): 404–9.

Gigerenzer, Gerd. 2008. *Rationality for Mortals: How People Cope with Uncertainty.* Oxford: Oxford University Press.

Gigerenzer, Gerd, Peter M. Todd, and the ABC Research Group. 1999. *Simple Heuristics That Make Us Smart.* Oxford: Oxford University Press.

Gilovich, Thomas, Dale Griffin, and Daniel Kahneman, eds. 2002. *Heuristics and Biases. The Pyschology of Intuitive Judgment.* Cambridge: Cambridge University Press.

Gish, Duane T. 1978. *Evolution? The Fossils Say No!* 3rd ed. San Diego: Creation-Life.

Godfrey-Smith, Peter. 1991. "Signal, Decision, Action." *Journal of Philosophy* 88 (12): 709–22.

Guthrie, Stewart. 1993. *Faces in the Clouds. A New Theory of Religion.* New York: Oxford University Press.

Haley, Kevin J., and Daniel M. T. Fessler. 2005. "Nobody's Watching? Subtle Cues Affect Generosity in an Anonymous Economic Game." *Evolution and Human Behavior* 26 (3): 245–56.

Haselton, Martie G., David Nettle, and Paul W. Andrews. 2005. "The Evolution of Cognitive Bias." In *The Handbook of Evolutionary Psychology*, edited by David M. Buss, 724–46. Hoboken: John Wiley and Sons.

Henrich, J., and F. J. Gil-White. 2001. "The Evolution of Prestige—Freely Conferred Deference as a Mechanism for Enhancing the Benefits of Cultural Transmission." *Evolution and Human Behavior* 22 (3): 165–96.

Hertwig, R., and G. Gigerenzer. 1999. "The 'Conjunction Fallacy' Revisited: How Intelligent Inferences Look Like Reasoning Errors." *Journal of Behavioral Decision Making* 12 (4): 275–305.

Hilton, Denis. 2002. "Thinking About Causality: Pragmatic, Social and Scientific Rationality." In *The Cognitive Basis of Science*, edited by Peter Carruthers, Stephen P. Stich, and Michael Siegal, 211–31. Cambridge: Cambridge University Press.

Hood, Bruce M. 2008. *Supersense: Why We Believe in the Unbelievable.* San Francisco: HarperOne.

Humphrey, N. K. 1976. "The Social Function of Intellect." In *Growing Points in Ethology*, edited by P. P. G. Bateson and R. A. Hinde, 303–17. Cambridge: Cambridge University Press.

Kelemen, Deborah. 1999a. "The Scope of Teleological Thinking in Preschool Children." *Cognition* 70 (3): 241–72.

———. 1999b. "Why Are Rocks Pointy? Children's Preference for Teleological Explanations of the Natural World." *Developmental Psychology* 35 (6): 1440–52.

———. 2004. "Are Children 'Intuitive Theists'? Reasoning About Purpose and Design in Nature." *Psychological Science* 15 (5): 295–301.

Kelemen, Deborah, and C. Di Yanni. 2005. "Intuitions About Origins: Purpose and Intelligent Design in Children's Reasoning About Nature." *Journal of Cognition and Development* 6 (1): 3–31.

Kelemen, Deborah, and E. Rosset. 2009. "The Human Function Compunction: Teleological Explanation in Adults." *Cognition* 111 (1): 138–43.

Kitcher, Philip. 2008. "Science, Religion and Democracy." *Episteme* 5 (1): 5–18.

Lombrozo, Tania, Deborah Kelemen, and D. Zaitchik. 2007. "Inferring Design—Evidence of a Preference for Teleological Explanations in Patients with Alzheimer's Disease." *Psychological Science* 18 (11): 999–1006.

Lyons, Linda. 2005. "Paranormal Beliefs Come (Super)Naturally to Some." *Gallup*, November 1. http://www.gallup.com/poll/19558/Paranormal-Beliefs-Come-SuperNaturally -Some.aspx.

Matzke, Nicholas. 2010. "The Evolution of Creationist Movements." *Evolution: Education and Outreach* 3 (2): 145–62.

Mayr, Ernst. 1982. *The Growth of Biological Thought. Diversity, Evolution, and Inheritance.* Cambridge: Harvard University Press.

Mekel-Bobrov, N., S. L. Gilbert, P. D. Evans, E. J. Vallender, J. R. Anderson, R. R. Hudson, S. A. Tishkoff, and B. T. Lahn. 2005. "Ongoing Adaptive Evolution of Aspm, a Brain Size Determinant in Homo Sapiens." *Science* 309 (5741): 1720–22.

Mercier, Hugo, and Dan Sperber. 2011. "Why Do Humans Reason? Arguments for an Argumentative Theory." *Behavioral and Brain Sciences* 34 (2): 57–74.

Moore, David W. 2005. "Three in Four Americans Believe in Paranormal." *Gallup*, June 16. http://www.gallup.com/poll/16915/Three-Four-Americans-Believe-Paranormal.aspx.

Morris, Henry M. 1974a. *Scientific Creationism*. Gen. ed. San Diego: Creation-Life.

———. 1974b. *The Troubled Waters of Evolution*. San Diego: Creation-Life.

Newman, G. E., F. C. Keil, V. A. Kuhlmeier, and K. Wynn. 2010. "Early Understandings of the Link between Agents and Order." *Proceedings of the National Academy of Sciences of the United States of America* 107 (40): 17140–45.

Newport, Frank. 2010. "Four in 10 Americans Believe in Strict Creationism." *Gallup*, December 17. http://www.gallup.com/poll/145286/Four-Americans-Believe-Strict-Creationism.aspx.

Oda, Ryo, Yuki Niwa, Atsushi Honma, and Kai Hiraishi. 2011. "An Eye-Like Painting Enhances the Expectation of a Good Reputation." *Evolution and Human Behavior* 32 (3): 166–71.

Perry, George H., Nathaniel J. Dominy, Katrina G. Claw, Arthur S. Lee, Heike Fiegler, Richard Redon, John Werner, Fernando A. Villanea, Joanna L. Mountain, Rajeev Misra, Nigel P. Carter, Charles Lee, and Anne C. Stone. 2007. "Diet and the Evolution of Human Amylase Gene Copy Number Variation." *Nature Genetics* 39 (10): 1256–60.

Phillips, Webb, Maya Shankar, and Laurie R. Santos. 2010. "Essentialism in the Absence of Language? Evidence from Rhesus Monkeys (*Macaca Mulatta*)." *Developmental Science* 13 (4): F1–F7.

Pigliucci, Massimo, and Maarten Boudry. 2011. "Why Machine-Information Metaphors Are Bad for Science and Science Education." *Science & Education* 20 (5–6): 453–71.

Pinker, Steven. 1997. *How the Mind Works*. New York: W. W. Norton.

———. 2002. *The Blank Slate: The Modern Denial of Human Nature*. London: Penguin Books.

Rosset, E. 2008. "It's No Accident: Our Bias for Intentional Explanations." *Cognition* 108 (3): 771–80.

Ruse, Michael. 2003. *Darwin and Design. Does Evolution Have a Purpose?* Cambridge, MA: Harvard University Press.

Samarapungavan, A., and R. W. Wiers. 1997. "Children's Thoughts on the Origin of Species: A Study of Explanatory Coherence." *Cognitive Science* 21 (2): 147–77.

Saxe, R., J. B. Tenenbaum, and S. Carey. 2005. "Secret Agents: Inferences About Hidden Causes by 10-and 12-Month-Old Infants." *Psychological Science* 16 (12): 995–1001.

Scott, Eugenie C. 2009. *Evolution vs. Creationism: An Introduction.* 2nd ed. Westport, CT: Greenwood Press.

Shtulman, Andrew, and L. Schulz. 2008. "The Relation between Essentialist Beliefs and Evolutionary Reasoning." *Cognitive Science* 32 (6): 1049–62.

Slingerland, Edward, and Maciej Chudek. 2011. "The Prevalence of Mind–Body Dualism in Early China." *Cognitive Science* 35 (5): 997–1007.

Sloman, S. A. 1996. "The Empirical Case for Two Systems of Reasoning." *Psychological Bulletin* 119 (1): 3–22.

Sperber, Dan. 1996. *Explaining Culture: A Naturalistic Approach.* Oxford: Blackwell.

Sperber, Dan, F. Clement, C. Heintz, O. Mascaro, H. Mercier, G. Origgi, and D. Wilson. 2010. "Epistemic Vigilance." *Mind & Language* 25 (4): 359–93.

Tentori, K., N. Bonini, and D. Osherson. 2004. "The Conjunction Fallacy: A Misunderstanding About Conjunction?" *Cognitive Science* 28 (3): 467–77.

Tooby, John, and Leda Cosmides. 1992. "The Biological Foundations of Culture." In *The Adapted Mind: Evolutionary Psychology and the Generation of Culture*, edited by J. Barkow, J. Tooby, and L. Cosmides, 19–136. New York: Oxford University Press.

———. 2005. "Conceptual Foundations of Evolutionary Psychology." In *The Handbook of Evolutionary Psychology*, edited by David M. Buss, 5–67. Hoboken: John Wiley and Sons.

Tversky, Amos, and Daniel Kahneman. 1974. "Judgment under Uncertainty: Heuristics and Biases." *Science* 185 (4157): 1124–31.

———. 1983. "Extensional Versus Intuitive Reasoning: The Conjunction Fallacy in Probability Judgment." *Psychological Review* 90 (4): 293–315.

Yu, Fuli, R. Sean Hill, Stephen F. Schaffner, Pardis C. Sabeti, Eric T. Wang, Andre A. Mignault, Russell J. Ferland, Robert K. Moyzis, Christopher A. Walsh, and David Reich. 2007. "Comment on 'Ongoing Adaptive Evolution of Aspm, a Brain Size Determinant in Homo Sapiens.'" *Science* 316 (5823): 370b.

20

Werewolves in Scientists' Clothing

Understanding Pseudoscientific Cognition

KONRAD TALMONT-KAMINSKI

Most pseudoscientific beliefs have very little in common with real scientific beliefs. Although they might fool at first glance because they claim the authority of science or bear similarities to scientific claims, pseudoscientific beliefs instead should be understood in one of two ways: as versions of supernatural claims that have taken on guises more fitting to the modern world or as claims drawing their motivation from such beliefs. To better illustrate this basic picture, I use an approach that focuses on the cognitive and cultural mechanisms that produce pseudoscientific beliefs and the evolutionary processes that likely shaped those mechanisms. This does not require giving up on the epistemic considerations, but it does mean putting them into the proper context.

The method pursued in this chapter is to relate pseudoscientific beliefs to the cognitive picture that Robert McCauley (2010, 2011) has put forward of the relationships between theology and popular religion on one hand, and science and commonsense beliefs on the other. The effect is to show that the fundamental difference between science and pseudoscience is found in the way they relate to beliefs that humans find intuitively attractive. Science, unlike pseudoscience and the other kinds of beliefs McCauley considers, does not seek to maintain agreement with those "maturationally natural" beliefs but instead investigates their shortcomings.

In both its topic and its underlying theoretical assumptions, this chapter is very closely connected to those written by Stefaan Blancke and Johan De Smedt, and John Wilkins (see chapters 19 and 21, respectively, in this volume). It shares the view that human cognition is best understood in terms of bounded rationality theory, originally put forward by Herbert Simon (1955) and developed by Gerd Gigerenzer (2000) as well as by William Wimsatt (2007). The Blancke and De Smedt chapter is doubly relevant in so far as it explores the cognitive explanation of pseudoscientific beliefs that underpins much of the picture that is developed in this chapter.

Mechanisms of Nonscience

The supernatural is very commonly defined in opposition to science. This is not the path taken in this chapter, however. Defining the supernatural (or, indeed, the pseudoscientific) in terms of scientific knowledge risks falling foul of Hempel's dilemma (Hempel 1969), for it requires that we either reference *current* scientific knowledge or *future* claims that scientific inquiry, carried to its endpoint (if this is even possible), would prove correct. Going on current scientific knowledge, however, we may well end up deeming pseudoscientific the claims that future scientific inquiry would legitimize. Indeed, this outcome is unavoidable once we recognize, as we must, that scientific inquiry has not reached its endpoint. To avoid this problem, we can hold that beliefs are pseudoscientific whenever an idealized, final science would not support them. But then, we virtually guarantee that much of what we consider best in today's science will turn out to be pseudoscientific. And we cannot even know which parts since we cannot know what finalized scientific inquiry will reveal until we get there.

Beyond this dilemma, much is wrong with thinking about either supernatural or pseudoscientific claims primarily in terms of their relation to scientific claims:

- It appears to treat science as primarily characterized in terms of the claims that are justified by it—a view of science that may be partly motivated by science textbooks, but that has little to do with scientific practice and that immediately puts science on par with any belief system—including supernatural ones. Viewed in this way, the conflict between scientific and antiscientific worldviews is reduced to a matter of picking different ontologies. Such an understanding of science (or the supernatural/pseudoscientific, for that matter) is singularly lacking in insight.

- It does not necessarily distinguish between supernatural and pseudo-scientific claims since both conflict with scientific knowledge. Indeed, it does not necessarily distinguish between such claims and all other claims that do not accord with science. After all, the claim that electrons and protons have the same weight runs counter to what science tells us, yet it would be hard to argue that it is a supernatural claim.
- It fails to tell us anything substantive about supernatural/pseudoscientific claims. In particular, it gives us no insight into why it is that those claims have proved as difficult to expunge as they have.

In short, this way of thinking is singularly unhelpful in getting us to understand the phenomenon in question. As such, it is better to abandon it. Before we do, however, it is useful to consider why this way of approaching the issue has proved fruitless. The reason, I argue, is that it treats both scientific and antiscientific explanations in abstraction, separate from the psychological and cultural processes that produce them. In this, it harks back to a positivist view of science as characterized in terms of a decontexualized inductive logic, instead of as a social endeavor carried out by organized groups of boundedly rational agents—the view that traces back to the work of Herbert Simon (1955) and that is pursued here (see also Talmont-Kaminski 2009b, 2012, forthcoming).

The alternative approach can be motivated by a seemingly simple question that Pascal Boyer (2001) has put forward: why do people believe the particular supernatural claims that they do accept? Or, we could just as well ask, why do people believe in particular pseudoscientific claims? After all, supernatural beliefs do not present a random gamut of scenarios but, instead, usually share many similarities. These questions refocus the issue of what such beliefs are as an investigation of the mechanisms that are responsible for their appearance and stabilization. The ultimate aim is to characterize the claims in terms of the causal processes that produce them. This, in turn, should make it possible to understand under what conditions human cognitive systems produce beliefs that, in an important sense, are irrational rather than rational. And this strikes me as a particularly worthwhile project, even if this chapter goes only a small way toward that goal.

In practice, this way of approaching the question of what the supernatural and the pseudoscientific are entails potentially drawing on a variety of disciplines that seek to explain human behavior using a range of scales—neurophysiological, through psychological, all the way up to the cultural. Evolutionary theory, applied at both the genetic and the cultural levels, is the

overarching framework for the approach taken up here. This is becoming the
norm in many investigations of human behavior, in fact. Vitally, taking up this
approach does not mean abandoning epistemological questions, but requires
reconsidering them in the context of the processes that produce actual beliefs
rather than in an abstracted fashion divorced from the details of human cognitive systems.

Modestly Counterintuitive Agency

The cognitive byproduct account of supernatural beliefs and practices developed by Boyer and others is probably the most widespread at this time and
seeks to explain these phenomena in terms of biases produced by the idiosyncrasies of the human cognitive system (for review, see Bulbulia 2004). For example, humans appear to be overly sensitive to cues of the presence of other
agents in their vicinity (Guthrie 1993). Thus, people returning home late at
night often imagine the presence of shadowy figures when there are none.
This oversensitivity is presumably highly adaptive, given that the cost of re-
acting to a nonexistent threat is much lower than that of failing to spot a threat
that is real (Haselton and Nettle 2006). It has been argued, however, that such
instances of imagining the presence of nonexistent agents may lead to the postulation of the presence of agents whose supernatural abilities allow them to,
for example, disappear when more closely investigated (Barrett 2000).

Originally put forward in the context of discussing religious beliefs, the
cognitive byproduct account is coming to be seen as inadequate when dealing
with the complexities of religious traditions, leading to the proposal of dual-
inheritance accounts that combine it with approaches that treat religions as
prosocial cultural adaptations (Talmont-Kaminski 2009a, 2012, forthcoming;
Atran and Henrich 2010). The cognitive byproduct account is much more
successful, however, when it comes to such beliefs as superstitions and, indeed, pseudoscientific explanations. This is because these beliefs most often
have not been recruited to systematically serve any function and, therefore,
generally do not require consideration in terms of cultural adaptation.

In answer to his question—why do people believe the particular supernatural claims that they do accept?—Boyer and others who argue for the cognitive byproduct account draw attention to the properties that supernatural
claims the world over tend to share. McCauley (2010) focuses on two. The
first is that supernatural beliefs usually give agency a much more fundamental
role in the functioning of the universe than do scientific explanations:

Scientific abstemiousness concerning intentional agents and their putative actions is to be contrasted with religions' pervasive recruitment of theory of mind and appeals to agent explanations. (McCauley 2010, 253)

While McCauley makes the point regarding religions, it is as true of supernatural claims in general. The obvious example of this difference is the contrast between evolutionary theory and creationism—while evolutionary explanations are based on the processes of blind selection working over millions of years, creationist explanations fundamentally rely on the postulation of purposeful actions undertaken by a supernatural agent. It is highly instructive to consider pseudoscientific beliefs that offer explanations that in some way compete with the evolutionary and creationist accounts. These "paleocontact" accounts typically involve stories of extraterrestrial species influencing the development of life on Earth. Thus, for example, Zecharia Sitchin, in his *The 12th Planet*, wrote about a species from the planet Nibiru that genetically engineered humans to work as slave labor for them. Or, to give another example, Erich von Däniken claimed in *The Chariots of the Gods?* that extraterrestrials constructed many (if not all) of the great prehistoric structures on Earth such as Stonehenge, the Easter Island statues, and the drawings in the Nazca Desert in Peru, in the process greatly influencing the development of human culture. Similar to supernatural explanations and in contrast to scientific explanations, these pseudoscientific accounts tend to fundamentally rely on the actions of (extraterrestrial) agents. Quite relevantly, many of those proposing paleocontact scenarios claim that world religions trace back to contact with such extraterrestrial agents—another example of an agent-based pseudoscientific explanation where the scientific approach taken in this chapter (as well as in other research in cognitive science of religion) is to look to evolutionary and cognitive explanations.

It would be incorrect to claim that all pseudoscientific claims lend agency such a central role. One significant exception is Immanuel Velikovsky's catastrophist pseudohistory presented in *Worlds in Collision* and other books. The past popularity of Velikovsky's account probably cannot be explained in the same cognitive terms as those applied to many other pseudoscientific claims. The book appears to be significantly motivated by interest in showing that biblical accounts of plagues were historically accurate, at least to some degree. It may seem that the most that can be said is that there is a strong tendency for pseudoscientific claims to be alike to supernatural claims in that they place agency at the center of the picture of reality they propose. We will see, however, that it is ultimately possible to formulate a stronger claim concerning pseudoscientific beliefs.

The second property that McCauley considers is the degree to which pseudoscientific accounts fit with people's intuitive ontology. As he goes on to explain (2010, 245), mundane supernatural claims are only modestly counterintuitive (see also Boyer 2001), making it easy for people to make inferences using them; scientific claims, however, typically run radically counter to what people generally expect and require extensive reflective reasoning to be understood and appreciated. Thus, ghosts may not have a physical body but, nonetheless, are believed to be comprehensible in terms of typical belief-desire folk psychology. It is modern neuropsychology that actually presents a picture that differs in much more fundamental ways from folk psychology.

Just as in the case of reliance on agent-based explanations, pseudoscientific accounts tend to have much more in common with supernatural accounts than with scientific ones. The point is clear when we consider the extraterrestrial agents central to the pseudoscientific accounts discussed above. These aliens are represented as having extraordinary abilities that allow them to fundamentally alter the course of the development of human life, much like the gods and spirits of many religious traditions. These abilities might not necessarily be impossible from the point of view of science, but they are counterintuitive from the point of view of common sense—the attraction of von Däniken's idea of alien architects being that it is hard to imagine that prescientific peoples were capable of such feats as building the Egyptian pyramids, seemingly necessitating the postulation of counterintuitive agents. At the same time, the decisions made by the extraterrestrials are typically explained in terms of commonsense notions of beliefs and desires—when Sitchin writes about the aliens creating humans, the factor motivating their actions appears to be nothing more than the desire to avoid physical labor. Again, this is much as was the case with gods and spirits.

Similarly to Velikovsky's avoidance of calls for agency, however, there are examples of pseudoscientific belief systems that have gone a long way toward requiring extensive reflective reasoning. The basics of astrology are only modestly counterintuitive, with the thought that the heavens should reflect human events holding significant intuitive attraction for human reasoners. On this basis, however, professional astrologers have built up an extensive pseudoscience that calls for complex calculations in order to construct horoscopes—professional astrology standing in something like the relation to popular astrology as theology in relation to popular religion. The radically counterintuitive aspect of this practice comes to the fore in their justification for their ability to formulate horoscopes that connect the actions of people, whom they hold to be able to exercise their free will, with the predetermined movement of the planets. In particular, the astrologers hold that the connection between the two is not to be understood as causal—which leaves the question of how the proposed connection is maintained if humans have free will. The overall picture is every bit as radically counterintuitive as the Calvinist doctrine of predestination. Popular belief in astrology is maintained only because there is no need to learn and agree with the abstruse pseudophilosophical claims of professional astrologers in order to have a conception of the significance of astrological predictions.

The deep similarities between supernatural beliefs and the majority of pseudoscientific claims are instructive in light of the surface similarities between pseudoscientific claims and those put forward by science. While eschewing reference to ghosts or other traditionally supernatural entities and relying on entities that prima facie fit with scientific knowledge, most pseudoscientific claims exhibit a profound similarity with supernatural claims. In the context of a cognitive byproduct account, the fundamental similarities between supernatural and pseudoscientific beliefs invite the conclusion that both kinds of beliefs are the byproduct of the same cognitive mechanisms: the human predilection for taking the intentional stance in the case of the preference for agent-based explanations and the relative ease of using modestly counterintuitive representations. Indeed, the chief similarities between these kinds of beliefs suggest that pseudoscientific beliefs (or, at least, some of them) perhaps ought to be thought of as a subset of supernatural beliefs—a conclusion that those who put forward such beliefs would probably find less than comforting.

Further evidence for the fundamental connection between supernatural and pseudoscientific beliefs is provided by examples of religions based on pseudoscientific beliefs, the most infamous being Ron Hubbard's Scientology. The claim that many of the modern world's problems can be traced back to the genocide of billions of individuals millions of years ago by the Galactic Confederacy is a narrative pulled straight from the pages of second-rate science fiction, yet it has all the hallmarks of typical supernatural accounts that McCauley considers. The Galactic dictator Xenu and the spirits of the murdered extraterrestrials are at the center of the narrative, ensuring that it has both the property of focusing on agency and of postulating modestly counterintuitive entities. Indeed, Hubbard's account presents the ancient extraterrestrial civilization as very similar in numerous respects to that of 1960s' America, making it particularly easy for people to make inferences about the agents he postulates. It should be noted that, even so, the Scientologist "truth" is revealed only to individuals who have already made a very significant commitment to Scientology and, therefore, are motivated to accept the story of Xenu. It is only known more generally thanks to ex-Scientologists who have been willing to reveal this particular secret to the broader, and much more skeptical, public.

The cause of the similarity between supernatural and pseudoscientific beliefs can be understood as analogous to the reason for the similarity between animals from radically different lineages that have come to occupy the same environmental niche. Thus, for example, ichthyosaurs, which were an ocean-dwelling species of dinosaur, looked very similar to tuna as well as to dolphins—the similarities between them being explained by their need to make their way through water. In the case of supernatural and pseudoscientific beliefs, the niche occupied by them is created by the idiosyncratic nature of human cognitive systems that leads to cognitive byproducts that have particular characteristics and that appear reliably across a wide range of conditions that humans find themselves in. The entities that populate pseudoscientific accounts may have their origin in science, but they have undergone significant change to fit the supernatural niche, with the result that they have come to look a lot like supernatural beliefs, even if the latter draw their content from very different cultural reference points.

Looking Back to Intuitions

The central point of McCauley's discussion of the difference between religion and science is that in one fundamental respect science is a lot more like theology than it is like popular religion, while popular religion is a lot more like commonsense beliefs in this respect. The difference is that religious and commonsense beliefs are produced by what McCauley (2010, 2011) calls "maturationally natural" cognitive systems, while theological and scientific claims require a great degree of further intellectual development and reflection because of their radically counterintuitive content. This difference, along with the degree to which the particular claims tend to squander agent-based explanations, allows McCauley to plot science, religion, theology, and commonsense beliefs on a simple two-by-two table.

It is particularly enlightening to consider what adding pseudoscience to the table tells us. Given the points that have been made previously, the obvious pigeon hole for a lot of pseudoscience is with popular religion and other supernatural claims. However, as has already been pointed out, not all pseudoscientific beliefs fit into that place in the table. While Scientology does share the traits of traditional religions, professional astrology is much more akin to theology, and Velikovsky's stories might even have to be put into the same pigeon hole as properly scientific accounts.[1] Without denying the significance of the cognitive approach to understanding all these phenomena, it does show that in so far as we wish to understand why—for example—pseudoscientific claims should not have the same epistemic status as scientific ones, we do have to go beyond the cognitive basis of these beliefs. McCauley, quite clearly, agrees with this assessment given the lengths he goes to explain the different epistemic status of theology and science. While a fuller consideration of these issues will have to wait until the final section of this chapter, it is instructive to point out that the two traits of scientific explanations that McCauley identifies are implicit in the naturalist stance that science is normally seen as adopting with regard to ontological claims. In particular, the cognitive approach that both McCauley and I are pursuing is probably most in line with provisory methodological naturalism (Boudry, Blancke, and Braeckman 2010)—a view that sees the basic scientific naturalist commitments as the fallibilist product of a long process of scientific reflection. Science does not presume that agency does not play a central role in how the universe functions—this is just something that science has discovered over time, despite the degree to which this thought runs counter to what people naturally assume. This interplay between normative epistemic considerations and the cognitive picture is exactly the optimal approach suggested at the beginning of this chapter.

Introducing pseudoscience into McCauley's table makes clear a further point. It breaks up the neat symmetry McCauley might be thought to have set up between the left- and right-hand sides of the table. As things stand, it might seem that while theology is the product of reflection on people's maturationally natural religious beliefs, science gets its start from a type of reflection on commonsense explanations.

Of course, as McCauley makes clear, even here the symmetry is not perfect. While science soon breaks free of commonsense beliefs; "[t]heology, like Lot's wife, cannot avoid the persistent temptation to look back—in the case of theology to look back to popular religious forms." (McCauley 2011, 228). Unlike science, which has its own justification, theology gets its raison d'être from the existence of popular religion.

At the same time, neither popular religion nor commonsense beliefs necessarily owe much to the more reflective sets of practices. Thus, popular religion pays little heed to theology as revealed by research into theological incorrectness (Slone 2004). Theists may be able to reproduce theologically correct dogma when explicitly required to, but they seem to operate with much simpler and less counterintuitive supernatural beliefs than those condoned by theology. Of course, it would be possible to talk about something quite similar—a scientific incorrectness, perhaps—in the case of popular understanding of phenomena that science has explained. The obvious example is that even though in many societies the majority of people will claim that they believe in Darwinian evolution, most of them would not be able to characterize it even in broadest terms. Instead, many would produce something more akin to a Lamarckian account, which gives the endeavors of individual agents a much more central place. According to Lamark those endeavors directly lead to changes in the next generation, rather than affecting it indirectly and in a limited fashion through changes in the incidence of particular genes.

Pseudoscientific claims present us with an interesting addition to this picture in that, at least on the outside, they wear conceptual cloth originally spun by the scientists. Many of the basic concepts necessary to express the idea of extraterrestrial agents traveling to Earth millions of years ago to influence the progress of evolution, for example, were originally made meaningful in the context of scientific research—even though some of them never rose above the level of conceptual possibilities there. Yet, as already discussed, the similarities between pseudoscientific and scientific explanations are mostly skin deep.

It seems that scientific concepts surprisingly easily devolve into pseudoscientific concepts given the right conditions. Many scientists whose work came to be referred to in newspapers or popular magazines have painful personal experiences of this process. Quantum physics is one area of science that has become infamous for the numerous pseudoscientific interpretations it has given rise to, with the likes of Deepak Chopra popularizing claims that actually have little in common with the original science. Indeed, the radically counterintuitive nature of scientific concepts might render them particularly suitable for pseudoscientific misunderstanding since it makes understanding them correctly so difficult. It seems that whereas theological beliefs are what results from reflection on popular religious beliefs, pseudoscientific beliefs are what one gets when scientific beliefs are allowed to erode away from the lack of necessary reflection.

Recognizing this difference between theology and science leads to two further points. The first is just how fragile science is from a cognitive point of view. While McCauley makes this observation, considering the example of pseudoscience emphasizes that it is not just that science requires social institutions to continue developing, but that it probably requires them for scientific beliefs not to devolve in the pseudoscientific ones that are just so much more natural for humans. Even with the existence of numerous research institutions and universal education, public understanding of scientific claims regularly has more in common with pseudoscience. Without such institutions it seems unlikely that scientific concepts could survive for long.

The second point is that scientific beliefs do not necessarily provide the best example of a contrast class to theology. As has already been observed, science owes little to commonsense beliefs. In this, it is unlike theology, which relies on the beliefs of popular religion in two regards: by retaining them as the subject of its reflection, as well as by having its motivation depend on people's commitment to them. Much more similar to theology in both these regards is traditional philosophy in its relationship to commonsense beliefs. Intuitions play a vital role in both regards when it comes to traditional philosophy. They provide the raw material that philosophy attempts to analyze rationally through careful reflection, and typically act as the ultimate justification of the views that traditional philosophers have proposed. As we will see, these similarities between philosophy and theology point to a very important difference between science and theology.

It should be noted that naturalized philosophy has a very different relationship to commonsense beliefs. Similarly to science, it sees no advantage in referring back to the beliefs that are intuitive to human cognizers. Breaking free of commonsense intuitions is justified in part by the extensive evidence that science has provided for the shortcomings of commonsense beliefs in general and intuitions in particular (Nisbett and Ross 1980), the specific implications for philosophical methodology having been explored by Bishop and Trout (2005). Instead of looking back to commonsense beliefs, naturalist philosophy takes scientific claims as the reference point for the further reflection it engages in.

Conclusion

Having drawn out the implications that considering the cognitive basis of pseudoscientific beliefs leads to, it is time to show how they help to come to grips with the issues this chapter opened with:

- Why are pseudoscientific (as well as supernatural) beliefs so hard to eliminate?
- What is the difference between supernatural and pseudoscientific beliefs?
- What is the difference between pseudoscientific and scientific beliefs?

The easiest to deal with is the question of why it is that pseudoscientific beliefs are so difficult to counter. Many of them rely on the same cognitive byproducts that lend plausibility to supernatural beliefs. This means that they will likely remain attractive so long as human cognitive systems produce those byproducts. Without needing to buy into the whole of memetics, it can be seen that human cognitive systems provide these kinds of beliefs with a ready environment in which to prosper. Eliminating individual pseudoscientific beliefs is only likely to allow others to take their place. The history of superstitions provides some evidence for this claim. Folklorists conclude that individual superstitions tend to remain popular for a limited amount of time measured in decades rather than centuries. However, as old superstitions disappear, new ones tend to fill their place (Roud 2006). Similarly, getting rid of individual pseudoscientific beliefs is only likely to lead to new ones becoming popular, much in the same way that eliminating certain species often only leads to other species quickly invading that particular environmental niche.

As we saw, the difference between pseudoscientific and supernatural beliefs is, for the most part, little more than skin deep. The supernatural beliefs may get their content from commonsense beliefs while the pseudoscientific beliefs are usually dressed up in scientific garb. This does not substantially alter how they interact with human cognitive systems, however. All that it may do is render pseudoscientific beliefs somewhat more attractive in the context of modern cultures that hold scientific knowledge in great regard but have limited actual understanding of it—cultural systems constrain what concepts can be acquired (Sørensen 2004).

Having said that, it does appear that pseudoscientific beliefs may interact successfully with human cognitive systems in a greater variety of ways than those used by supernatural beliefs. It is not clear to what degree this is just a matter of the ways those different sets of beliefs are classified—the difference between popular religion and theology is stressed a lot more than that between popular and professional astrology. Even so, it might be possible to find clear examples of pseudoscientific beliefs that avoid excessive reference to agent-based explanations and run profoundly counter to human intuitions, thereby sharing those traits with properly scientific claims—Velikovsky's views being a potential example I earlier suggested. The question with such pseudoscientific beliefs is how it is that they hold sufficient attraction to remain viable, given that they cannot straightforwardly rely upon the cognitive byproducts that supernatural beliefs find support in, while lacking the kind of support that properly scientific claims have. This suggests a way of thinking about pseudoscientific beliefs that is based on the cognitive picture we have been examining and develops out of the analysis provided by McCauley.

Pseudoscientific beliefs need not all involve moderately counterintuitive agents, but many do; and those that do not draw their strength from that well. While both Velikovsky's claims and professional astrology fall outside of the box typically occupied by supernatural and pseudoscientific beliefs, they both call on the beliefs that are found in that box to find sufficient motivation to make them attractive. In the case of Velikovsky, the motivation is to provide existing religious beliefs with pseudoscientific interpretations that are attractive in modern culture. In the case of professional astrology, it is to reconstruct naïve astrological beliefs in a more logical fashion. This is much the same kind of relationship as that between theology and popular religion in so far as theology would hold little or no interest were it not for the cognitive attraction of the popular religious views. But, as has been already noted several times, science does not seek its motivation in unreflective beliefs, be they commonsense or supernatural. Instead, it finds justification in the way it ties its claims to empirical evidence, which is intended to be independent of the idiosyncrasies of human commonsense beliefs. By divorcing itself from those beliefs, science is therefore different even from the potential pseudoscientific accounts that might fall inside the same box as it does on McCauley's table. Pseudoscientific beliefs find their motivation, directly or indirectly, in the cognitive byproducts that human cognitive systems produce—science looks to standards of evidence that are significantly different from those that come intuitively to humans. It is in this, ultimately, that the difference between the two lies.

This chapter started with the suggestion that focusing on the content of pseudoscientific as opposed to scientific beliefs would not lead to a deep understanding of the difference between them. The alternative pursued here focuses on the cognitive basis for pseudoscientific as opposed to scientific reasoning. This has revealed that while there are significant dissimilarities between the content of the two kinds of beliefs science and pseudoscience produce, the reason for the difference between them is ultimately found in the disparate attitudes they take in relation to human maturationally natural beliefs. The dissimilarities between pseudoscientific and scientific beliefs reflect that deeper and more profound difference.

Drawing the difference between science and pseudoscience in this way is important in that it helps to bring out very clearly what is special about science. This is particularly vital given that any cognitively informed account of science must recognize that the cognitive basis for science is the same as that for commonsense beliefs as well as supernatural or pseudoscientific beliefs. Human cognitive mechanisms underpin all these phenomena, the differences between them lying in the details of how those mechanisms are used in each case. Science is already hobbled in that it is undertaken by boundedly rational humans. To build on this basis, rather than be trapped by it, it must be free to explore conceptions that are not maturationally natural to our minds.

Acknowledgments

Science is a cooperative endeavor. Some of this is indicated through citations. However, many intellectual exchanges are informal. The list of people who have influenced this chapter's content in one way or another is very long, and I trust I will be forgiven if I mention only a few names here. Massimo Pigliucci and Maarten Boudry, the editors of this volume, gave me numerous insightful comments on the first draft. I am also grateful to the many International Association for the Cognitive Science of Religion members at the 2011 IACSR meeting in Boston with whom I discussed ideas that made their way into this chapter. Finally, Bob McCauley good-naturedly assisted me at various stages of work on this chapter, and I am very much in debt to him.

NOTE

1. While not making extensive use of folk psychology, one highly relevant respect in which Velikovsky's story is tied to commonsense beliefs is in its use of folk physics. This point was suggested to me by Maarten Boudry and would place Velikovsky's account in the easy-to-think but agency-eschewing corner of McCauley's figure.

REFERENCES

Atran, Scott, and J. Henrich. 2010. "The Evolution of Religion: How Cognitive By-products, Adaptive Learning Heuristics, Ritual Displays, and Group Competition Generate Deep Commitments to Prosocial Religions." *Biological Theory* 5 (1): 18–30.

Barrett, Justin L. 2000. "Exploring the Natural Foundations of Religion." *Trends in Cognitive Sciences* 4 (1): 29–34.

Bishop, Michael A., and J. D. Trout. 2005. *Epistemology and the Psychology of Human Judgment*. New York: Oxford University Press.

Boudry, Maarten, Stefaan Blancke, and Johan Braeckman. 2010. "How Not to Attack Intelligent Design Creationism: Philosophical Misconceptions About Methodological Naturalism." *Foundations of Science* 15 (3): 227–44. doi:10.1007/s10699–010–9178–7.

Boyer, Pascal. 2001. *Religion Explained: The Evolutionary Origins of Religious Thought*. New York: Basic Books.

Bulbulia, Joseph. 2004. "The Cognitive and Evolutionary Psychology of Religion." *Biology & Philosophy* 19 (5): 655–86. doi:10.1007/s10539–005–5568–6.

Gigerenzer, Gerd. 2000. *Adaptive Thinking: Rationality in the Real World*. Oxford: Oxford University Press.

Guthrie, Stewart. 1993. *Faces in the Clouds*. New York: Oxford University Press.

Haselton, Martie G., and Daniel Nettle. 2006. "The Paranoid Optimist: An Integrative Evolutionary Model of Cognitive Biases." *Personality and Social Psychology Review* 10 (1): 47–66. doi:10.1207/s15327957pspr1001_3.

Hempel, C. 1969. "Reduction: Ontological and Linguistic Facets." In *Philosophy, Science, and Method: Essays in Honor of Ernest Nagel*, edited by Sidney Morgenbesser, Patrick Suppes, and Morton White, 179–99. New York: St. Martin's Press.

McCauley, Robert N. 2010. "How Science and Religion Are More Like Theology and Commonsense Explanations than They Are Like Each Other: A Cognitive Account." In *Chasing Down Religion: In the Sights of History and Cognitive Science*, edited by P. Pachis and D. Wiebe, 242–65. Thessaloniki: Barbounakis.

———. 2011. *Why Religion Is Natural and Science Is Not*. New York: Oxford University Press.

Nisbett, Richard, and Lee Ross. 1980. *Human Inference: Strategies and Shortcomings of Social Judgment*. Englewood Cliffs, NJ: Prentice-Hall.

Roud, Steve. 2006. *The Penguin Guide to the Superstitions of Britain and Ireland*. London: Penguin.

Simon, Herbert. 1955. "A Behavioral Model of Rational Choice." *Quarterly Journal of Economics* 69:99–188.

Sitchin, Zecharia. 1976. *The 12th Planet*. New York: Stein and Day.

Slone, D. Jason. 2004. *Theological Incorrectness: Why Religious People Believe What They Shouldn't*. Oxford: Oxford University Press.

Sørensen, Jesper. 2004. 'Religion, Evolution, and an Immunology of Cultural Systems." *Evolution and Cognition* 10 (1): 61–73.

Talmont-Kaminski, Konrad. 2009a. "Effective Untestability and Bounded Rationality Help in Seeing Religion as Adaptive Misbelief." *Behavioral and Brain Sciences* 32 (6): 536–37.

———. 2009b. "The Fixation of Superstitious Beliefs." *teorema* 28:1–15.

———. 2012. *In a Mirror, Darkly: How the Supernatural Reflects Rationality*. Lublin: UMCS Press.

———. Forthcoming. *Religion as Magical Ideology: How the Supernatural Reflects Rationality*. Durham: Acumen.

Velikovsky, Immanuel. 1950. *Worlds in Collision*. London: Macmillan.

Von Däniken, Erich. 1969. *Chariots of the Gods?* London: Souvenir Press.

Wimsatt, William. 2007. *Re-engineering Philosophy for Limited Beings: Piecewise Approximations to Reality*. Cambridge, MA: Harvard University Press.

21

The Salem Region

Two Mindsets about Science

JOHN S. WILKINS

People believe silly things and known falsehoods for all kinds of reasons, ranging from cognitive deficits, groupthink, stereotyping, and cognitive dissonance. The issue I address here is not why people believe things that are false; other researchers have already advanced many causes and hypotheses for that (see, e.g., Peirce 1877; Shermer 1997). Instead, it is this: why do educated people who thoroughly understand their own scientific or technological fields continue to adopt positions that are contrary to our best science?

PhDs in science-related fields who promote antiscience-establishment, antiscience, or pseudoscience agendas are very often engineers, dentists, surgeons, or medical practitioners. While this does not mean that all members of these professions or disciplines are antiscience, of course, the higher frequency of pseudoscientific belief among them indicates what I call the "deductivist mindset" regarding science itself. Opposing this is the "inductivist mindset," a view that philosophers since Karl Popper have deprecated. Roughly, a deductivist tends to see problems as questions to be resolved by deduction from known theory or principle, while the inductivist sees problems as questions to be resolved by discovery. Those who tend toward a deductivist mindset may find results that conflict with prior theoretical commitments unacceptable. The deductivist tends to be a cognitive conservative, and the inductivist a cognitive progressive. The conservative mindset more often

leads to resentment about modernism and hence about certain scientific results, or so I argue in this chapter.

Highly educated and trained scientists are no more immune to the effect
of their cognitive dispositions than are other experts. Being at one end of
the reasoning spectrum or the other—deductivist pole or inductivist pole—
disposes even educated people to reject some scientific conclusions as false
and accept false propositions as true, despite the evidence for and success
of accepted claims within science. To some extent, both poles represent legitimate approaches within real science. But it may be that some people are
inclined, by virtue of their modes of reasoning, to remain at one extreme—the
conservative or the radical—rather than to range along the spectrum of available scientific views.

Why Are People Opposed to Science?

Science is the process of learning how the world works. It would seem that
this is something all reasonable people should approve of and take seriously.
As the eighteenth-century bishop Joseph Butler noted in a different context,
"Things and actions are what they are, and consequences of them will be what
they will be: why then should we desire to be deceived?" (Butler 1726) We
should, in principle, accept the results of our best science simply because we
have no alternative, nothing against which we can test science that has a better warrant.[1] There may be other sources of knowledge, such as revelation or
intuition or moral faculties, but these are not sources of knowledge about the
natural world. If there is a knowledge claim about biology, physics, or psychology, for example, the best and only authority on such matters is the best
science that we presently have. In what follows, I take as given that our best
science is our best knowledge of nature.

And yet, a large number of people—apparently rational and educated
people in other respects—do not accept the best science as authoritative.
They instead look for what they see as alternatives to scientific views. Why
is this? Why do reasonable people often refuse to accept that, for example,
climate change is caused by human power generation and industry, that vaccination is a cheap and relatively safe medical prophylactic that has no causal
connection with autism, and that the diversity of life is due to a process of
evolution that has occurred over millions, indeed, billions, of years? There are
a plethora of explanations on offer, ranging from deficits in human-evolved

cognitive psychology, poor education by scientists and educators, and manipulation of opinion by vested interests running "astroturf" campaigns (fake grassroots movements funded and run by the tobacco and chemical industries and more recently the oil and coal industries, see Oreskes and Conway 2010). To varying degrees, these are all plausible accounts in some respects, but even taken together they fail to explain why pseudoscience is as common a problem as it is, especially among the more industrialized and developed nations that pride themselves on their progress in education and public communication of science.

Some commentators presume that antiscience proponents are irrational or suffer from some cognitive deficit. Given that some proportion of the population will act irrationally or have cognitive deficits, which can be as simple as an inability to accurately estimate risks due to anxiety (Gasper and Clore 1998) or social factors (Johnson and Tversky 1983), it follows that this will often be true. Indeed, it follows that on some topics or concerns, we are all irrational. But if nobody manages to be completely rational because of deficits of this kind, rationality becomes either an unattainable idea that has no explanatory power or we must redefine rationality. We need to conceive of rationality as a humanly achievable state, with all the attendant fallibility and limitations that entails.

Rational action models have long been used to explain economic, social, and conceptual behaviors. While almost nobody now would suggest that the "rational actor" theory is fully explanatory, as a first approximation, a rationality model sets up a background against which we can identify deviations and deficits. And even the nature of rationality is something we can investigate if we presume very roughly that people act in their own interests. For example, recent work by Skyrms (2001) develops Jean-Jacques Rousseau's idea that Hobbesian self-interest can lead to different outcomes of cooperation in the "stag hunt" case, depending on the local conditions.[2] So assumptions of rationality have a utility independent of work done on irrational deficits. In fact, we may find that many phenomena we had thought not to be rational are a form of bounded rationality. I have argued previously that this is true of ordinary creationists, who are making boundedly rational decisions as to what beliefs to adopt based on limited information and reflective opportunities (Wilkins 2011). Because of the prior social heuristic that Gerd Gigerenzer and colleagues (1999) call "follow the good," it is boundedly rational to accept, *ceteris paribus,* what leading figures in your community believe, in part because in believing it, they are not dead yet. The general assumption that agents are rational actors need not presume they are fully rational. We can also distinguish between a "strong" rational actor model, in which every choice of belief is rational, and a "weak" rational actor model, in which only the choices based on the leading principles held by the reasoner are rational.

There are two main reasons to assume a weak rational actor model to begin our deliberations. One is the problem of confirmation bias, and the other is the problem of demonization. Confirmation bias is obvious: if we presume that those who adopt a certain position are suffering from a deficit of some kind, then every case in which we find someone of those views who does suffer a deficit (and there must be some) will be taken as confirmation of the presumption. But this has no more force than concluding that all brilliant scientists have massive corpus callosum connective brain tissue because Albert Einstein did.

The demonization of people holding unscientific beliefs is a bigger problem. First of all, it is contraindicated by personal experience: many people find that antiscience advocates are often intelligent, educated, and clever individuals in other respects. Moreover, demonizing too easily permits proscience advocates to wash their hands of their opponents by assigning them to a lost cause. This neither deals with the problem of irrationality nor has any long-term utility for the advancement of science in society. Finally, it is rude and uncivil. We should not assume people are stupid simply because we think their views are silly, especially if, as I have asserted, all of us have rational deficits of some kind or another (and that it is a mark of the self-unaware to assert that they do not). All of us should assume we have made mistakes, including silly ones, and seek to find them out.

So in what follows, I argue that we should presume as a first approximation that antiscience views result from rational decisions being made in contexts and with dispositions that lead to suboptimal outcomes. In some ways, this is rather like natural selection settling on suboptimal solutions (Wilkins 2008). Assuming that antiscience proponents are being rational, then their unscientific beliefs may be suboptimally trapped on cognitive developmental "peaks." If we find that bounded rationality explains the phenomena, that is excellent, for now we can employ the same rational dispositions to combat these suboptimal outcomes. If we find that it fails, then that too is a result worth knowing, and one that sets up the problem for further research to identify the actual irrational dispositions, rather than presuming them to exist.

Mindsets and Bounded Rationality

As in my previous work on creationism (Wilkins 2011), I appeal to the ideas of Herbert Simon and Gerd Gigerenzer and colleagues on bounded rationality (Simon 1972, 1981, 1986, 1997; Gigerenzer et al. 1999; Gigerenzer 2000; Gigerenzer and Selten 2001; Todd and Gigerenzer 2003). Here, "rationality" is bounded by several things: first we all act under uncertainty. Second, we all have limited time and resources to devote to reasoning. Third, we all have limited information. Each of these limits our ability to reason about the world. A major failure of "rational man" theory in economics was that it assumed no limits in any of these aspects of reasoning. Moreover, Gigerenzer's group has argued that there are a number of simple heuristics, presumably handed to us by evolution, that we use for social inferences, and by analogy we might expect there are also natural heuristics of the same kind.[3] What matters is that, as rational agents, we are bounded by circumstance and capacity.

This is not necessarily the consensus view in cognitive psychology. However, on these matters there is no consensus, so I am not committing, I think, a *petitio* if I take this to be consonant with my experience in dealing with antiscience advocates over the past quarter century. Moreover, Gigerenzer's and colleagues' approach is well founded with considerable evidentiary support. Alternative explanations include cognitive biases, cognitive illusions and heuristics, false belief and reasoning, counterfactual reasoning, functional neurobehavioral anatomy of false beliefs, delusion and confabulation, brain damage and developmental disorders, distorted memory, persuasion, and neurotheology (including mystical experiences and spiritual, religious, and psi beliefs).[4] Broadly speaking, false beliefs are given developmental dispositional explanations in which some deficit or failure to act normally is adduced. These include social psychological explanations, in which one's belief set is derived from social influences, and heuristic and logical explanations, in which prior heuristics or logical limitations cause invalid or unsound inferences. Examples of developmental deficit explanations include the pathological neuroanatomical accounts of delusions and confabulation; of social accounts arguments regarding tendencies to conform or seek to advance one's status by adhering to the dominant belief set; and of the heuristic approach, the views advanced by Gigerenzer and colleagues. Often, accounts are of more than one kind. There is no need to exclude other explanations *tout court*, but I believe that we tend to overlook that there is a *normal* variation on all metrics in a large population, and so at least some false belief will turn out to be normal. Some false belief may even act to drive further investigation and learning socially.

Antiscience, Pseudoscience, Contested Science

The term "antiscience" is not new (Ashby 1971). It referred originally to intellectual critiques of science by leftist and new age critics in the late 1960s and early 1970s (Laing 1969a, 1969b; Roszak 1969). Here I take it to be any view that sets itself against the best science of the day. There is a similar phenomenon, pseudoscience, in which a set of views that are not arrived at through scientific investigation are dressed up in scientific-appearing terminology, form, publications, and organizations. Intelligent Design (ID) is a case in point: no science has been done to arrive at the ID conclusion, and its mathematics and terminology are designed to mislead uninformed readers into thinking that the view has scientific merit. Similar phenomena and movements have appeared since the sciences first evolved. Astrology, homeopathy, theosophy, and even socialism (Engels 1892) have all been dressed up as science. There are extensive treatments of pseudoscience (Hines 1988; Aaseng 1994; Shermer 1997; Curd and Cover 1998; Bauer 2001; Frazier 2009; Pigliucci 2010; Smith 2010), although many are polemic rather than dispassionate treatments, understandably. Often, though, the term is used to dismiss alternative theories in a field such as psychology (Blum 1978; Lilienfeld et al. 2003; Lilienfeld et al. 2008), education (Warnick et al. 2010), or archeology (Feder 1990). Sometimes this is warranted, but other times it is a rhetorical ploy within the science.

Contestations of a theory or research program as "pseudoscience" often arise when there is limited or no consensus in the discipline, or when competing disciplines are addressing similar material or problems. It is common for scientists who object to a theory or research program, or even an entire discipline, to insult that project by calling it "pseudoscience." Similar claims were made about Charles Darwin's theories during his lifetime, and others such as Einstein, Stephen Jay Gould, and Sigmund Freud (see Cioffi, chapter 17, in this volume) have all received the label or some similar term like "nonscience," "unscientific," and the like. Debates over what counts as scientific in a discipline are common, particularly over methodological matters. For example, the taxonomy wars in biological systematics have focused on questions of methodology, philosophical approaches, and special techniques (Dupré 2001; Hull 2001; Will et al. 2005). Simply because something is called "antiscience" or "unscientific" does not imply that it actually is. There is a vague border between such accusations *within* the science and *between* science and nonscientific discussions. At one point in the history of climatic studies, it was feasible to think that global warming was an unscientific hypothesis, or that it was not human caused, without being, ipso facto, unscientific. However, that time has passed now, and to make the assertion contrary to all evidence and scientific modeling is to be antiscientific no matter what the qualifications of the speaker are (see Prothero, chapter 18, in this volume).

Educated Antiscience Advocates

Certain Internet forums devoted to discussing creation science and ID present a "hypothesis" known as the "Salem Hypothesis" (SH) (after Bruce Salem, who first mentioned it). It runs roughly like this:

> An education in the engineering disciplines forms a predisposition to [scientific creationist] viewpoints.[5]

The SH generated a lot of debate in these forums, with many engineers defending their profession by pointing out that they are hardly more likely to be creationists than people from any other discipline. The hypothesis evolved over time into weak and strong versions:

> Weak: In any evolution versus creation debate, a person who claims scientific credentials and sides with creation will most likely have an engineering degree.
> Strong: An education in the engineering disciplines forms a predisposition to creation/ID viewpoints.

The strong SH is difficult to defend. If there is a tendency for, say, creationists claiming scientific credentials to be engineers, it need not translate into any statistically significant difference in the levels of creationism among engineers in general, since populations can vary quite a lot without that signifying any deep difference. On the other hand, some professions *do* show a strong tendency toward antiscience. For example, a study done at Monash University in Australia, at one of the biggest medical schools in that country, showed that roughly 40 percent of medical freshmen were creationists; what is more, six years of medical study did not change that proportion significantly (Short 1994)! A similar study of evolutionary biology students at Capetown University showed the same result (Chinsamy and Plagányi 2008).

Many explanations of this phenomenon have been given. It represents an apparent paradox that education does not shift false ideas. We tend to think that education is just about changing false ideas and beliefs into true or warranted ones. Yet many educated people believe things that are simply false and are often highly resistant to correction (Kahan, Jenkins-Smith, and Braman 2011). Why? Some accounts assert the cultural contingency of beliefs. American exceptionalism and fundamentalism is one explanation given, for example, in the citations offered by Chinsamy and Plagányi (2008), as well as the claim that students are inadequately prepared for their tertiary studies by prior education. However, the widespread cross-cultural nature of these results indicates that such explanations are probably not sufficient. For example, Australian education teaches evolution at secondary school very well indeed (see, for example, the excellent textbook Huxley and Walter 2005), and while fundamentalism is not nearly as ubiquitous in Australian society as it is in the US "Bible Belt," it is still common in Australian polity and social makeup. Sociological explanations may partly account for the spread of fundamentalism, but they cannot be the whole story, or else graduates would still consider geocentrism and the miasma theory of disease to be true. Granted, I have encountered a Cambridge philosophy PhD who insisted that disease was caused by moral failure rather than germs (due to his neo-Platonist views), but that is hardly representative even of Cambridge.

The weak SH, on the other hand, is a statement about the reference class of educated people who believe things contrary to their education (in the original, engineers believing in creationism; but we can generalize). Why do educated people believe antiscience when they have been taught science or their field relies on the veracity and reliability of science? The following letter to an electrical engineering professional newsletter is an exemplar of the weak SH:

> Naturalistic evolution is the antithesis to engineering. Engineers understand that complex structures are intelligently designed, not the product of random variations. Engineers should be the first to recognize that a highly complex optimized structure, like the human eye or ear (not to mention the intricacies of individual cells), is not likely the result of mere time + chance + natural selection.[6]

This style of reasoning is not uncommon in the author's experience. It is that form of inference presented by Cleanthes (pt. 1) and then rebutted by David Hume's spokesman, Philo, in the *Dialogues Concerning Natural Religion* (pt. 5):

> [Cleanthes:] The curious adapting of means to ends, throughout all nature, resembles exactly, though it much exceeds, the productions of human contrivance; of human designs, thought, wisdom, and intelligence. Since, therefore, the effects resemble each other, we are led to infer, by all the rules of analogy, that the causes also resemble; and that the Author of Nature is somewhat similar to the mind of man, though possessed of much larger faculties, proportioned to the grandeur of the work which he has executed. By this argument a posteriori, and by this argument alone, do we prove at once the existence of a Deity, and his similarity to human mind and intelligence.
>
> [Philo:] Were this world ever so perfect a production, it must still remain uncertain, whether all the excellences of the work can justly be ascribed to the workman. If we survey a ship, what an exalted idea must we form of the ingenuity of the carpenter who framed so complicated, useful, and beautiful a machine? And what surprise must we feel, when we find him a stupid mechanic, who imitated others, and copied an art, which, through a long succession of ages, after multiplied trials, mistakes, corrections, deliberations, and controversies, had been gradually improving? Many worlds might have been botched and bungled, throughout an eternity, ere this system was struck out; much labour lost, many fruitless trials made; and a slow, but continued improvement carried on during infinite ages in the art of world-making.

Cleanthes represents the design-first inferential style that Philo critiques on behalf of Hume and much modern philosophy since. It is, however, clear that the tendency to argue by analogy from human mentation and dispositions to the physical world is an old one (arguably one that goes back to Socrates, according to Sedley 2007), and it bespeaks a psychological and cognitive disposition, often in that context titled "anthropomorphism" or "design stance" or "teleological reasoning." That such dispositions exist is not controversial. That they are not overcome among the educated, even in fields in which the best scientific theories have disposed of them, is intriguing.

The weak SH seems to generalize outside evolutionary biology as well.

We find biochemists and virologists who reject the pathogenic causes of disease. We find educated geologists and statisticians who dispute anthropogenic global warming. It appears that there are some mindsets, ways of belief formation, that occasionally supersede and transcend epistemic commitments, and that lead to beliefs that are critical of scientific knowledge for reasons other than the merely scientific, whether the affected professionals themselves realize it or not.

Modes of Thought

To explain this tendency of contrariness for extrascientific reasons, I propose that we consider people's belief formation as the end result of distributions of cognitive dispositions along several axes. That is to say, every population of cognizers, including the educated, tends to be arrayed along a distribution curve for each independent aspect of cognition. There is a similarity here with treatments on authoritarianism and conservatism in political psychology (Feldman and Stenner 1997; although see Martin 2001 for a cautionary discussion; Jost 2003). Jost and colleagues discern several variables, in decreasing order of significance: death anxiety; dogmatism-intolerance of ambiguity; openness to experience; uncertainty tolerance; needs for order, structure, and closure; integrative complexity; fear of threat and loss; and self-esteem. But in the case of reasoning about science, I conjecture that the variables are more directly epistemic.

If we conceive of the conceptual space in which inferential styles may be located as a phase space of n dimensions, my hypothesis is that several such dimensions stand out as likely important, in particular: essentialist thinking, resistance to novelty, deductive bias, and authority bias. Let us consider these in turn. If they are variables, then they must have polar contrasts, so we can get an idea of how they might operate by examining these variables (fig. 21.1).

Essentialist thinking involves setting up one's reference classes by taking a singular definition of a class or kind and adopting a binary inclusion-exclusion approach to phenomena. The contrast to this is exemplary thinking, which means taking an exemplary case or specimen and aggregating phenomena around it. Exemplary thinking is akin to what Wittgenstein (1968) called "family resemblance," or as it is regarded in biological systematics, clustering, although it was first described by William Whewell (1840).

The dimension of resistance to novelty denotes the individual's disposition to adopt novel ideas from the surrounding culture, including the culture

Modes of belief formation: a phase space

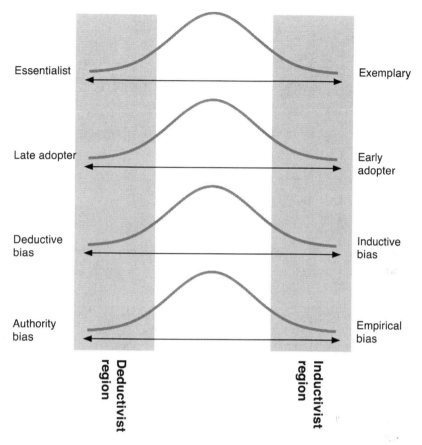

Figure 21.1

of a scientific or technical discipline. In simple terms, it measures whether an individual is an early adopter, a late adopter, or a modal adopter. Late adopters tend to resist novel ideas that they were not, in general, enculturated into when they developed their professional belief set. We might think of this as the "undergraduate effect": what one learns first tends to be more deeply entrenched in one's overall belief development, much as an earlier developmental process in biology affects the downstream phenotype in ways that are hard, if not impossible, to reverse. Many scientists assert as statements of faith things they learned in their freshman year and that they have not since needed to revise in the light of empirical experience. The more closely related to their own specialty a belief is, however, the less doctrinaire they are.

Deductive bias is my name for a tendency to treat scientific inference as deduction from axioms of "theory" or "what science knows." There is a historical reason for this in many cases—some people take Popperian critical rationalism as a foundation for the practice of their science—but just as many people have never heard of Popper and yet they find deductive reasoning agreeable. The polar contrast is of course inductive thinking, which among many philosophers is unnecessarily deprecated (see below). Deductive thinkers tend not to reason ampliatively, and so they find such reasoning distasteful and suspect.

Finally, authority bias is the degree to which an individual might defer to authority even when she is familiar with information and evidence that would contradict authoritative claims. For example, a creationist might defer to the authority of a religious leader or text despite working in the science that undercuts that source's teachings. A classic example is the Australian geologist Andrew Snelling. Despite working on mining and petrogeology that covers millions of years, and publishing work on that basis, Snelling nevertheless asserts that the earth is only a few thousand years old, as (his interpretation of) the Bible states (Ritchie 1991).

These four (probably) independent variables form a phase space in which we may identify regions representing different types of cognitive styles, or mindsets. One region will tend to be the tail of these four distributions toward the lower bound: someone whose dispositions are typically essentialistic, conservative, deductive, and authority deferent. Another region will be the other end of the distributions: someone who is exemplar-based in his taxonomies, novelty seeking, inductive, and evidence-based in his belief choices. And there will be a field of possible states of all the combinations of distributions. I am particularly interested in the first of these two regions, which I call the Salem Region.

A Salem Region occupier will tend to rely on prior knowledge and accept the truth of science at some particular time, most likely the time of his education (or rather, the time at which the textbook writers learned the science since textbooks and teaching tend to lag behind the cutting edge of science). He will tend to rely on authorities both within and without science, and will resist anything that challenges the consensus or cultural verities accepted by his community, for values of "community" that might include social, religious, and political as well as professional ones. He will permit external influences to override the consensus of the professional science to which he appeals or applies his values.

Examples of Salem Regionism include, most obviously (since I am basing my account on them) creationists and ID proponents, but also global warming skeptics, antivaccinationists, opponents of scientific medicine, and various conspiracy theorists. The latter reject scientific and ancillary disciplinary explanations of events as being done by some secret group that is feared, such as the 9/11 "Truther" movement, John F. Kennedy assassination conspiracies, and the like. No amount of evidence shifts these people's beliefs, even when they are regarded as technically informed and have good reputations in other fields, such as medicine, engineering, or even philosophy of science.

The Salem Region leads to a distrust of novelty; this in turn can lead to stances that are like "frozen accidents."[7] For example, initial distrust of vaccination at the time of Edward Jenner led to a tradition of opposing vaccination in the United Kingdom (Alfred Russel Wallace was such an opponent). Once a tradition like this is in play, it can be almost impossible to eliminate it long after the time for any reasonable opposition has passed. People who are in the Salem Region will then tend to follow authorities of their community (here, the community of antivaccination, not the community of medical science), maintaining and even extending the reach of the antiscience. While it is untrue that the Planck Principle, that theories die with their proponents, applies (Levin et al. 1995), the inverse is equally unfortunately true: old ideas are readily and stubbornly passed on to progeny and confederates. The lack of receptiveness to novelty of the Salem Region will ensure that some people will continue to resist modern ideas.

Resistance to Modernism

And this is the crucial matter: modernism. There will always be a number of individuals in a population who are relatively more conservative, deductivist, and so forth, no matter what the educational status of the reference class is, and who will therefore fear any novelty or change from traditional views; they will fear the modern. While education does seem to ameliorate dislike of the modern, it does not eliminate it (Lindeman et al. 2011). Since the 1960s, and earlier in the European traditions, the modern has been attacked by intellectuals, for reasons ranging from the justified to the absurd. There have been attacks on medicine, psychiatry, physics, neurology, and even germ theory, by otherwise educated individuals acting to "correct" modernist biases. "Mod-

ernism" is, of course, so protean a notion as to be almost meaningless. It usually means some aspect of the recent culture that the opponents dislike. And
one would be seriously misled if it were not accepted that people using the
label, or something like it ("scientific progress" or "enlightened thinking"),
have promoted social policies in immoral and often inhumane ways (eugenics
being the most obvious). However, a large part of the resistance to science has
to do with a disposition that mistrusts the modern no matter what it is or how
it has played out, a *ressentiment* of the modern, to appropriate the Nietzschean
term. Opposition to genetically modified organisms, for example, is a mixture
of justified criticism (about corporate ownership and control) and unjustified
fear of how genes might be shared, and in large part relies on a curiously ignorant understanding of genetics.

It would be tempting to set up a contrast of the good guys versus the bad
guys here, as Nietzsche did. The Salem Region occupiers are not, however,
bad guys in and of themselves, nor are those who in the opposite region necessarily bright or enlightened defenders of the good. Life is not arrayed into
white and black hats. Moreover, there will always be a Salem Region because
it is defined in relative terms. Today's Salem Region may represent the views
that a thousand years ago would have been regarded as dangerously radical.
In fact, it may for a given field like biology or psychology represent the radical
novelties of less than a generation ago.[8] Like the Overton Window in politics
(recently appropriated by Tea Party writers[9]), these issues slide and shift, and
there are no absolute positions of conservative or radical stance. Moreover,
the interplay between Salem Regionists and other areas of the scientific cultural enterprise is part of the development of science and culture itself, since
conflict as well as agreement drives science productively. But on any given
issue, some fraction of Salem Regionists will adopt an antiscience position.

That said, there is a contrasting region, just in virtue of the geometry of the
phase space. I will call people falling into that region Inductivists, for want of
a better term. The inductivist mindset does not rely on prior knowledge and
prizes discovery. The reasoning they use is ampliative, often to the point of
speculation that offends the Salemists.[10] They are empiricists and treat theory
as an outcome, rather than a determinant, of investigation. They are not es-
sentialists in their classifications, but take exemplary cases and organize phenomena around them, as Whewell said. Inductivists use consilient reasoning
and do not tend to respect dogma, scientific or otherwise. Authority matters
only when the question is not contested (that is, when the authority suffices
to bracket other concerns for now), and it is to be subordinated to data. They

are early adopters, who take an epistemic bet that a novel hypothesis has a chance of paying off.

Moreover, we must be careful not to assume that someone is an Inductivist in all matters because they are in some, and the same will be true for other regions of the phase space. Individuals can shift and hold distinct dispositions in different domains. To the degree that these dispositions may be innate, we might expect a correlation across all stances on different issues, but given that biological dispositions are modulated by developmental environments, people may be conservative in one domain and radical in another, and so on, depending on the environmental factors they encountered at critical periods of their conceptual development. So, rather than assessing *individuals* as Salem Region occupiers or Inductivists, we should instead assess *acts of reasoning and belief formation* in this way. It is the mindset that occupies the region relative to a given issue, not the person, although we may expect the person to be more or less consistent in her disposition.

Deductivism in Philosophy

Deductivism in the philosophy of science is the view that ampliative reasoning is not possible, or not justified, or that discovery is fortuitous. Inductivism is widely regarded by deductivists as a dead horse, or a patch of grass where there used to be a dead horse. To what extent is this an outcome of Salem Region occupancy by some philosophers? Since Hume we have known that induction may not be justified by deductive reasoning because there must always be some missing premise that the world is regular, what Hans Reichenbach called a "straight rule" (Reichenbach 1949; Salmon 1991). Since John Stuart Mill was revived in the early twentieth century, this has been a widely held opinion among English-speaking philosophers, aided and abetted by Popper and his followers. However, induction still appears to be something that scientists actually do, whether under the guise of "inference to the best explanation" (Lipton 1990, 1991) "consilience" (Wilson 1999), or just as "induction" (Kornblith 1993; Heit 2000; Achinstein 2010).

The induction-blindness exhibited by some philosophers of science is somewhat perplexing. We may be missing a straight rule to justify induction, but why do some philosophers leap from that to the conclusion that induction is never justified? I suspect it may be because some are disposed toward deductivism and distrust discovery as an inferential process. Others may find deductivism distasteful and yet be led to that conclusion for philosophical reasons; but the sorts of assumptions that feed into those arguments, such as the deductivist assumption itself (that everything must be justified as a sound deductive argument), may have entered philosophical debate by one of someone who lived in the Salem Region, possibly Mill. Popperian demarcationism relies on the deductivist assumption, for if we do not have a single, clear, and universal criterion for distinguishing science from nonscience, how can it be that science is a real thing or different from any pseudoscience? That sort of straightforward answer is not, however, available to us. The world is not demarcated on the basis of (essentialistic, be it noted) definitions, and yet, as Edmund Burke, no stranger to authority himself, once noted: "Though no man can draw a stroke between the confines of day and night, yet light and darkness are upon the whole tolerably distinguishable" (Burke 1876).

Science is recognizable if not definable. The psychological need that some have for a definition is, I believe, due to these cognitive predispositions. This, of course, in no way prejudices the philosophical arguments on the matter.

Conclusion

If the Salem Region and the Inductivist Region represent distinct mindsets, how may we apply this knowledge to the problem of educated antiscientific advocates? This depends on what the problem is seen to be. If the problem is that there are antiscientific stances being taken, then the social manipulation of the population of educated people must include normalizing recent science so that the issues of competing authority sources, late adoption dispositions, and essentialism do not arise regarding it. Basically, if no tradition of opposition has arisen, then we may change the dramatic narrative of popular debate and the media so that it is not science that inflames the attention of the cognitive predispositions of the Salem Regionist. However, when antiscience traditions have evolved, this will not work. No matter how we introduce those with the right predispositions to science, some aspects of it will always be seen as controversial and threatening given the right stimuli.

Acknowledgments

I am enormously grateful to Jocelyn Stoller for an extensive introduction to the psychological literature that I have not yet managed to assimilate, and for discussions. I also thank Paul Griffiths for many discussions over the years and for pointing me to Peter Todd's work and hence Gerd Gigerenzer's project, and Glenn Branch and Joshua Rosenau for papers, ideas, and chats. The editors gave full, close, and constructive criticism of earlier drafts. Of course, my errors are self-generated, even in the light of these helpful individuals.

NOTES

1. As sometimes noted, for example by the performer Tim Minchin, if alternative science worked, then we'd call it science. The standards by which we judge something to be science are debated, but include reliability, success at prediction and explanation, extension into novel fields, and so forth. If we had a rival claim for acceptance from a nonscientific source, say clairvoyance, we would have no reason to accept it if the claim contradicted the virtues of science. If, on the other hand, the claims of clairvoyance had these virtues, we would have to incorporate them into science, as an example of our best knowledge.

2. Skyrms has revived a problem of Rousseau's using the example of a cooperative hunt for large game (stags), where cooperation has an expected payoff that is rational to pursue unless an individual hunter is offered a more immediate chance to catch a rabbit. Whether it is rational to behave cooperatively or individually depends crucially on the immediate conditions of each hunter. This shows us that rational self-interest need not result in a universal solution. It is rational to aid in hunting the big game until small game becomes available.

3. Gigerenzer et al. do not address natural heuristics or, as I call them in my 2011 work, ecological heuristics. They are likely to be closely related, both in functionality and phylogenetically, however.

4. I am deeply indebted to Jocelyn Stoller, a neurobiological and learning consultant, for much help covering and synthesizing the literature on false belief.

5. "Salem Hypothesis," *RationalWiki*, last modified May 21, 2012, http://rationalwiki .org/wiki/Salem_Hypothesis 15.

6. "Evolution Debate Engages Readers on Both Sides of the Argument," *EETimes*, February 20, 2006, http://www.eetimes.com/electronics-news/4058608/Evolution-debate-engages-readers-on-both-sides-of-the-argument.

7. A "frozen accident" is a historical event that is not necessary but that, having occurred, constrains the future. For example, the QWERTY keyboard was instituted to stop key strikes from colliding in manual typewriters. It now has no purpose other than that most typists use it. A similar example is driving on the left in the British Commonwealth; there is no benefit to this other than colluding with everyone else also driving on the left (see Lewis 1969).

8. Like the British Liberal MP, Sir William Vernon Harcourt, who famously declared in 1894 that "we are all socialists now." Similar sayings are attributed to Richard Nixon: "we are all Keynesians now." Given how US politics has changed, these are the radicals of the past now.

9. The Overton Window is the sliding of extremes to the right in political discourse, so that views once seen as mildly conservative are now seen as radically leftist. Ironically, Tea Partiers use it to assert that the extremes have shifted to the left.

10. Examples are Georges Cuvier's attack on Jean Baptiste Lamarck and Richard Owen's attack on Charles Darwin, for exceeding the limits of observation in theorizing. For this mindset, any kind of reasoning in a contentious domain that is hypothetical is unjustified.

REFERENCES

Aaseng, N. 1994. *Science versus Pseudoscience*. New York: F. Watts.

Achinstein, P. 2010. "The War on Induction: Whewell Takes on Newton and Mill (Norton Takes on Everyone)." *Philosophy of Science* 77 (5): 728–39.

Ashby, E. 1971. "Science and Antiscience." *Nature* 230: 283–86.

Bauer, H. H. 2001. *Science or Pseudoscience: Magnetic Healing, Psychic Phenomena, and Other Heterodoxies*. Urbana: University of Illinois Press.

Blum, J. M. 1978. *Pseudoscience and Mental Ability: The Origins and Fallacies of the IQ Controversy*. New York: Monthly Review Press.

Burke, E. 1876. *Select Works*. New ed. Oxford: Oxford University Press.

Butler, J. 1726. *Fifteen Sermons Preached at the Rolls Chapel, etc.* London: J. & J. Knapton.

Chinsamy, A., and É. Plagányi. 2008. "Accepting Evolution." *Evolution* 62 (1): 248–54.

Curd, M., and J. A. Cover. 1998. *Philosophy of Science: The Central Issues*. New York: W. W. Norton.

Dupré, J. 2001. "In Defence of Classification." *Studies in History, Philosophy, Biology, and Biomedical Science* 32 (2): 203–19.

Engels, Friedrich. 1892. *Socialism: Utopian and Scientific*. London: Swan Sonnenschein.

Feder, K. L. 1990. *Frauds, Myths, and Mysteries: Science and Pseudoscience in Archaeology*. Mountain View, CA: Mayfield.

Feldman, S., and K. Stenner. 1997. "Perceived Threat and Authoritarianism." *Political Psychology* 18 (4): 741–70.

Frazier, K. 2009. *Science under Siege: Defending Science, Exposing Pseudoscience*. Amherst, NY: Prometheus Books.

Gasper, K., and G. L. Clore. 1998. "The Persistent Use of Negative Affect by Anxious Individuals to Estimate Risk." *Journal of Personality and Social Psychology* 74 (5): 1350–63.

Gigerenzer, Gerd. 2000. *Adaptive Thinking: Rationality in the Real World*. New York: Oxford University Press.

Gigerenzer, Gerd, and R. Selten. 2001. *Bounded Rationality: The Adaptive Toolbox*. Cambridge, MA: MIT Press.

Gigerenzer, Gerd, Peter M. Todd, and ABC Research Group. 1999. *Simple Heuristics that Make Us Smart*. Oxford: Oxford University Press.

Heit, E. 2000. "Properties of Inductive Reasoning." *Psychonomic Bulletin & Review* 7 (4): 569–92.

Hines, T. 1988. *Pseudoscience and the Paranormal: A Critical Examination of the Evidence*. Buffalo: Prometheus Books.

Hull, D. L. 2001. "The Role of Theories in Biological Systematics." *Studies in History and Philosophy of Science Part C: Studies in History and Philosophy of Biological and Biomedical Sciences* 32 (2): 221–38.

Huxley, L., and M. Walter. 2005. *Biology an Australian Perspective*. Melbourne: Oxford University Press Australia.

Johnson, E. J., and A. Tversky. 1983. "Affect, Generalization, and the Perception of Risk." *Journal of Personality and Social Psychology* 45 (1): 20–31.

Jost, John T., Jack Glaser, Arie W. Kruglanski, and Frank J. Sulloway. 2003. "Political Conservatism as Motivated Social Cognition." *Psychological Bulletin* 129 (3): 339–75.

Kahan, Dan M., Hank Jenkins-Smith, and Donald Braman. 2011. "Cultural Cognition of Scientific Consensus." *Journal of Risk Research* 14 (2): 147–74.

Kornblith, H. 1993. *Inductive Inference and Its Natural Ground: An Essay in Naturalistic Epistemology*. Cambridge, MA: MIT Press.

Laing, R. D. 1969a. *The Divided Self: An Existential Study in Sanity and Madness*. London: Tavistock.

———. 1969b. *Self and Others*. 2nd rev. ed. New York: Pantheon.

Levin, S. G., et al. 1995. "Planck's Principle Revisited: A Note." *Social Studies of Science* 25 (2): 275–83.

Lewis, David K. 1969. *Convention: A Philosophical Study*. Cambridge, MA: Harvard University Press.

Lilienfeld, S. O., et al. 2003. *Science and Pseudoscience in Clinical Psychology*. New York: Guilford Press.

———. 2008. *Navigating the Mindfield: A Guide to Separating Science from Pseudoscience in Mental Health*. Amherst, NY: Prometheus Books.

Lindeman, M., et al. 2011. "Core Knowledge Confusions Among University Students." *Science & Education* 20 (5): 439–51.

Lipton, P. 1990. "Contrastive Explanation." *Royal Institute of Philosophy Supplements* 27 (1): 247–66.

———. 1991. "Contrastive Explanation and Causal Triangulation." *Philosophy of Science* 58 (4): 687–97.

Martin, J. L. 2001. "The Authoritarian Personality, 50 Years Later: What Questions Are There for Political Psychology?" *Political Psychology* 22 (1): 1–26.

Oreskes, N., and E. M. Conway. 2010. *Merchants of Doubt: How a Handful of Scientists Obscured the Truth on Issues from Tobacco Smoke to Global Warming*. New York: Bloomsbury.

Peirce, Charles Sanders. 1877. "The Fixation of Belief." *Popular Science Monthly*, November, 1–15.

Pigliucci, Massimo. 2010. *Nonsense on Stilts: How to Tell Science from Bunk*. Chicago: University of Chicago Press.

Reichenbach, Hans. 1949. *The Theory of Probability, an Inquiry into the Logical and Mathematical Foundations of the Calculus of Probability*. Berkeley: University of California Press.

Ritchie, A. 1991. "Will the Real Dr Snelling Please Stand Up?" *Skeptic* 4 (11): 12–15.

Roszak, Theodore. 1969. *The Making of a Counter Culture: Reflections on the Technocratic Society and Its Youthful Opposition*. Garden City, NY: Anchor Books.

Salmon, W. C. 1991. "Hans Reichenbach's Vindication of Induction." *Erkenntnis* 35 (1): 99–122.

Sedley, D. N. 2007. *Creationism and Its Critics in Antiquity*. Berkeley: University of California Press.

Shermer, Michael. 1997. *Why People Believe Weird Things: Pseudoscience, Superstition, and Other Confusions of Our Time*. New York: W. H. Freeman.

Short, R. V. 1994. "Darwin, Have I Failed You?" *Trends in Ecology & Evolution* 9 (7): 275.

Simon, H. A. 1972. "Theories of Bounded Rationality." In *Decision and Organization: A Volume in Honour of Jacob Marschak*, edited by C. McGuire and R. Radner, 12:161–76. Amsterdam: North-Holland.

———. 1981. *The Sciences of the Artificial*. Cambridge, MA: MIT Press.

———. 1986. Theories of Bounded Rationality. *Decision and Organization: A Volume in Honour of Jacob Marschak*, 2nd ed, edited by C. B. McGuire and R. Radner, 2:161–76. Minneapolis: University of Minnesota Press.

———. 1997. *Models of Bounded Rationality*. Cambridge, MA: MIT Press.

Skyrms, B. 2001. "The Stag Hunt." *Proceedings and Addresses of the American Philosophical Association* 75 (2): 31–41.

Smith, J. C. 2010. *Pseudoscience and Extraordinary Claims of the Paranormal: A Critical Thinker's Toolkit.* Malden, MA: Wiley-Blackwell.

Todd, P. M., and Gerd Gigerenzer 2003. "Bounding Rationality to the World." *Journal of Economic Psychology* 24:143–165.

Warnick, J. E., et al. 2010. *Educational Policy and Practice: The Good, the Bad and the Pseudo-science.* New York: Nova Science.

Whewell, William. 1840. *The Philosophy of the Inductive Sciences: Founded Upon Their History.* London: John W. Parker.

Wilkins, John S. 2008. "The Adaptive Landscape of Science." *Biology and Philosophy* 23 (5): 659–71.

———. 2011. "Are Creationists Rational?" *Synthese* 178 (2): 207–18.

Will, K. W., et al. 2005. "The Perils of DNA Barcoding and the Need for Integrative Taxon-omy." *Systematic Biology* 54 (5): 844.

Wilson, Edward O. 1999. *Consilience: The Unity of Knowledge.* London: Abacus.

Wittgenstein, Ludwig. 1968. *Philosophical Investigations.* Oxford: Basil Blackwell.

22

Pseudoscience and Idiosyncratic Theories of Rational Belief

NICHOLAS SHACKEL

I take pseudoscience to be a pretense at science. I will not rehash the difficulties in demarcating science. Pretenses are innumerable, limited only by our imagination and credulity. As Stove points out, "numerology is actually quite as different from astrology as astrology is from astronomy" (Stove 1991, 187). We are sure that "something has gone appallingly wrong" (Stove 1991, 180) and yet "thoughts . . . can go wrong in a multiplicity of ways, none of which anyone yet understands" (Stove 1991, 190).[1] Often all we can do is give a careful description of a way of pretending, a motivation for pretense, a source of pretension. In this chapter, I attempt the latter. We will be concerned with the relation of conviction to rational belief. I shall be suggesting that the question of whether an inquiry is a pretense at science can be, in part, a question over the role of conviction in rational belief, and that the answer is to be found in the philosophical problem of the role of values in rational belief.

The Borders of Science and Pseudoscience

Pseudoscientists seek to be taken seriously for the same reason that scientists claim our attention, that the propositions of a rigorous and rational science are more worthy of belief than the common run of opinion. But why do they wish to be taken seriously in this way? Certainly in some cases, cases of outright fakery and deceit, there is some interest of theirs that they think will be served by exploiting our credulity. These people do not think that they are scientists, they just want you to think they have the imprimatur of science. Others are just engaging fools who fool themselves and others. Yet others may have a very strong and deep need to bolster certain beliefs and must find whatever tools they can to do it. These errors are important but have been widely analyzed and do not manifest the problems of philosophical interest in this chapter. My concern here is not with the worst of pseudoscience but with the best. Equally, my concern is not with the best of science but with, if not the worst, with ways in which it can go and has gone bad.

Acquaintance with the history of science inclines me to think that science has grown out of practices that were confusions of what we would now call science and pseudoscience,[2] distinctions that we can make with hindsight but that were not clear to the practitioners of the time. For them, the issues were obscure and so were the surrounding methodological and philosophical questions. Of course, the nonscientific parts were not at that time exactly pretenses since no one knew better, nor were they motivated by wanting the prestige and authority that science now has. Rather, what we had were sincere inquirers wandering around in the borders of science and pseudoscience, sometimes getting lost and sometimes finding something.

The borders are still inhabited. The pseudoscientists who are of greatest philosophical interest are reasonable and sincere pretenders at science: trained inquirers who appear to want to know and who, if they pretended well enough, might even cease to be pretenders and become the real thing. Also of interest are those who are the real thing but who are becoming pretenders. Something about the borderers makes us wonder whether they really *do* want to know. We see the appearance of inquiry but we detect the portents of ideological conviction in the propositions pursued and in the manner of pursuit. The inquiry and the form it takes may arrive wreathed in the philosophical glories of a rooted conviction. We must wonder whether conviction is driving the inquiry where it wills.

Yet trained inquirers are often thus driven by strongly held conviction. In concurring with Stove above and in contributing to this book, I am expressing my convictions. It will perhaps be no surprise to the reader if my destination here is consonant with my convictions. Is being a pseudoscientist borderer just having the *wrong* convictions then? Well maybe, but something more is required than a dogmatic yes. Maybe all convictions are wrong! If not, then we need to know the basis for distinguishing right and wrong convictions.

Science is hard, and we are still learning how to do it. Sometimes we find our way, but which route the discipline of empirical method indicates depends on philosophical assumptions. Hence the room for the borderer's philosophical defense of an inquiry. Among those assumptions are what amount to theories of rational belief, and to take a route is to accept, if only tacitly, such a theory. These theories are the philosophical assumptions that bear on whether conviction is an illicit input to science. Hence I think that knowing the proper relation of conviction and rational belief will help in distinguishing science and pseudoscience at the borders.

Conviction

In general, our convictions are assemblies of deeply intertwined factual and evaluative beliefs that are important to us and hang together for us. Despite being beliefs, their importance to us makes it unclear to what extent conviction is disciplined by truth. Things would be simpler if we could reject conviction outright. Yet conviction is not simply a bad thing. It is often necessary in order for us to persist in a hard task despite difficulties and setbacks. When shared it creates affiliations and alliances of the most reliable kind and thereby furnishes not only the practical benefits that flow from trust safely placed but a good more highly prized still, the feeling of being with like minds, of being at home and at one with others.

It is, then, perhaps not surprising that conviction is common in us, indeed, that it is often something of which we are proud. Yet with the good comes the bad. We parade our convictions and demand submission to them. We indoctrinate our children in our convictions and think it right to do so. We cast creeds before strangers and know the unbeliever by his pause. Conviction can be ideological and fanatical. There is often something that we *want* to be true, that we are sure it matters a great deal that it *be* true, and that we are so sure *is* true that denial is heresy and deniers heretics to be anathematized and cast out.

There are important empirical questions in the middle of this, puzzles over the muddle of belief, desire, and self-delusion that we inhabit. Granted, for example, the apparent role of belief as information carrier and the practical value of information, how could we ever end up with a psychology that subverts that role and turns it to other purposes? Economists and psychologists have brought to our attention the general importance of self-deceit, signaling, persuasion, and commitment strategies in which belief entrained to something other than truth can have a role (Spence 1973; Hechter 1987; Bulbulia 2007; Caplan 2007; Rasmusen 2007; Trivers 2011). Presumably there is some trade-off between informational loss and practical gain on which, if selection pressure *can* bear, it *will* have borne.

Whilst I look forward to having good empirical answers to those questions, and I think such answers will be a valuable contribution to our understanding of pseudoscience, they are not my concern here. I am concerned with the rights and wrongs of conviction because I think this will tell us something about science and pseudoscience, will articulate and distinguish something they share and something that goes wrong in the case of pseudoscience. That is to say, I am not concerned with the causal explanation of how we come to conviction but with the evaluation of its role and our views of such evaluation.

So for our purposes, it is the normative link between conviction and belief that is crucial. People who have convictions are by and large convinced also that their conviction is what *ought* to be believed. This is what can lead pseudoscientists to think that if science says otherwise it must be in error, and since science is our organized project of inquiry we need to reformulate it to give the right answers. This can strike us as back to front, but is it? Scientists have sometimes had convictions at odds with science and have rightly reformulated science as a result. To determine whether this is just luck in being right, we need to know what ought to be believed and what relation what ought to be believed has to science.

Ethics of Belief

Well, what ought we to believe? Presumably, whatever the determinants of right belief determine to be right to believe. But what are those determinants, how widely do they range, how do they determine the rightness of belief, and relative to what do they determine the rightness of belief? These questions are the central questions of the ethics of belief.[3]

A normative principle that many find intuitive is that what you ought is whatever is rational. So a theory of rational belief is commonly taken to be the formal answer to what you ought to believe. Substantive disputes can then be conducted in terms of disputes over substantive theories of rational belief.

A traditional answer in this line has been to say that you ought to believe in accord with reasons rather than, for example, with emotion or faith. When put like that, however, it has drawn the response that "the heart has its reasons of which reason knows nothing" (Pascal 1670, XVIII) and that faith has its reasons too. In recent literature (e.g., Meiland 1980; Heil 1983; Haack 1997; Stanley 2005), this conflict manifests in the controversy over whether practical considerations play a role in what we ought to believe—practical considerations such as the loyalty owed a friend or the better outcome that having a certain belief may secure—or whether only theoretical considerations count.

The distinction between practical and theoretical considerations can be cashed out in a number of ways. Sufficient for our purposes is to note that theoretical considerations are, in broad, purely truth-directed or truth-conducive considerations; I say "in broad" in order to include internalist notions such as consciously accessible principles of inference and evidence as well as externalist notions such as reliably based beliefs.[4] To cut a long story short, we are going to use "evidence" to cover the purely theoretical considerations and "values" to cover the practical considerations. I use the latter because they are essentially ethical considerations, in the broad sense of the term concerned with those things constitutive of a worthwhile life.[5] We are going to call believing as and insofar as the purely theoretical considerations determine to be correct believing in accordance with the evidence.

Evidentialism and Pragmatism

Strict evidentialists (such as Conee and Feldman 2004) hold the stringent position that you ought and ought only to believe in accordance with the evidence. At the other extreme, pure pragmatists hold that only values count. Pure consequentialism can end up here; for example, we could understand Sidgwick ([1906] 1981, bk. 3, chap. 14) as taking this position.[6] Between these extremes, we can distinguish positions that are relatively more or less evidentialist, more or less pragmatist. Because this can be done in two ways, distinguishing the kinds of values and distinguishing the ways values interact with evidence to determine right belief, there is a complex range of positions available for specific substantive theories in the ethics of belief.

In terms of the kinds of values, an evidentialism with a minimal concession to pragmatism would be an axiology that confines the relevant values to the value of knowledge or virtuous belief, or to some notion of epistemic utility that is not evidential yet still purely epistemic.[7] A purely hedonistic axiology, on the other hand, would take us to an extreme pragmatism.

In terms of the ways values and evidence interact, a minimal concession to pragmatism would be for values and evidence to have entirely independent roles in determining what ought to be believed: values determine *which* propositions are worth believing or disbelieving and that is the entirety of their role; whether we ought to believe or not is *then* fixed by the evidence. We could call this pure factorism. This is an appealing position and is a natural retreat for the strict evidentialist. From pure factorism we can move to impure factorism, where whilst the roles of evidence and values remain distinct, they are not wholly independent. For example, perhaps the values can break evidential ties or determine starting places when evidence cannot. Positions in which values and evidence are both taken to have the same role *qua* reasons in determining what ought to be believed leave factorism behind, and varieties of such positions will fall out of different accounts of how reasons "add up," whether they are commensurable, comparable, incomparable, whether they can silence or exclude one another, and so on.[8]

Science and the Ethics of Belief

It is tempting to think that addressing pseudoscience does not require us to take a detour through the ethics of belief.[9] Are not pseudoscientists as committed as scientists to taking the aim to be truth? So the truth is what ought to be believed and the issue is simply how pseudoscientists pursue this in a distorted way! It is certainly true that the literature on pseudoscience has presupposed something like this, and for that reason authors in this area have also a tacit and unanalyzed presupposition of strict evidentialism.[10] I think it would be fair to say that strict evidentialism has also been a presupposition of the public face of science.

Philosophical analysis can seem to be a matter of raising questions that do not need asking. This is especially the case when, as is common in philosophy, at the end of the analysis no definitive answer is offered. To some extent that will be the case here. However, it seems to me that there is a kind of idiosyncrasy in pseudoscience that can only be made sense of by investigating why we have these presuppositions and whether they are well or ill founded. So I am aiming to show an aspect of pseudoscience that has been neglected. That it is neglect rather than correct peremptory dismissal depends partly on showing the presuppositions to be less well founded than is presumed and partly on the extent to which such a dismissal poses problems also for understanding science. I am now going to sketch briefly some weaknesses of the presuppositions before we turn to what is for us a central issue: whether pragmatism is avoidable. My answer will be that it is not, and we will then turn to considering the consequences.

The nature of science is itself in dispute. For a start, instrumentalist philosophers of science such as van Frassen (1976) would not agree that science aims at the truth. Whether science has a single aim and whether that aim is truth is also questioned by scientists, some of whom explicitly reject talk of truth in science and regard truth as an unscientific concept best left for philosophers to waffle about. Instead they abjure the term and wish only to discuss models and their uses. Turning to pseudoscientists, they too may state their aim for science in other terms and sometimes explicitly in terms of conviction, for example:

> The Discovery Institute . . . talks about a strategy to "defeat scientific materialism" and "reverse the stifling dominance of the materialist worldview, and to replace it with a science consonant with Christian and theistic *convictions*." (IDEA 2011, my emphasis)

Taking truth as the aim of science faces obvious problems. Truth cannot be an aim like a target because when we see the truth we have already attained our aim. If someone was just lucky at guessing the truth, then on that basis they would count as a scientist; but obviously that is not right. So achieving the truth is not sufficient for science. Nor is it necessary. Scientists are not simply given the truth and what is meant by "scientific truth" need not even be true! Rather, what may make a doctrine scientific is that it is the output of a rational inquiry, thereby being what is rational to believe, which is to say, what ought to be believed.

Even if we accept that truth is the aim of science, the nature of truth is up for dispute. The assumption that this aim takes us swiftly to strict evidentialism depends on assuming a correspondence theory of truth.[11] But if semantic antirealism is true, truth is something like warranted believability.[12] In that case the aim of science *just is* whatever ought to be believed. The fact that truth has been thus defined in normative terms means that values can now count as truth-conducive considerations, but that is a merely verbal victory for evidentialism. Evidentialists want to exclude values from counting toward what ought to be believed, and if values count in this way the pragmatists win. So truth being the aim does not imply that we can ignore the ethics of belief.

Furthermore, some scientists, perhaps especially in the social sciences (e.g., Sampson 1978; Rabinow and Sullivan 1979), reject the notion of facts that are independent of values. In such a case, even correspondence truth would be relative to values, and so not even a correspondence theory of truth guarantees that we can ignore the ethics of belief.

Finally, science is respected for having a wider social and moral significance in being a source of what is worthy, or more worthy, of belief. It is exactly this respect that pseudoscientists wish to have. But that wider social and moral significance is a matter of taking science to be a source of what ought to be believed. So the dispute over pseudoscience depends in part on the ethics of belief.

Some Degree of Pragmatism Unavoidable

I am now going to argue that strict evidentialism is false, which implies that any true theory of rational belief must have a pragmatic element. The essential problem is that arguments for strict evidentialism fail by failing to attend to an important distinction in kinds of normativity and to a correlate distinction in what we are talking about when we are talking about rationality.

It is regrettable that our terminology is so congested here. To get clear we must distinguish two uses of the word "rationality." The first is the sense we used above, in which we took a theory of rationality to answer the question of what ought to be believed.[13] The second is the sense that characterises our mentality and agency. When I need to be careful in distinguishing these senses, I shall speak of *normative* rationality for the first sense and *intrinsic* rationality for the second.

A central kind of argument for strict evidentialism is to formulate a theory of normative rational belief in terms of the intrinsic rationality of belief (e.g., Adler 1999, 2002), that is, the correctness of a mental state whose role in the mental economy is to represent the world as in fact being a certain way. This apparently offers a short path to evidentialism. For example, one can argue that if belief were to be otherwise guided, then the rational economy would have to have some *other* state that was purely truth directed in order to keep track of how things are, thereby to determine whether believing how things are is practically better or not. But in that case, belief would be otiose, since its intrinsic role in the rational economy is played by that other state.

Thus a putative short path to evidentialism, but a path taken at the cost of evading the question. Yes, in one sense of "ought,"[14] the sense that here expresses what is correctly responsive to the role of belief in a system of mental states constituting rational agency, you ought to believe in accordance with the evidence. But then, what the ethics of belief is asking is whether what is intrinsically rationally correct to believe is what ought to be believed, in precisely the sense of "ought" that goes beyond mere intrinsic rational correctness and advances on what is right.[15] After all, it may be that the greedy nephew who wants his aunt's fortune ought to poison her in the first sense,[16] but not in the second. So the whole question over strict evidentialism is whether what is intrinsically rational is what is normatively rational. Absent some *further* argument, this kind of approach to strict evidentialism merely assumes what it was supposed to prove.

An assumption of the argument just considered is that belief is truth directed, that truth is the *aim* of belief. Prima facie there is something right in this thought (although there are outright rejections in the literature, e.g., Rosenberg 2002; Steglich-Petersen 2006; Gluer and Wikforss 2009). Finding and delimiting what is right has proved to be harder than it might at first appear (Velleman 2000; Wedgwood 2002; Steglich-Petersen 2006; Engel, forthcoming). A natural approach is to think that being true is necessary and sufficient for being what you ought to believe. Necessity is appealing but fails if, for example, you can have strong enough evidence for a falsehood that you ought to believe it. Sufficiency fails more clearly since presumably it is permissible not to believe the myriad trivial truths even if you had the capacity to do so. The latter difficulty is usually addressed by a clause requiring significance (which can include significance for further inquiry, but then in the end must be grounded in the significance of that inquiry). What is significance if not a practical consideration? Certainly, significance may sometimes be the value of knowledge or the value of excellence in belief, but in appealing to these we have gone beyond purely theoretical grounds into the ethical value of theoretical goods: "Knowledge is valuable because knowledge *of certain matters* adds so importantly to the flourishing of one's life individually, and of life in community" (Sosa 2010, 189).

Another path to evidentialism is to take standard answers in epistemology to the question of justified belief to be answers to our question. But consider the telling qualification at the end of this remark: "another kind of normative fact—epistemic facts . . . concern what we ought to believe, provided that our beliefs are aimed at the truth" (Shafer-Landau 2006, 226). The provision is accurate and significant, and leaves clear room for the broader question. Furthermore, just because the deontological vocabulary deployed in epistemology is the same as that deployed in ethics it does not mean that the normativity in play answers to the ethics of belief. Alston, for example, uses some of the difficulties that arise if we assume it does to argue that "deontological justification is not epistemic justification" (Alston 1988, 293). That is to say, despite the normativity of epistemic justification *sounding* like a notion correlative to the normative concerns of the ethics of belief, it is not. My suggestion for understanding this disjunction is that the normativity of the epistemic facts discussed in epistemology is correctness rather than directivity; the issue over whether what is correct is also right is not much considered and is often assumed without argument.

Whatever the obscurity in the use of deontological notions within epistemology, whether they are to be taken as merely a loose analogy to their use in ethics or whether they are to be taken full bloodedly, when we come to the ethics of belief we have come precisely to the place where analogy is laid aside and identity assumed. The ethics of belief is where we ask what ought to be believed, in precisely the same sense as in ethics we ask what ought to be done. In so doing, we move to the widest or deepest or most fundamental normative perspective, the perspective not of correctness but of directivity, marked by the directive sense of "ought," and are looking for the final and complete answer that takes everything into account. This is sometimes expressed as the question of what, all things considered, we ought to do or believe.[17]

On occasion it has been argued that the answer to any such question is determined once we know what is morally right (e.g., Prichard 1912), but that is clearly a further question. So identifying the senses of "ought" in the questions of what ought to be done or believed, absent a prior commitment to morality overriding, is not to identify right belief with moral belief. It is merely getting clear which question is being addressed in the ethics of belief. When we have determined what is prudent, we have still more work to do before we have answered what ought to be done; likewise, when we have determined what is correctly in accordance with the evidence, we have still more work to do before we have answered what ought to be believed. We have to determine whether what is intrinsically rational is also what is right.

The clarity of thus distinguishing what is intrinsically rationally correct from what is right can be obscured when we define reasons *in terms of* what ought to be believed, thereby returning to the normative sense of "rational." Once again we can be tempted to evade the difficulty. If, having defined reasons as the determinants (whatever they are) of what ought to be believed, we now identify them with the determinants of the intrinsic rationality of belief (evidence), we have taken a question-begging shortcut to evidentialism from the other direction. Pragmatists can similarly evade the difficulty by staunchly affirming reasons of the heart and faith. Instead, given this definition, having identified what is right to believe with what is normatively rational to believe, now the disagreement between evidentialists and pragmatists has to return to where we started. Deciding to call the determinants of what we ought to believe "reasons" has not advanced us one bit since we must still consider whether such reasons include theoretical considerations alone or include practical ones as well.

We can now see that the move from the normative to the intrinsic sense of rational belief is also a move between two correlative senses of "ought," the "directive" sense that attributes the normativity of what is right and the "correctness" sense that attributes the normativity of correctness alone. Insufficient marking of this division within normativity can lead us to assume that which was to be proved. We can do this because taking the argument through a truism that to be rational is to act and believe in accordance with reasons leads us to miss the shift in senses of "rationality." When we start at the *intrinsic* rationality of belief and identify the intrinsic reasons derived on that basis with normative reasons, we have begged the question against pragmatism. And if we start at *normative* reasons and derive the rationality of belief, we only get back to *intrinsic* rationality of belief if we started by assuming that normative reasons are evidence.

The argument I have just given is not conclusive, but I think it makes clear that the burden is on the strict evidentialist to advance beyond the mere assumption of the identity of normative rationality and intrinsic rationality. Absent some good argument to that effect, I think we have to give up on strict evidentialism. I do not know of such arguments: I think some degree of pragmatism is unavoidable.

Source of Pretension

So now we can draw the threads together to characterise the source of pretension that it has been my purpose to describe. We are concerned with the role of conviction in rational belief in aid of understanding the borders between science and pseudoscience. Recall that we are not concerned with the pseudoscience of fakes, fools, and fanatics and hence are not interested in blind conviction. Pseudoscientists of the most problematic philosophical kind are those who are sincere, reasonable, scientifically trained, driven by conviction in a way similar to scientists, and who seek to defend their pretense at science as rational inquiry.

Essentially, the source of pretension is that, despite the lip service paid to strict evidentialism, there is a tacit subscription by inquirers to pragmatism, a subscription driven by conviction and leading to the deployment of conviction in inquiry. The tacit pragmatism allows that there is a proper role for conviction, the public evidentialism rules it out, and discomfort at this conflict forestalls explicit philosophical inquiry into the proper and improper roles for conviction. Because conviction results from and is expressive of our values, we do not renounce it. Partly because its role is underanalyzed, it has freest rein wherever there is obscurity in inquiry. The upshot is that conviction has a significant role in inquiry and also (as has been long recognized) a potentially corrupting role in inquiry. The question is what to do about it.

If strict evidentialism is true, the answer is straightforward. Under strict evidentialism any intrusion of values and convictions into science looks only peculiar and irrational. Rather, scientists must be purely disinterested and dispassionate inquirers. A problem here is that many scientists are neither disinterested nor dispassionate inquirers. So if strict evidentialism is true, we can make no distinction between scientists who are driven by their convictions and pseudoscientists who are driven by *their* convictions: neither are doing science. Still, that might be right, and we might just need to train the conviction out. There are, however, reasons to be uncomfortable with that conclusion, reasons independent of the argument above against strict evidentialism.

The irrationality of conviction has been a doctrine of the philosophy of science since Bacon[18] and remained largely unquestioned until Polanyi (1966) pointed out the importance of tacit knowledge and Kuhn (1970) convinced us that philosophy of science must attend carefully and more respectfully to what scientists actually do. These made the doctrine look false even when we sustain the distinction between the context of discovery and the context of justification. Yes, sometimes it is irrational, but sometimes it isn't, so what makes the difference? How does a theory of rational belief countenance conviction?

Pragmatic theories of rational belief, theories that countenance some role for values, can allow the intrusion of values and conviction without necessarily marking them irrational. Such theories, in distinguishing proper and improper roles for values, distinguish proper and improper roles for convictions, and may thereby distinguish the convictions of scientists from those of pseudoscientists. The kinds of theories of rational belief surveyed above are very different in their import for the proper influence of values, and this is in part why what one person sees as illicit influence another may see as required. What would be needed in any particular case is an analysis of just what roles convictions are playing in a particular inquiry, what that entails for the role of values in that inquiry, and what kind of theory of rational belief countenances such a role for values. Absent knowledge of the true theory of rational belief, an issue that is not likely to be settled any time soon, any such analysis brings with it the possibility of extensive and deepening philosophical dispute.

A New Area of Work for the Analysis of Pseudoscience

A full analysis of this source of pretension depends on knowing which principles are the true principles of rationality. Certainly we have some knowledge here, but less than we would like, and far less in the specific area that has been our focus. The literature on the ethics of belief has explored the role of values, but their import for the epistemology and practice of science is barely discussed. Strict evidentialism has been the assumption in the literature on pseudoscience, and because it has been the assumption the source of pretension that I have sought to bring into focus here has been neglected through being dismissed as merely irrational. Since strict evidentialism is probably false, to advance the analysis we need new work on the import of pragmatic theories for understanding the rational role of conviction in science. Getting this right will, I think, help us determine new markers of science and pseudoscience.

Once we see that we cannot defend strict evidentialism, we must concede that we may have been begging some questions against sophisticated pseudoscientists. Furthermore, by failing to take on directly those elements of their defenses that are grounded in explicit or implicit theories of the role of values in rational belief, our prophylaxis has been less effective. In evading the issue of the role of values, the convictions of truth wanters such as I am may have struck pseudoscientists as mere prejudice, and they may have resented our attitude as bigoted.

A further point that becomes evident is that pseudoscience is closer to us than the easy examples we like to reject. We can see powerful convictions and accompanying bigotries lurking in almost any area of science with strong practical import, and we see it on both sides of controversies. There is no shortage of ideologically driven true believers in economics, social science, psychology and climate science. Perhaps in part because these disciplines study complex systems, which are by their nature obscure, conviction has an especially free rein here.

The range of strategies available for the philosophical defense of pseudo-science is wider than has been previously considered. The fact that pragmatic theories of rational belief are defensible means they can be deployed in defending a program of inquiry, and controversy over that inquiry can be addressed by moving to the controversy over those theories. The fact that pragmatic theories of rational belief have some tacit currency in our general intellectual lives and have some intuitive appeal when explored is in part an explanation of the extent to which both scientists and pseudoscientists engage in pragmatically tinged philosophical defenses of their projects. It will seem right to be motivated by conviction in defending their projects, and when their convictions drive them to thinking that science is in need of reform, they will have to articulate that reformed science on the basis of some theory of rational belief, a theory that licenses the role their convictions are playing. Hence do inhabitants of the borders rationalize their activities in the borders. The question is whether they are doing it in the psychiatric or the success sense. Developing a better understanding of the rational role of conviction will help us here.

As Quine pointed out,[19] sophisticated systems of belief have immense capacity for resisting unwanted change in one area by making changes elsewhere. A theory of rational belief itself is one such area. Put these elements together and we have the materials for the self-enclosed systems of inquiry and belief that we can fall into. Because the same materials are deployed in our open enquiries and the differences are subtle and philosophically disputable, we can hardly be surprised that the borders of science and pseudoscience are inhabited. For these reasons, distinguishing science and pseudoscience can in part be a matter of distinguishing the nature of an implicit theory of rational belief.

Insofar as we do not know the true principles of rationality, we are unable straightforwardly to distinguish true and false theories of rational belief. What I think we can often distinguish are philosophically defensible theories from idiosyncratic theories. Even when we cannot fully specify the principles being transgressed, there comes a point at which we recognize a theory as being bent ingeniously and entirely to preserving the precious propositions.

The literature on pseudoscience has focused on idiosyncrasy in the treatment of the evidential aspects of rational belief. We have not spent much time analyzing idiosyncrasies in the treatment of values. The fact that values have some role means that the ways theories of rational belief can go wrong are more varied than the literature has tended to address. If we cannot simply dismiss the influence of values as distortion, then we have to tease out the range of potentially legitimate influences and point out idiosyncrasies where they can be identified. Further complications arise when we consider the interaction of values and evidence as warranted by some theories.

Indeed, I think work on such interactions will illuminate some puzzles about pseudoscience. One problem here is that the distortions are sometimes so bizarre and obviously wrong that it is very hard to understand why anyone should ever end up with them on a purely evidential basis. If values have a proper input to rational inquiry, it is easier to understand how mistakes could be made and lead to improper inputs. For example, it is hard for us now to understand the acceptance of eugenic "science." When, however, we consider the values that scientists held, it makes more sense. The belief was that science was not ethically neutral, but that it was on the side of progress for humanity. From there the belief in "scientifically" engineering the biological progress of humans was not so far, at least rhetorically speaking.[20]

Two Examples of Idiosyncrasy in the Treatment of Values in Theories of Rational Belief

Plainly there is a program of work here that I cannot possibly undertake, or even advance much, in the space remaining. What I am going to do is illustrate idiosyncrasy over the role of values in rational belief by a couple of examples and then conclude.

The first example is a matter of equivocation on "values." When I introduced the distinction between values and evidence, I stipulated that I meant ethical values, in the broad sense of ethical, that is, those things constitutive of worthwhile life. Sometimes, however, when speaking of values in this context, people speak of epistemic values. But the latter term is ambiguous over the crucial distinction around which I have organized this discussion, the distinction between values and evidence. On the one hand, epistemic values might be about what questions are important, or about the value of knowledge or excellence in believing. As such, they are ethical values. On the other hand, they might be epistemic standards constituting truth-conducive methodology and the intellectual virtues accompanying it such as open-mindedness, curiosity, intellectual rigor, and diligence. All of these I subsumed under evidence. But my distinction between values and evidence is not a distinction between the normative and the nonnormative, it is a distinction within normativity. The issue between strict evidentialists and pragmatists is a question over the nature of the normativity had by evidence as such, whether it is intrinsically directive or not. By failing to make this distinction, the interference in the role of evidence by values can be rationalized on the grounds that values are ineliminable from inquiry. But of course, the whole issue is not whether epistemic value, that is, what I have subsumed under evidence, is ineliminable but whether ethical value is ineliminable. I agree that value is ineliminable, but this argument to that conclusion is just a cheat, and worse than a cheat, it is a source of idiosyncratic theorizing about science.

Notoriously, Kuhn was criticized as an irrationalist on this basis.[21] Although Kuhn denied his work had this import, his work was understood by others in precisely this vein. For example, there are social scientists who aim at using the outputs of their research to advance a political agenda. Some use biased methodologies and reject criticism by denying the existence of an impartial basis from which to criticize their methodology. The reason for the no-impartial-basis claim is the fact that people are not ethically neutral, they have ethical commitments and interests (i.e., ethical values), but clearly that does not entail the impossibility of epistemic impartiality as a constraint on methodology.

The second example is a well-known difficulty of empirical inquiry, which can be an occasion for the action of values in inquiry. It is impossible to conduct an inquiry with a completely open mind. There are infinitely many hypotheses that might be entertained, and for any hypothesis there are infinitely many kinds of data that might be relevant to confirming or rejecting it. How then should we cut them down? Presumably we want some principled way of discerning hypotheses worth considering and relevant kinds of data. Unfortunately, there are infinitely many hypotheses that might be entertained for which principles are correct and infinitely many kinds of data that might be relevant to choosing the principles. So the problem recurs, and recurs at every level. This difficulty appears in many guises; for example, Popperians (e.g., Bartley 1964) got into a difficulty of circularity or regress in trying to defend critical rationalism (is it itself falsifiable or up for defeat by criticism?) and more recently in the framing problem in artificial intelligence. It has its roots in any attempt to formulate rationality in terms of rules, where presumably one needs rational rules to choose the rules, and so on (e.g., see Brown 1988, chap. 2). Fodor offers an engaging description of our difficulties here and concludes that "it strikes me as remarkable . . . how regularly what gets offered as a solution of the frame problem proves to be just one of its formulations" (Fodor 2008, 121).

If Fodor is right (see his remarks about Kyburg and Laudan, pages 117 and

119), strict evidentialism is floundering with this problem. The philosophical obscurity in epistemic standards resultant from this irresoluble regress leaves room for the appeal to values. Hence we meet with Kierkegaardian claims that rationalism (by which he meant something like strict evidentialism) is on a par with Christian faith because to be a rationalist is also to make a leap of faith. Insofar as I have suggested that some concession to pragmatism is unavoidable, I am conceding that there is a normative truth lurking here. Setting aside the question of just what it is, there is at least an empirical truth in play. What steps in and cuts short the regress are convictions, often the very same convictions that set us off on the inquiry in the first place. Depending on exactly how and on what basis convictions step in, this might be defensible by pragmatism. But, of course, there are worries here, a worry about path dependence and a worry about the intrusion of values overstepping whatever role is proper to them.

Taking the latter first, any such overstepping may distort our inquiries somehow or other. In particular, it is plausible that conviction can result in loading the dice to get the number we want, that we set out in a way we present as being neutral but which in fact fixes our destination before the inquiry is even under way. So a recognizable feature of ideological conviction is its power to make the wanted truths the centre around which everything else will be arranged. Heterodox hypotheses are declared heresy, awkward facts declared taboo. Methodologies that will lead in the desired direction and produce the desired evidence are deployed. Epistemic standards that will warrant the desired answers are articulated and their application to evidence is selectively arranged to afford the desired proofs.

The worry about path dependence is deeper than the worry about the improper intrusion of values, and in a sense subsumes that worry. An improper intrusion of values requires a distinction with their proper role whereas path dependence might result from the propriety or otherwise of intrusion depending on the values started from. To put it more finely, whilst there may be some purely formal constraints on the propriety of roles for values, that might be as far as it goes. Granted that there are a variety of substantive values that constitute the good for persons, and that as a result the good for different persons is itself various, respecting the formal constraints on the basis of different substantive values may result in proper but opposing answers to one and the same nonnormative factual question.

That is not an upshot I find at all congenial. Truth wanters are inclined to think that there cannot be proper but opposing answers to one and the same nonnormative factual question.[22] It is, however, in a loose sense a recognizable feature of our experience. Pursuing disagreements frequently moves on to disagreements about methodology, about how evidence counts and in the end about what matters, which is to say, about values. What at the outset seems to be a straightforward and resolvable dispute about facts ends up in a convoluted and frustratingly irresoluble dispute about values.

Conclusion

I set out to offer a description of a source of pretension to science. This description has been necessarily very broad in order that I could illustrate the nexus of conviction, value, and theories of rationality that constitute the particular source of pretension of interest. I think we are familiar with the existence of this source but have not attended to it much in a philosophical way, tending to see it as only a psychological source of distortion. Yet it has played a role in science as well as in pseudoscience, and so to dismiss entirely anything whose origins include it may be to rule out of science things we wish to rule in. Insofar as its manifestation has been philosophically considered, it has been through the analysis of idiosyncratic treatments of evidence in pseudoscience, and that is itself very important. Indeed, we should see those analyses as illuminating a wider tendency to subscribe to idiosyncratic theories of rational belief. I have suggested that the role of value in rational belief has been neglected and that as a consequence our understanding of this source has been cruder than it needs to be. In particular, if some variety of pragmatism is true, we cannot simply dismiss this source as irrationalism. In so doing, we have neglected an entirely distinct range of theory in which idiosyncrasy can manifest.

If we are to advance our understanding of the borders of science and pseudoscience, we need better analyses of the proper and improper roles of values and convictions in rational belief. My own convictions are that the truth is paramount and that as a consequence we are required to permit the widest ranging and most open inquiries, however obnoxious we find them. Taboo is forbidden. Heresy must be tolerated. Despite these convictions, I do not think that strict evidentialism can be defended. Features that are ethical in the broadest sense, that is to say, features determinative of worthwhile lives, must be among the determinants of right belief. Absent such features, whilst there may be much that is correct or incorrect to believe, and whilst correctness of belief is belief in accordance with evidence, there would be nothing we ought to believe, in the directive sense of "ought," because it would not matter what we believed. Instead I seek to defend factorism of some kind. I answer the toughest challenges to the requirement to believe the truth on the basis that insofar as beliefs are relevant to settling hard practical and ethical conflicts, only the truth is a neutral ground; hence this defense is based in impartiality but allows some retreat from truth when partiality is properly in play. It seems to me that such an account, while denying strict evidentialism as the true theory of what ought to be believed, makes evidentialist *practice* right for the most part.

I would be very disturbed if something beyond factorism were true. The upshot of granting values a role that results in proper path dependence seems to be either that living in illusion can be rational or that reality is relative; it is literally *your* reality, as your values make it. My conviction is that this cannot be true: it is, however, what some pseudoscientists seem to believe.

NOTES

1. Stove's target is not pseudoscience alone but includes philosophical horrors and the problem of our lack of what he calls a nosology (a classification of diseases) of thought.

2. See, e.g., Dobbs (1975).

3. It is an error to mistake the ethics of belief for epistemology. The thought that the output of theories in epistemology settles the ethics of belief is sometimes assumed dogmatically. To do so is to miss vital questions over the normative status of such theories and the nature of the normativity referred to when such theories are expressed in normative terms. For example, a causal theory of knowledge may be a purely positive theory. If such a theory identifies the justification of the belief with being caused by the fact believed, there is a question over whether this is *really* a normative use of the term "justification." If it is, there is a further problem in explaining how being caused by a fact makes something normative and in what way is it normative.

4. This water is muddied somewhat in some discussions of deontological notions of internalist justification for belief (Alston 1988), but even then, rarely do contra-truth practical considerations feature as justifiers. Rather, the worry is whether the combination of the ought-implies-can principle and belief being outwith voluntary control undermine responsibility, and thereby the possibility of justification, for belief. We shall return below to the status of deontology in the justification of belief.

5. See Williams (1985, chap. 1) for this notion of the ethical and why it is broader than morality.

6. Moore (1903, 85), at least, interprets him in this way.

7. Whether there is any such thing is unclear. Exploration of consequentialist epistemology intended to be analogous to consequentialism as virtue epistemology is to virtue (such as

Percival 2002) have tended to leave the nature of epistemic utility undefined whenever they have gone beyond taking truth as the good.

8. These metanormative issues have been discussed more deeply in ethics than in epistemology (e.g., Raz [1975] 1999; Dancy 2000; Broome 2006; Schroeder 2007).

9. The next few sections draw on material developed in Shackel (n.d.).

10. These were my presuppositions too.

11. In which truth is correspondence to the facts. The most recent versions are given in terms of truth makers rather than facts.

12. Strictly speaking, I have moved a bit swiftly here since usually it is defined in terms of warranted assertibility, and then we have to work on the route to belief. That would be an unilluminating technical journey for our purposes.

13. This is the same sense as when we take it to answer the question of what we ought to do.

14. A sense that I have called the correctness sense of "ought" (see Shackel 2004, chap. 2).

15. A sense that I have called the directive sense of "ought" (see Shackel 2004, chap. 2).

16. The correctness of instrumental rationality. Cf. Kant's hypothetical imperative.

17. And in the latter case especially it is sometimes held that it is without answer because the question is without meaning (see Feldman 2000, 694).

18. See his four idols (Bacon [1620] 1994, bk. 1, Aphorisms 38–44), especially of the cave and of the theater.

19. E.g., "Any statement can be held true come what may, if we make drastic enough adjustments elsewhere in the system" (Quine 1951, §6).

20. This is not to deny other and more obnoxious inputs to the acceptance of eugenics.

21. See, for example, the shifting to and fro between epistemic and ethical values whilst appearing to give them the same or very similar roles in the big shifts in science (Kuhn 1970, 185–6).

22. Modulo such complications as possessing different evidence, the possibility of justified belief in a falsehood, and so on.

REFERENCES

Adler, Jonathan Eric. 1999. "The Ethics of Belief: Off the Wrong Track." *Midwest Studies in Philosophy* 23:267–85.

———. 2002. *Belief's Own Ethics*. Cambridge, MA: MIT Press. http://www.netLibrary.com/urlapi.asp?action=summary&v=1&bookid=74721.

Alston, William P. 1988. "The Deontological Conception of Epistemic Justification." *Philosophical Perspectives*, 2:257–99.

Bacon, Francis. (1620) 1994. *Novum Organum*. Peru, IL: Open Court.

Bartley, William Warren. 1964. *The Retreat to Commitment*. London: Chatto and Windus.

Broome, John. 2006. "Reasons." In *Reason and Value: Essays on the Moral Philosophy of Joseph Raz*, edited by P. Pettit et al., 29–55. Oxford: Oxford University Press.

Brown, Harold. 1988. *Rationality*. London: Routledge.

Bulbulia, Joseph. 2007. "The Evolution of Religion." In *Oxford Handbook of Evolutionary Psychology*, edited by R. I. M. Dunbar and L. Barrett, 621–36. Oxford: Oxford University Press.

Caplan, Bryan Douglas. 2007. *The Myth of the Rational Voter: Why Democracies Choose Bad Policies*. Princeton, NJ: Princeton University Press.

Conee, Earl Brink, and Richard Feldman. 2004. *Evidentialism: Essays in Epistemology*. Oxford: Oxford University Press. http://www.loc.gov/catdir/toc/fy046/2004300585.html.

Dancy, Jonathan. 2000. *Practical Reality*. Oxford: Oxford University Press.

Dobbs, Betty Jo Teeter. 1975. *The Foundations of Newton's Alchemy, or, the Hunting of the Greene Lyon*. Cambridge: Cambridge University Press.

Engel, Pascal. Forthcoming. "In Defence of Normativism About the Aim of Belief." In *The Aim of Belief*, edited by Timothy Chan, 65–107. Oxford: Oxford University Press.

Feldman, Richard. 2000. "The Ethics of Belief." *Philosophy and Phenomenological Research* 60 (3): 667–95. http://www.jstor.org/stable/2653823.

Fodor, Jerry A. 2008. *Lot 2: The Language of Thought Revisited*. Oxford: Clarendon Press. http://dx.doi.org/10.1093/acprof:oso/9780199548774.001.0001.

Gluer, Kathrin, and Asa Wikforss. 2009. "Against Content Normativity." Mind 118 (469): 31–70. http://mind.oxfordjournals.org/content/118/469/31.abstract.

Haack, Susan. 1997. "The Ethics of Belief Reconsidered." In *The Philosophy of Roderick M. Chisholm*, edited by R. M. Chisholm and L. E. Hahn, xviii. Chicago: Open Court.

Hechter, Michael. 1987. *Principles of Group Solidarity*. Berkeley: University of California Press.

Heil, John. 1983. "Believing What One Ought." *Journal of Philosophy* 80:752–64.

IDEA 2011. "FAQ: Isn't Intelligent Design Just a Movement Trying to Push a Political Agenda?" *Intelligent Design and Evolution Awareness Centre*. Accessed December 8, 2012. http://www.ideacenter.org/contentmgr/showdetails.php/id/1188.

Kuhn, Thomas S. 1970. *The Structure of Scientific Revolutions*. 2nd ed. Chicago: University of Chicago Press.

Meiland, Jack W. 1980. "What Ought We to Believe? Or the Ethics of Belief Revisited." *American Philosophical Quarterly* 17 (1): 15–24.

Moore, George Edward. 1903. *Principia Ethica*. Cambridge: Cambridge University Press.

Pascal, Blaise. 1670. *Pensees De M. Pascal Sur La Religion Et Sur Quelques Autres Sujets*. Paris: Chez Guillaume Desprez.

Percival, Philip. 2002. "Epistemic Consequentialism." *Proceedings of the Aristotelian Society, Supplementary Volumes*, 76:121–68. http://www.jstor.org/stable/4106966.

Polanyi, Michael. 1966. *The Tacit Dimension*. Garden City, NY: Doubleday.

Prichard, Harold Arthur. 1912. "Does Moral Philosophy Rest on a Mistake?" *Mind* 21 (81): 21–37. Reprinted in *Moral Obligation: Essays and Lectures by H. A. Prichard*, edited by W. D. Ross (Oxford: Oxford University Press, 1968), to which page references refer.

Quine, Willard van Orman. 1951. "Two Dogmas of Empiricism." *Philosophical Review* 60:20–43.

Rabinow, Paul, and William M. Sullivan. 1979. "The Interpretive Turn: Emergence as an Approach." In *Interpretative Social Science: A Reader*, edited by Paul Rabinow and William M. Sullivan. Berkeley: University of California Press.

Rasmusen, Eric. 2007. *Games and Information: An Introduction to Game Theory*. 4th ed. Malden, MA: Blackwell. http://www.loc.gov/catdir/toc/ecip0620/2006029009.html.

Raz, Joseph. (1975) 1999. *Practical Reason and Norms*. 2nd ed. Oxford: Oxford University Press.

Rosenberg, Jay. 2002. *Thinking About Knowing*. Oxford: Clarendon Press.

Sampson, E. E. 1978. "Scientific Paradigms and Social Values: Wanted—A Scientific Revolution." *Journal of Personality and Social Psychology* 36:1332–43.

Shafer-Landau, Russ. 2006. "Ethics as Philosophy: A Defence of Ethical Non-Naturalism." In *Metaethics after Moore*, edited by T. Horgan and M. Timmons, 209–32. Oxford: Clarendon Press. http://www.loc.gov/catdir/toc/ecip0518/2005023277.html.

Schroeder, Mark Andrew. 2007. "Weighting for a Plausible Humean Theory of Reasons." *Nous* 41 (1): 138–60.

Shackel, Nicholas. 2004. "On the Obligation to Be Rational." PhD diss., University of Nottingham.

———. n.d. "Ethics of Belief, Normativity, and Pragmatism."

Sidgwick, Henry. (1906) 1981. *The Methods of Ethics*. Indianapolis: Hackett.

Sosa, Ernest. 2010. "Value Matters in Epistemology." *Journal of Philosophy* 107 (4): 167–90.

Spence, A. Michael. 1973. "Job Market Signalling." *Quarterly Journal of Economics* 87 (3): 355–74. http://dx.doi.org/10.2307%2F1882010.

Stanley, Jason. 2005. *Knowledge and Practical Interests*. Oxford: Oxford University Press. http://www.oxfordscholarship.com/oso/public/content/philosophy/9780199288038/toc.html.

Steglich-Petersen, Asbjørn. 2006. "No Norm Needed: On the Aim of Belief." *Philosophical Quarterly* 56 (225): 499–516.

Stove, David Charles. 1991. "What Is Wrong with Our Thoughts? A Neo-Positivist Credo." In *The Plato Cult and Other Philosophical Follies*, 179–205. Oxford: Blackwell.

Trivers, Robert. 2011. *Deceit and Self-Deception: Fooling Yourself the Better to Fool Others*. London: Allen Lane.

van Frassen, Bas C. 1976. "To Save the Phenomena." *Journal of Philosophy* 73 (18): 623–32.

Velleman, James David. 2000. "On the Aim of Belief." In *The Possibility of Practical Reason*, 244–81. Oxford: Clarendon Press.

Wedgwood, Ralph. 2002. "The Aim of Belief." *Nous* 16:267–97.

Williams, Bernard. 1985. *Ethics and the Limits of Philosophy*. London: Fontana.

23

Agentive Thinking and Illusions
of Understanding

FILIP BUEKENS

Think of how puzzling a dream is. Such a riddle does not have a solution. It in-
trigues us. It is as if there were a riddle here. This could be a primitive reaction.
—Ludwig Wittgenstein (1982, sec. 195)

Dedicated to Jacques van Rillaer

The Hermeneutics of Stomach Rumblings

The *Annals of Improbable Research* recently reported how psychoanalysts
were able to find meaning in stomach rumblings. The technical name for these
sounds is borborygmi, and they are produced by the contraction of muscles in
the stomach and intestines of animals and humans. In 1984 Christian Müller
of the Hôpital de Cery in Prilly, Switzerland, published 'New Observations
on Body Organ Language' in the journal *Psychotherapy and Psychosomatics*
(Müller 1984). Müller presented a 1918 essay by a certain Willener, who, in
Müller's words, 'conclude[d] that the phenomenon generally known as borborygmi must be regarded as crypto-grammatically encoded bodily signals
that could be interpreted with the help of [special] apparatus" (1984, 125).
Combining "electromesenterography with Spindel's alamograph, and in ad-

dition the use of digital transformation for a quantitative analysis of the curves via computer" (Müller 1984, 17), Müller claimed to have gained access to their meaning:

> The presence of a negative transference situation was not difficult to deduce from the following sequence: "Ro . . . Pi . . . le . . . me . . . Lo. . . ." The following translation is certainly an appropriate rendering: "Rotten pig, leave me alone." (Müller 1984, 17)

Marc Abrahams, who reports these "observations" in the *Annals of Improbable Research*, adds that a Montreal psychoanalyst, whose name shall not be mentioned here, published several "apparently quite serious papers about the psychoanalytic significance of borborygmi." The title of one of these papers was "Borborgymi as Markers of Psychic Work During the Analytic Session: A Contribution to Freud's Experience of Satisfaction and to Bion's Idea About the Digestive Model for the Thinking Apparatus" (Da Silva 1990).

Muller's hoax is instructive because he could unproblematically present the investigation of gut sounds in intentional terms: there is talk of meaning and understanding, of translation into English of the gut language, of grammatically encoded signals, and the unavoidable (and quite disturbing) hidden message.[1] Freud announced his "discovery" of the meaning of dreams using exactly the same concepts: "at that time I learnt how to translate the language of dreams into the forms of expression of our own thought—language" (Freud 1953, 7:15).

As for borborygmi, serious science would try to find the underlying mechanisms that produce such sounds, and which states of mind and external circumstances (nervousness before a talk or an exam, for example) activate dispositions of the mechanisms (Bechtel and Richardson 1993). Progress in our knowledge of the proximal and distal enabling conditions of manifest mental and bodily phenomena can be made, for example, when we discover low-level neurological and physical mechanisms that implement manifest capacities and dispositions, or when we discover systematic patterns in external causes of those sounds. But in Müller's hoax, such empirical questions are simply bypassed. There is, in Müller's description, a familiar narrative that runs from sounds produced by the stomach and the intestines to an interpretation (Müller's 'translation'). The narrative does not require conjectures about underlying mechanisms or how in nonpsychoanalytic settings similar rumblings could occur. The reason is simple: knowledge of underlying mechanisms or external triggering causes would be irrelevant in the hermeneutic project of finding *meaning* in them, and the explanandum already anticipates that meanings will be found because the rumblings are described as "signs." (The story illustrates Frank Cioffi's claim that psychoanalysis is best described as a pseudohermeneutics; see Cioffi 1998a.) Moreover, if the mechanisms were at work in nonpsychoanalytic settings, why would sounds produced in *those* settings *not* have meaning?[2] While it is not logically impossible that dreams could have hidden meanings, there is nothing in Freud's technique—the free associations—that would suggest that it generates plausible evidence for that claim, and his descriptions of the underlying mechanisms that produced the dreamwork were not based on independent evidence but recapitulated, in quasi-mechanistic terms, the agentive explanations he had provided from behind the couch.

The Intentional Strategy and Illusions of Understanding

Müller's hoax illustrates the familiar criticism that psychoanalysis, qua hermeneutic method, is capable of understanding every anthropological phenomenon, an objection first formulated in 1901 by Wilhelm Fliess, Freud's one-time friend and collaborator, but also by Ludwig Wittgenstein, and further developed by contemporary critics like Bouveresse (1995), Cioffi (1998b), and Crews (2006).[3] Müller's piece reveals important components of explanatory projects that rely on the intentional strategy: by applying the intentional strategy to objects or systems that are not candidates for that type of explanation, *illusions of understanding* emerge. When the intentional strategy is systematically applied to natural facts, a more or less systematic "theory" seems acceptable because such explanations are accompanied by the feeling of thereby having understood the facts. We are constitutively inclined to apply the intentional strategy to ourselves and other persons; and when we successfully apply it, a gratifying sentiment signals that we now have understood an action, an attitude, or an emotion. "You have grasped its meaning, you are in touch with the other, you follow her" is what the presence of the epistemic feeling suggests to you (Turner 2010).[4] Since there are no *natural* halting points for explanations, the function of the sentiment is to signal that you have "found" a satisfactory explanation given your explanatory goal (understanding a person's actions), but the sentiment itself hardly indicates that you have found the *true* explanation (Trout 2007, 567). I suggest this feature of intentional explanations is a psychological feature of explanations in general—it is the feeling of epistemic satisfaction that marks a significant moment and halts the explanatory process (Gopnik 1998).[5] But this sense of understanding is, as Trout (2007, 566) puts it, "a common, but routinely unreliable index of intellectual achievement." Psychological satisfaction need not indicate real insight. This also explains why, once meaning was found in dreams (or gut sounds) by giving them an agentive explanation, the further and urgent questions—How were these rumblings produced? Was there a pattern in the external causes?— were thought to be irrelevant. The feeling of understanding that accompanies good explanations suggests that further inquiries are superfluous.

An intentional explanation gives us the agent's reasons—usually a combination of a belief and a desire—and it creates a feeling of having understood what she did: we see how the reasons we constructed could also make sense of ourselves when performing that action under a suitable intentional description. A *cui bono* question seemed pertinent.[6] The intentional strategy is perfectly legitimate when actions of real agents figure on the explanandum side. Artifacts can be understood in terms of the purpose or goal with which they were designed. The 'design stance' is the intentional stance, but now applied to the mind of a real or virtual designer of an object: 'What's the function of this object?' is in that context the pertinent question. Müller's hoax works well because he plays out exactly that point: the gut sounds are encoded messages, cast in a language that can be interpreted. There is a hidden agent whose actions, with the help of Spindel's alamograph, we can make sense of. The explanandum, under an intentional description, cries out for an intentional explanation. But did the explanandum *deserve* to be described in intentional terms?

The term "intentional stance" was introduced by Daniel Dennett (1987, 1992). Our social nature being shaped by evolution, we make sense of each other in order to arrive at mutual understanding, and this with the goal of coordinating our actions, beliefs, intentions, and desires when realizing shared projects:

> The intentional stance is the strategy of interpreting the behavior of an entity (person, animal, artefact, whatever) by treating it as if it were a rational agent who governed its "choice" of "action" by a consideration of its "beliefs" and "desires." . . . The basic strategy of the intentional stance is to treat the entity in question as an agent, in order to predict—and thereby explain, in one sense— its actions or moves. (Dennett 1996, 27)

In making sense of others, we attribute large patterns of contextually appropriate true beliefs and reasonable desires; false beliefs (we all have false beliefs, and the world can always lead us astray) are identified on a background of shared truths (Davidson 1984). The genealogy and function of the intentional strategy that characterizes interactions between our kinds of mind can be explained in evolutionary terms, including the fact that we sometimes have useful *misbeliefs* (McKay and Dennett 2009). The intentional strategy is also the capacity to apply purposes and functions to things that are the products of minds like us, for not only human actions but also their products are involved in coordinating our actions and behavior. For artifacts, teleological reasoning ("reverse engineering") is perfectly legitimate, and since we are constantly surrounded by artifacts, the question "what is the purpose of this or that object (sign, signal)?" is often acutely relevant.

The intentional strategy is likely a naturally evolved explanatory strategy, but when not appropriately bridled, the strategy that is so successful in interpersonal interactions with a clearly defined explanandum ("What is she doing?" "What do you mean?") is prone to be applied to events and processes it does not apply to—from natural objects (by children), to dreams (by Freud), and even to genes (by some overenthusiastic evolutionary psychologists). My claim (which is certainly not original) is that overapplication of the intentional or agentive strategy can leave unwanted and potentially damaging traces in scientific theories (Dennett 1991; Kelemen 2004; Davies 2009b).

For starters, we have a natural tendency, which Jean Piaget (1929) first observed in very young children, to reason about natural phenomena—including mechanisms that explain human intentionality, life, and evolution—in terms of purpose and intention-based concepts (Kelemen 2004). As humans, we are primed to detect agency in phenomena around us, which makes evolutionary sense because our environment was full of predators, preys, mates, and rival conspecifics. It matters who (or what) is what, which involves the ascription of intentions and goals to the creatures that surround us. According to Deborah Kelemen (2004), children are "promiscuous teleologists," and even for adults who are aware of the illusion it is not easy to withhold intentional strategies in explanations of natural phenomena.[7] Despite the acknowledgment that intentional and agentive concepts can play a useful heuristic role in contexts of discovery of natural phenomena, and often help us see the point of a scientific model because they work as powerful metaphors that allow us to see things as "constructed according to a plan," illusions of understanding are lurking because the original psychological sense of understanding is vitally linked to agentive explanations. As with visual or cognitive illusions, the fact that one can be fully aware of being subjected to illusions of understanding in specific contexts does not automatically immunize one against the illusion and its psychological after-effects.[8]

When children give reasons for why rocks are pointy ("they are pointy so that animals wouldn't sit on them"), they assign a teleological function to physical processes (Kelemen 2004, 296). Keil (1992) and Kelemen (2004), among others, have further explored these findings, and some of them have become important components of cognitive and evolutionary explanations of the development of theistic explanations (Boyer 2001). But the same cognitive phenomenon may also play a role in explaining why the less controversial question of whether higher animals have a "theory of mind" was and still is hotly disputed. The issue has been on the agenda ever since David Premack and Guy Woodruff (1978) explored it in their famous article "Does the Chimpanzee Have a 'Theory of Mind'?" (see Corbey 2005 for an overview of the ensuing scientific and moral controversies). Premack and Woodruff used the term "theory of mind" because they took the system of inferences that yield social understanding to be about states that are not directly observable, and because it was used to make predictions about the behavior of others. But what made (and makes) research on social cognition in animals controversial, in a sense that is relevant for our purposes, is that we are, given our natural tendency to think teleologically and to see intentions and purposes where there are not any, prone to ascribe intentional states to animals and to explain their behavior in intentional terms. Initially, the tide was with those who argued that research on social cognition in animals showed that there was no special barrier between humans and apes, and that we should not indulge in the idea of human uniqueness, which, it was said, is itself a product of culturally transmitted theological, political, and metaphysical beliefs about ourselves and our place in nature. But it remains plausible to hold that the enormous gap between humans and higher animals is underestimated due to the fact that the "theory of mind" of animal researchers tends to produce false positives: it is "because we can't help but see and interpret (animal) behavior through the lens of our own theory of mind . . . that we may be seeing more than is actually there. Perhaps we're simply reading into their behaviors by projecting our own psychology into theirs" (Bering 2011, 29). While it remains undecided whether we are unique in being able to explain behavior in terms of the rational and norm-governed interplay between mental states, no one doubts that we are "exquisitely attuned to the unseen psychological world" (Bering 2011, 33).

Another reason the intentional strategy is of the utmost importance in social interaction is that I myself have something to gain when my beliefs, desires, and intentions are properly understood by others. Survival requires not only that I can predict how others act, but that I give others evidence for correctly predicting *my* behavior. As Morton (2003, 27) points out, by acting in regular ways we create conditions under which we give others an understanding of our *own* motives and desires. The desire to understand is therefore not independent from the desire to be understood. Both capacities in fact may make use of the same modules in the brain. But the idea that others understand you, that you are surrounded by others who are able to predict your behavior, also gives rise to the feeling that there is *always* someone out there who understands you and "sees" what you are about to do. This is, as Boyer (2001) and others stress, exactly the capacity ascribed to deities. And consider the idea "that Reason rules the world, and that world history has therefore been rational in its course." This was once an extremely serious philosophical idea (the quote is from G. W. F. Hegel, and served as a motto for Brian Keeley's 1999 paper on epistemic flaws in conspiracy theories).[9] If history were indeed the unfolding of a rational process, as Hegel seems to have thought, then it is almost unavoidable to locate the root of that thought in the idea that someone is indeed watching over *us* (humanity as a whole) and keeping track of our moves. We'd better be understood correctly by Reason, for only if we are correctly understood will Reason make space for the realization of our plans.[10]

What renders these projects problematic are the not always easily recognizable remnants of the intentional strategy in their central concepts. The hypothesis is that owing to the weight of the selective importance placed on the agentive strategy, we cannot help but postulate intentions, beliefs, and purposes behind events and objects.[11] The following projects are commonly recognized as containing residues of agentive thinking and are sources of illusions of insight:

- The Freudian unconscious (McMillan 1996; Cioffi 1998a, 1998b)
- The "message" of dreams in Freudian psychoanalysis (Bouveresse 1995; Cioffi 1998)
- Conspiracy theories (Keeley 1999; Coady 2003; Buenting and Taylor 2010)
- Narrative understanding and meaning-of-life discourse (Velleman 2003; Strawson 2004)

- Intelligent Design (Boyer 2001)
- The self as a hidden agent (Hume [1739–40] 1888; Metzinger 2003)
- Form-giving powers in biology (Davies 2009b)
- Hegel's *Weltgeist* and history as a guided process (Godfrey-Smith 2009; Ferguson 2011)
- The explaining of coincidences; moral explanations of accidents; "God's plan with my life" (Dawkins 1998)

More contentious examples include:

- Purposive thinking and normative accounts of biological function (Davies 2009a, 2009b)
- The representational mind and Cartesian theater conceptions of the mind (Dennett 1987)
- Selfish genes in biology (Godfrey-Smith 2009)

In the rest of this chapter, I focus on the more contentious examples.

Normative Functions, Homunculi, Hegel's Weltgeist, and Strange Coincidences

Consider biological theories that work with a normative concept of function. Paul Sheldon Davies (2009a, 2009b) holds that the power of the concept of purpose when thinking about living things is extremely intuitive: we have the intuition that living creatures, or at least parts of them, are "supposed" to fulfill functions. But this may be more or less a direct consequence of our psychological constitution, reinforced by a cultural environment that explicitly promotes the idea that nature was designed (Davies 2009b, 104). But even when you abandon these intuitions and fully recognize the full force of the Darwinian argument, the idea of living things, their organs, and their external features having purposes does not let go easily. Davies documents how the German physiologist and anthropologist Johann Friedrich Blumenbach (1752-1840) tried to model the apparent purposiveness of living things by positing a nonmechanical form-giving power on the basis of what appeared to be scientifically solid methods. Just as Isaac Newton had insisted that one could accept the existence of gravity as a fundamental force on the ground that it explains the observed phenomena, even though he had no account of its basic constituents, Blumenbach could hold the view that it was rational to accept archetypal, form-giving powers on the plausible ground that they (and only they) could explain what is distinctive of living things. Similar ideas could be found in Henri Bergson's 'élan vital.'

Davies draws an interesting connection between the views of Blumenbach and those of William Paley. The latter explained the apparent design in nature by its origin in a creative God: Blumenbach too introduced an "agent-like source of creative power that is difficult to square with a naturalistic world view" (Davies 2009a, 131). As Davies puts it, "The emergence and perpetuation of living forms, according to Blumenbach, comes from a nonmechanical power potent enough to cause the perpetuation of living forms, though nothing about the actual workings of this power is ever revealed to us" (Davies 2009a, 132). Davies points out that while it may be true that Blumenbach did not *intend* his theory of formative powers to be a step outside the natural order, and while there are some superficial analogies with the Newtonian explanation, the differences between them are stark: Blumenbach formative powers were entirely unmechanical and unexplained, which amounted to the positing of a mystery. "Newton was not offering us a mystery; he was confessing his ignorance and placing his bets on further inquiry. Blumenbach, by contrast, was asking that we accept an unexplained force from the silent realm of the nonmechanical" (Davies 2009a, 133). The explanatory gap Blumenbach encountered was filled by a formative power that had distinctively agentive features: it motivated the mechanical parts of cells, it imposed a template of the species' form on the processes of growth and reproduction, and the products resembled the phenomena that explained them. Davies concludes that, just as Paley explains away all the difficulties by positing a creator, Blumenbach "dumps (the difficulties) into the lap of an unexplained nonmechanical power—a hermetically sealed mystery" (Davies 2009a, 133).

The unintended positing of a surrogate agent, a "center of command and control" (Davies 2009a, 135) in Blumenbach's theory, is clearly a remnant of the intentional strategy in a scientific theory; agents operating outside the mechanical realm and their interventions are utterly inconsistent with current mechanical explanations. (Note that in Paley's design argument the positing of a creator was clearly an intentional move.) Davies goes on to insist, plausibly in my view, that even given the standards of eighteenth-century biology, Blumenbach's ideas must have been controversial, and he shows how even more recent normative accounts of functions are "unacceptably conservative because [they are] a product of our psychological inclination to conceptualize objects as purposive even when they are devoid of purpose" (Davies 2009a, 135).

Are theories that give normative functions in biology center stage therefore worthless? Philosopher of biology Michael Ruse thinks not since they
have, in his view, an important heuristic value: "without the metaphor [of
design,] the science [of evolutionary biology] would grind to a halt, if indeed
it ever got started" (Ruse 2003, 139; see Dennett 2011 for a similar position).
But I suggest we should side with Davies on this point: the stubborn persistence of the concept of purpose, given its roots in the intentional strategy
and agentive explanations, and its explanatory role in our social lives, need
not entail that it is a concept we cannot do without in the scientific study of
biological phenomena. This would also explain why it requires a deep cognitive transformation to substitute the questions "why?" and "what for?" that
come so easily to mind with "how" questions that seek explanations in terms
of mathematical or statistical models of underlying mechanisms. While it is
true that human psychology was not selected to produce scientific models
of the world, it does not follow that advocates of design metaphors make the
right recommendation when they hold that such metaphors are merely harmless cognitive tools in contexts of discovery; the sense of understanding they
engender may halt inquiry rather than stimulate it.

A second example of traces of the agentive strategy and the misplaced psychological sense of understanding it can generate is found in theories about
beliefs as harboring *mental representations*. Cognitive theories about beliefs
and desires conceptualize mental states in ways that it becomes almost inevitable to introduce an internal agent who watches and intervenes in the passing show of beliefs and desires as they occur in the mind. The map theory of
beliefs, originally due to the British philosopher F. P. Ramsey, holds that beliefs are a self-contained model of the world and asks how a cognitive system
uses that model ("beliefs," Frank Ramsey held, "are maps by which we steer"
[Ramsey 1990, 146]). This immediately raises the question of who is *using* or
consulting the inner map? The map theorist must posit a homunculus who
links features on the map ("in the mind") with features in the world. The map
metaphor simply does not work unless one postulates a reader of maps. The
intentional strategy applied to the inner representational states creates a sense
of understanding because we (qua agents) *do* understand how we use maps
to locate and orient ourselves in the world. Replicating this familiar picture
in explanations of how internal representations work begs the question. As
William Ramsey puts it, "If ordinary notions of non-mental representations
are to form the basis for understanding representations in cognitive theories,
and if those ordinary notions always presuppose some sort of representation

user, then we need to provide some sort of account of representation use where the user isn't a full blown mind" (Ramsey 2007, 23). Not a full-blown mind perhaps, but something like a mind, something about which we cannot say much more than that it is postulated to do exactly that—consulting inner representations.

The "homunculus fallacy" occurs "when one attempts to explain what is involved in a subject's being related to objects in the external world by appealing to the existence of an inner situation which recapitulates the essential features of the original situation to be explained . . . by introducing a relation between the subject and inner objects of essentially the same kind as the relation existing between the subject and outer objects" (Evans 1985, 397). The root of the problem is the overwhelming explanatory force of the intentional strategy: it is very hard to avoid understanding the concept of an (inner) representation as not referring to entities consulted, interpreted, and evaluated by an inner agent, and it remains an open question whether the concept of representation itself, so central in cognitive accounts of the mind, is not the indirect product of the intentional strategy, now applied to the inner workings of the mind itself.

Philosopher Colin McGinn holds that "we do not need to presuppose beliefs about mental models in order for mental models to act as the machinery of belief . . . mental models do not need to be interpreted, they just need to be used" (McGinn 1989, 200). But who is the user, the agent herself, or an inner user? I am not suggesting here that representational theories of mind are false or worthless.[12] My claim is that representational theories of mind contain remnants of the intentional strategy; when the issues are pressed, it becomes very difficult to grasp how internal representations are supposed to do their work. Remnants of the intentional strategy figure at central nodes in the framework, not in some faraway corner where they can be easily discarded or dismissed as a rhetorical device or a heuristic tool. There was therefore a point in the behaviorist's argument against the use of mental images as explanatory constructs on the grounds that homunculi were posited, and the lesson to be learned was that the concept of a representation—qua theoretical concept, not one that figures in our natural psychology—was not an easy one to unpack without creating illusions of understanding. Behaviorism's reductionist stance toward the mental inadvertently freed its proponents from unfounded applications of the intentional strategy. Note that, if Cartesian skepticism has its roots in a representational conception of the mind, as philosopher John McDowell argues (McDowell 2008), then the natural pull toward skepticism about knowledge of the external world could be explained by the tendency to apply the intentional strategy to internal enabling conditions of our mental economy.

The problem I am trying to describe here is not that applying the intentional strategy can bring about false beliefs. We are prone to false beliefs anyway. The world is a capricious place, and we do not always update what we believe in view of new evidence. But some false beliefs at the center of an explanatory theory have their source in the agentive strategy itself. There are interesting speculations about what makes systematic false beliefs (or misbeliefs, as McKay and Dennett (2009) call them) beneficial for their owners, and whether we should systematically seek to correct those beliefs given that they may harbor important evolutionary advantages (McKay and Dennett 2009, 493). Are these cases where illusions of understanding are eventually adaptive? Is the false positives account sufficient ("false positives are generally harmless," it is claimed, but "being harmless" is obviously itself a purpose-relative concept). They may be a cost worth paying in everyday life (the presumption is, of course, that being possessed of different psychological attitudes would be far worse). But when the aim is to acquire knowledge of the animate and inanimate world, they lead us astray and should be avoided.

Agentive concepts may be deeply hidden within an explanatory story. But sometimes the intentional concepts are right there at the surface, at the center of the approach. The role of a *Weltgeist* in Hegel's philosophy illustrates our stubborn incapacity to live with the disturbing idea that significant and morally relevant actions and events in our lives are simply consequential to luck (Nagel 1979; Ferguson 2011), and it has often been noticed that the constitutive role of luck in the success of our action is almost unbearable (Strawson 1996; compare the idea that "luck swallows everything" with the psychological impact of the idea that determinism is true). Hegel's seductive narrative about the *Weltgeist* presents things as if they had to happen, and understanding itself is often thought to be a good thing: "a richly narrative outlook is essential to a well-lived life, to true or full personhood" (Strawson 2004, 428). But if such a narrative is correct, who is the director, who decided that this or that had to happen? Who wrote the script?

The Freudian unconscious has a lot in common with hidden agents purportedly at work in paranoid minds (see Farrell 1996 for an extensive analysis of paranoid thinking in Freud and psychoanalysis). But there are less idiosyncratic traces of agentive thinking in psychoanalysis. Freud also held that the solution to the analyst's problem of dream interpretation is entirely and unequivocally predetermined, so that he could present the meaning of a dream as a discovery, not a construction he himself imposes on something intrinsically meaningless:

> What makes [the analyst] certain in the end is precisely the complication of the problem before him, which is like the solution of a jigsaw puzzle. A colored picture, pasted upon a thin sheet of wood and fitting exactly into a wooden frame, is cut up into a larger number of pieces of the most irregular and crooked shapes. If one succeeds in arranging the confused heap of fragments, each of which bears upon it an unintelligible piece of drawing, so that the picture acquires a meaning, so that there is no gap anywhere in the design, and so that the whole fits into the frame—if all these conditions are fulfilled, then one knows that one has solved the puzzle and that there is no alternative solution. (Freud, "Remarks upon the Theory and Practice of Dream Interpretation," qtd. in Bouveresse (1995, 113))

The ease with which a dream message could be constructed in every version of psychoanalysis reveals the power and versatility of the intentional stance and the open character of the apparatus for applying it: under the (dubious) assumption that there is meaning in a dream, we will always end up finding meaning and purpose in it.

In *Unweaving the Rainbow,* Richard Dawkins explains how to take the sting out of coincidence by calculating the likelihood that it would happen anyway. He retells the story of Richard Feynman, who in a story published in 1998, recounts how Feynman's first wife died at 9:22 in the evening while the clock in her room was later found to have stopped at exactly that time. The striking coincidence suggests that the clock stopped for a reason, that some hidden (psychic?) force had decided to stop the clock at that time because the force "thought" that it was a significant moment. Again we see agentive explanations at work: we can easily mentally simulate a mind that makes these decisions to remind us of, say, the importance of the event (which itself presupposes that the psychic medium was sensitive to emotional events—but why should Feynman's wife have attracted the medium's special attention?). The poetic effect that remains, even if one immediately realizes that this was just a coincidence, may be a remnant of what we have to learn to resist: letting our intentional strategies generate an explanation that creates a pleasing illusion of understanding.

Agentive Explanations and Darwinism

The pull of agentive thinking figures in a critical line of argument by Peter Godfrey-Smith (2009) regarding replicator approaches in evolutionary thinking (Dawkins 1976; see Dennett 2011 for critical remarks). The replicator approach is "in many of its presentations, designed to mesh with an 'agential' way of thinking of evolution (in which) evolution is treated as a contest between entities that have purposes, strategies and agendas" (Godfrey-Smith 2009, 10).[13] The problem is obvious: the agentive perspective is an "uneasy mix of the metaphorical and the literal," and the agentive perspective "engages a particular set of concepts and habits: our cognitive tools for navigating the social world" (Godfrey-Smith 2009, 10). Darwinian explanations do not come naturally to us because, there is, as Godfrey-Smith points out, "a premium on compact schemata and models with which we can impose order on [nature]" (Godfrey-Smith 2009, 13).[14] And even in the absence of a role for an Intelligent Designer, a teleological mode of thinking seems unavoidable for our understanding of the mindless principles of evolution by random mutation and natural selection (Buller 1999a; Thompson 1995).[15]

But even staunch Darwinists tend to present their theories using agentive metaphors, and they often had to face misunderstandings as a consequence. The Dawkins/Midgley controversy in the past century was partly due to the fact that Dawkins, at strategic places in *The Selfish Gene* (the first and last pages of the book), connected in one fell swoop the concept of the selfish gene—intended to be a *technical* term designating a complex biochemical process—with issues pertaining to our manifest notions of egoism and altruism. The phrase "let us try to teach generosity and altruism, because we are born selfish" (Dawkins 1976, 3) suggests a *manifest* notion of selfishness, while the rest of the book explores genes as the ultimate level of selection; and in that theory, the gene's selfishness indicates that they only "care" about replication (note how difficult it is to even describe this idea without using agentive notions). Notice the irony: while Darwin's main achievement was to introduce a style of explanation that aimed at eliminating agentive thinking, some of his popular defenders—who present themselves as staunch opponents of Intelligent Design—reintroduced agentive thinking in biology via powerful but potentially misleading metaphors. But, as Godfrey-Smith adds, "the feeling that some particular way of looking at things yields understanding should not always be taken at face value, is not the end of the matter" (2009, 13; and see also Trout 2007).

Godfrey-Smith locates the allure of the selfish gene model in the agentive narrative that surrounds it. When positing an agent in an explanatory story, two explanatory schemata can be developed. The first he calls a *paternalist* scheme, where the explanation works because some large, benevolent agent is postulated. The agent intends that, as Godfrey-Smith puts it, "all is ultimately for the best" (2009, 144). Such explanations often postulate gods and spirits, but Hegel's notorious *Weltgeist* would also be a good example. The second is the *paranoid* scheme, where the explanation postulates small, hidden powers and agents. Freud's model of the mind (the unconscious as a realm full of forces that explain manifest emotions, but also hidden desires that ultimately explain the content of dreams), demonic possession narratives, and selfish genes and memes are prime examples, but the psychological appeal of such hypotheses "often far outruns their empirical warrant" (Godfrey-Smith 2009, 144). The force of the intentional strategy explains (but obviously does not justify) this tendency.

Godfrey-Smith argues that there is a style of thinking in biology that does not invite paranoia—"the kind of investigation when someone asks: suppose a population was like this, and such-and such a mutation happened, what would happen to that population?" (2009, 145). J. D. Trout makes a similar observation: "We can easily imagine—even picture vividly—an individual's search for prey, or their ultimate triumph in the struggle to mate. But it is much harder to imagine the actual dispersion of individual's traits within and between species" (2007, 579). Godfrey-Smith's preferred style of explanation does not invite the idea of hidden agents and genes as "ultimate beneficiaries." Concepts employed in mathematical models of evolution of a certain trait are thin theoretical ones ("population," "mutation") that have no ancestor life in our manifest image (they are, as Carnap once put it, "successor concepts"), and statistical correlations written out in complex mathematical models are presented in connotationless equations. Theoretical concepts and mathematical equations are all we need to explain how mutations occur and what their effects are, but it is not difficult to appreciate that except for the happy few who are fully immersed in the language of the theory and fully understand its models, such formulations do not speak to outsiders in terms *they* understand. Concepts like "selfishness," on the other hand, inevitably trigger or activate the intentional strategy and agentive concepts. The problem was never the selfishness of genes but the language and implications such metaphors generate: "*There are these tiny little strategists who keep us under control.*" Another difficulty in grasping theoretical concepts and mathematical models is that evolution is gradual, and processes that often extend over thousands of generations are extremely difficult for our kinds of mind to cognitively grasp. Both factors—the anonymity of the concepts and formulas, and the gradual character of evolution (a factor already noticed by Darwin as a source of resistance to the evolutionary hypothesis)—account for why bona fide evolutionary explanations freed from intentional concepts do not easily yield the kind of epistemic satisfaction we expect from understanding a phenomenon. The real explanation continues to feel like a nonexplanation.

Conclusion

I have recommended a form of "puritanism" about the intentional strategy (and Dennett's intentional stance). Trusting the feeling that we have encountered something with a purpose is not always mistaken, but when talk of purposes, intentions, and meanings are woven together into a coherent narrative, we should be wary of illusions of understanding that accompany misapplication of the intentional strategy.

Illusions are worse than local mistakes. As Charles Taylor put it, "we speak of illusion when we are dealing with something of greater substance than error, [it is] error which in a sense builds a counterfeit reality of its own. . . . Such illusions are more than errors in this sense: they are sustained by certain practices of which they are constitutive" (Taylor 1985, 54). Just as optical illusions persist even when you have full knowledge of the underlying mechanism of the illusion, so can cognitive illusions persist even when you know which underlying (and evolutionary adaptive) processes have led you astray. Illusions of understanding mislead you even when you are fully aware of their causes. The intrusion of intentional thinking and intentional concepts in scientific models may have a potentially dangerous side effect: exaggerated vigilance vis-à-vis intentional concepts can also, unintentionally, affect our natural confidence in the explanations in which agentive concepts figure correctly, thus inspiring revisionist and eliminativist proposals. Agentive explanatory strategies did not evolve to function as protoscientific theories, but their role in explanatory strategies that enhance human cooperation should not be discredited just for that reason.

Acknowledgments

Thanks to Maarten Boudry, Helen de Cruz, Johan Braeckman, and Wouter D'hooghe for comments on earlier versions of this paper. Special thanks to

Maarten Boudry and Massimo Pigliucci for inviting me to further reflect on the sources and effects of flawed thinking in science and philosophy.

NOTES

1. Psychoanalysts, of course, will deny that it was disturbing.

2. "The strongest reason for considering Freud a pseudo-scientist is that he claimed to have tested—and thus to have provided the most cogent grounds for accepting—theories which are either untestable or even if testable had not been tested. It is spurious claims to have tested an untestable or untested theory which are the most pertinent grounds for deeming Freud and his followers pseudoscientists (though pseudo-hermeneutic would have been a more apposite and felicitous description)" (Cioffi 2005).

3. See Buekens and Boudry (2011) for an explanation of why Freud's explanations were so successful.

4. Gopnik (1998) compares the satisfaction conveyed by an explanation with that of an orgasm.

5. Children not always appreciate that explanations come to an end. Does that mean that their epistemic sentiments ("now I understand") are not yet fully developed? Or that they find pleasure in seeking satisfaction of that sentiment because they have learned to appreciate it?

6. An important argument for the model approach in philosophy of science was that models produce understanding. Good models render the unfamiliar intelligible. Illusions of understanding emerge when one takes one's understanding of other persons as a model for understanding material processes and events in general.

7. Consider how William James experienced the San Francisco earthquake when he visited Stanford University in 1906:

> I personified the earthquake as a permanent individual entity. . . . Animus and intent were never more present in any human action, nor did any human activity ever more definitely point back to a living agent as its source and origin. All of whom I consulted on the point agreed as to this feature in their experience, "It expressed intention," "It was vicious," "It was bent on destruction," "It wanted to show its power." . . . For science . . . earthquake is simply the collective name of all the cracks and shakings and disturbances that happen. They are the earthquake. But for me the earthquake was the cause of the disturbances, and the perception of it as a living agent was irresistible. It had an overpowering dramatic convincingness.

William James, "On Some Mental Effects of the Earthquake," in *Memoirs and Studies* (New York: Longmans, Green 1911), 212–13 (qtd. in Cioffi 1998b, 95).

8. Michael Ruse argues that we cannot dispense with the metaphor of design in biology and we can, in contexts of discovery, exploit the illusion to good purposes. Paul S. Davies, who discusses Ruse's view, argues that precisely because we *know* that purposive thinking has its roots in misapplications of the intentional strategy, we should withhold such applications.

9. The full quotation: "The only thought which philosophy brings with it, in regard to history, is the simple thought of Reason—the thought that Reason rules the world, and that world history has therefore been rational in its course."

10. Keeley argues that conspiracy theorists are some of the last believers in an ordered universe: "By supposing that current events are under the control of nefarious agents, conspiracy theories entail that such events are capable of being controlled" (1999, 123).

11. According to Sellars (1963), this habit is not part of the manifest image itself; the agentive ideas are, according to him, part of the "original image." The manifest image is a modified version of the original, the modification consisting of a gradual pruning of agentive ideas ascribed to inanimate entities.

12. See Brooks (1991) and the recent wave of work on embodied cognition for some nonrepresentational conceptions of cognition.

13. Similar tendencies were criticized by D. J. Buller (1999a, 111), who connects this tendency with the Freudian legacy: "It is the Freudian legacy of the dynamic unconscious that tempts us . . . to internalize adaptive goals into the unconscious and then view them as the hidden driving force behind our behavior and reproductive success."

14. This echoes Wittgenstein's claim in the *Tractatus*: "Men have always had a presentiment that there must be a realm in which the answers to questions are systematically combined—a priori—to form a self-contained system" (Wittgenstein 1922, 5.451).

15. Shtulman (2006) argues that the influence of the intentional stance shows up in the understanding of evolutionary ideas, even in students who had extensive instruction in the theory.

REFERENCES

Abrahams, Marc. 2011. "The Psychoanalyst Says Your Gut Says . . ." *Annals of Improbable Research* 17:2, 17.

Bechtel, W., and Robert C. Richardson. 1993. *Discovering Complexity: Decomposition and Localization as Strategies in Scientific Research*. Princeton, NJ: Princeton University Press.

Bering, Jesse. 2011. *The God Instinct: The Psychology of Souls, Destiny and the Meaning of Life*. London: Nicholas Brealey.

Bouveresse, Jacques. 1995. *Wittgenstein Reads Freud: The Myth of the Unconscious*. Princeton, NJ: Princeton University Press.

Boyer, Pascal. 2001. *Religion Explained: The Evolutionary Origins of Religious Thought*. New York: Basic Books.

Brooks, R. A. 1991. "Intelligence without Representation." *Artificial Intelligence* 47:139–59.

Buekens, Filip, and Maarten Boudry. 2011. "Psychoanalytic Facts as Unintended Institutional Facts." *Philosophy of the Social Sciences* 42 (2): 239–69.

Buenting, J., and J. Taylor. 2010. "Conspiracy Theories and Fortuitous Data." *Philosophy and the Social Sciences* 40 (4): 567–78.

Buller, David J. 1999a. "DeFreuding Evolutionary Psychology: Adaptation and Human Motivation." In *Where Biology Meets Psychology: Philosophical Essays*, edited by V. G. Hardcastle, 99–114. Cambridge, MA: MIT Press.

———. 1999b. *Function, Selection, and Design*. Albany, NY: SUNY Press.

Cioffi, Frank. 1998a. *Freud and the Question of Pseudoscience*. Chicago: Open Court.

———. 1998b. *Wittgenstein on Freud and Frazer*. Cambridge: Cambridge University Press.

———. 2005. "Was Freud a Pseudoscientist?" *Butterflies & Wheels* (blog), November 9. http://www.butterfliesandwheels.org/2005/was-freud-a-pseudoscientist/.

Corbey, Raymond. 2005. *The Metaphysics of Apes*. Cambridge: Cambridge University Press.

Coady, David. 2003. "Conspiracy Theories and Official Stories." *International Journal of Applied Philosophy* 17 (2): 197–209.

Crews, Frederick C. 2006. *Follies of the Wise: Dissenting Essays*. Emeryville, CA: Shoemaker and Hoard.

Da Silva, G. 1990. "Borborgymi as Markers of Psychic Work During the Analytic Session: A Contribution to Freud's Experience of Satisfaction and to Bion's Idea About the Digestive Model for the Thinking Apparatus." *International Journal of Psycho-Analysis* 71 (4): 641–59.

Davidson, Donald. 1984. *Inquiries into Truth and Interpretation*. Oxford: Oxford University Press.

Davies, Paul S. 2009a. "Conceptual Conservativeness: The Case of Normative Functions." In *Functions in Biological and Artificial Worlds*, edited by U. Krohs and Peter Kroes, 127–46. Cambridge, MA: MIT Press.

———. 2009b. *Subjects of the World*. Chicago: University of Chicago Press.

Dawkins, Richard. 1976. *The Selfish Gene*. Oxford: Oxford University Press.

———. 1998. *Unweaving the Rainbow*. London: Penguin.

Dennett, Daniel. 1987. *The Intentional Stance*. Cambridge, MA: MIT Press.

———. 1991. *Consciousness Explained*. Boston: Little, Brown.

———. 1996. *Kinds of Minds: Towards an Understanding of Consciousness*. New York: Basic Books.

———. 2011. "Homunculi Rule: Reflections on Darwinian Populations and Natural Selection by Peter Godfrey Smith." *Biology and Philosophy* 26:475–88.

Evans, Gareth. 1985. "Molyneux' Question." In *Collected Papers*. Oxford: Oxford University Press.

Farrell, John. 1996. *Freud's Paranoid Quest: Psychoanalysis and Modern Suspicion*. New York: New York University Press.

Ferguson, Niall. 2011. *Virtual History*. London: Penguin Books.

Freud, Sigmund. 1953. *The Standard Edition of the Complete Psychological Works of Sigmund Freud*. Translated by James Strachey. London: Hogarth Press.

Godfrey-Smith, Peter. 2009. *Darwinian Populations and Natural Selection*. Oxford: Oxford University Press.

Gopnik, Alison. 1998. "Explanation as Orgasm." *Mind and Machines* 8:101–18.

Hume, David. (1739–40) 1888. *A Treatise of Human Nature*. Edited by L. A. Selby-Bigge. Oxford: Oxford University Press.

Keeley, B. L. 1999. "Of Conspiracy Theories." *Journal of Philosophy* 96:109–26.

Keil, Frank C. 1992. "The Origins of an Autonomous Biology." In *Minnesota Symposia on Child Psychology: Modularity and Constraints in Language and Cognition*, edited by M. R. Gunnar and M. Maratsos, 25:103–37. Hillsdale, NJ: Erlbaum.

———. 2004. "Are Children 'Intuitive Theists'? Reasoning about Purpose and Design in Nature." *Psychological Science* 15:295–301.

McDowell, John. 2008. "The Disjunctive Conception of Experience as Material for a Transcendental Argument." In *Disjunctivism: Perception, Action and Knowledge*, edited by Adrian Haddock and Fiona McPherson, 376–89. Oxford: Oxford University Press.

McGinn, Colin. 1989. *Mental Content*. Oxford: Blackwell.

McKay, Ryan T., and Daniel Dennett. 2009. "The Evolution of Misbelief." *Behavioural and Brain Sciences* 32:493–561.

Macmillan, Malcolm. 1996. *Freud: The Complete Arc*. Cambridge, MA: MIT Press.

Metzinger, Thomas. 2003. *Being No-One*. Cambridge, MA: MIT Press.

Midgley, Mary. 1985. *Evolution as Religion*. London: Routledge.

Morton, Adam. 2003. *The Importance of Being Understood: Folk Psychology as Ethics*. London: Routledge.

Müller, Christian. 1984. "New Observations on Body Organ Language." *Psychotherapy and Psychosomics* 42 (1–4): 124–26.

Nagel, Thomas. 1979. *Mortal Questions*. Cambridge: Cambridge University Press.

Piaget, Jean. 1929. *The Child's Conception of the World*. London: Routledge and Kegan Paul.

Premack, David, and Guy Woodruff. 1978. "Does the Chimpanzee Have a 'Theory of Mind'?" *Behavioral and Brain Sciences* 1 (4): 515–26.

Ramsey, Frank. 1990. *Philosophical Papers*. Edited by D. H. Mellor. Cambridge: Cambridge University Press

Ramsey, William M. 2007. *Representation Reconsidered*. Cambridge: Cambridge University Press.

Ruse, Michael. 2003. *Darwin and Design: Does Evolution Have A Purpose?* Cambridge, MA: Harvard University Press.

Sellars, Wilfrid. 1963. "Philosophy and the Scientific Image of Man." In *Science, Perception and Reality*. London: Routledge and Kegan Paul

Shtulman, A. 2006. 'Qualitative Differences Between Naïve and Scientific Theories of Evolution.' *Cognitive Psychology* 52:170–94.

Strawson, Galen. 1996. "Luck Swallows Everything." http://www.naturalism.org/strawson.htm.

———. 2004. "Against Narrativity." *Ratio* 17:428–52.

Taylor, Charles. 1985. "Interpretation and the Sciences of Man." In *Philosophy and the Human Sciences*. Philosophical Papers 2. Cambridge: Cambridge University Press.

Thompson, M. 1995. "The Representation of Life." In *Virtues and Reasons*, edited by R. Hursthouse, G. Lawrence, and W. Quinn, 247–97. Oxford: Oxford University Press.

Trout, J. D. 2007. "The Psychology of Scientific Explanation." *Philosophy Compass* 2 (3): 564–91. doi 10.1111/j.1747–9991.2007.00081.x.

Turner, Stephen. 2010. *Explaining the Normative*. Oxford: Polity.

Velleman, David J. 2003. "Narrative Explanation." *Philosophical Review* 112:1–25.

Wittgenstein, Ludwig. 1922. *Tractatus Logico-Philosophicus*. London: Routledge and Kegan Paul.

———. 1982. Last Writings on the Philosophy of Psychology. Vol. 1, *Preliminary Studies for Part II of Philosophical Investigations*. Oxford: Blackwell.

CONTRIBUTORS

Stefaan Blancke
University of Ghent
9800 Deinze
Belgium

Maarten Boudry
Department of Philosophy
   and Moral Sciences
Ghent University
B 9000 Ghent
Belgium

Sheralee Brindell
University of Colorado–Boulder
Louisville, CO 80027

Filip Buekens
University of Tilburg
5037 AB Tilburg
Netherlands

Frank Cioffi
University of Kent–England

Carol E. Cleland
University of Colorado–Boulder
Lafayette, CO 80026

Johan De Smedt
University of Ghent
9000 Ghent
Belgium

Evan Fales
University of Iowa
Iowa City, IA 52246

Barbara Forrest
Southeastern Louisiana State
Holden, LA 707443

Erich Goode
New York University
New York, NY 10012

Sven Ove Hansson
Royal Institute of Technology
16850 Bromma
Sweden

Jesper Jerkert
Royal Institute of Technology
SE-17075 Solna
Sweden

Noretta Koertge
Indiana University
Bloomington, IN 47401

James Ladyman
Bristol University
Bristol, UK

Martin Mahner
Center for Inquiry
D-64380 Rossdorf
Germany

Thomas Nickles
University of Nevada–Reno
Reno, NV 89509

Ronald L. Numbers
University of Wisconsin
Madison, WI 53703

Massimo Pigliucci
City University of New York
New York, NY 10075

Donald Prothero
Occidental College
La Cresenta, CA 91214

Michael Ruse
Florida State University
Tallahassee, FL 32303

Nicholas Shackel
Cardiff University
Oxford OX2 9ED
United Kingdom

Michael Shermer
Skeptic Society
Altadena, CA 91001

Konrad Talmont-Kaminski
Konrad Lorenz Institute
05-822 Milanowek
Poland

Daniel P. Thurs
New York University
New York, NY 10012

Jean Paul Van Bendegem
Vrije Universiteit
B-9000 Ghent
Belgium

John S. Wilkins
University of Sydney
Sydney, Australia

INDEX

AAAS (American Association for the Advancement of Science), 128, 130, 160
Aaronovitch, David, 92
abduction. *See* alien abduction
Académie des Sciences, 170
ACLU (American Civil Liberties Union), 210, 212–13
acupuncture, 206, 312–13
adaptation (biological), 235–36, 238, 274, 322, 363, 365, 384
ad hoc reasoning, 88, 90, 92, 109–10, 187, 189, 310; hypothesis, 12, 41
Adler, Alfred, 10–11, 16, 80, 325
agency detection, 272, 275, 365, 369, 443
alchemy, 13, 122, 126, 133, 138, 147, 172–75, 178
alien abduction, 159, 206
alternative medicine, 3, 5, 29, 94, 138, 206, 301, 305–19
Alvarez, Luis and Walter, 194–95, 200
analogy (in argumentation), 289, 291, 298–99, 405
analytic/synthetic distinction, 95, 108
anomalistics, 101, 175–77
anomaly, 87, 89, 157, 160, 162, 175–77, 188–89, 194–95, 200, 206, 217, 253, 261, 269, 337–38

anthropology, 66, 132–33, 272, 281, 446
anthropomorphism, 272, 275, 405
anthroposophy, 243, 309, 314, 317
antiscience, 5, 138, 140, 341, 382–83, 397, 399, 400–404, 409–10, 412
archeology, 35, 42, 402
argumentation, 5, 95, 237, 287–304
Aristotle, 12–13, 22, 34, 104–5, 115, 170, 172, 225, 231, 366
Arlow, Jacob, 333, 336
astronomy, 12–13, 24, 35–36, 61, 134, 151, 188–89, 250, 303, 352, 417; Ptolemaic, 13, 211
asymmetry of overdetermination, 191–92, 194, 196–201
atheism, 131, 251, 260, 356
Atlantis, 65, 138
atomism, 56
Atran, Scott, 272, 274–79
Aura, 161, 166
autism. *See* vaccine scare

Bacon, Francis, 14, 21, 83, 105–6, 428
bad science, 31, 45–49, 52, 81, 86, 101, 103, 110–12, 116, 136–37, 167, 203, 221, 251
Barrett, Justin, 272, 275–76, 280
Bauer, Henry, 175

Bayes: Bayesian reasoning, 15; causal
　　networks, 184, 201; induction, 195, 199;
　　logic, 187
behaviorism, 449
Behe, Michael, 258–59, 261, 267, 367
belief system, 113–14, 117–18, 382; pseudo-
　　scientific, 3, 89, 92, 146, 147–49, 152,
　　155, 158–59, 162, 165, 168, 343, 345,
　　351, 365, 368, 373, 386
Bergson, Henri, 447
Berlitz, Charles, 135
Bernstein, Jeremy, 150
bias, 58, 161, 216–18, 318, 332, 352, 355–56,
　　365, 372–73, 406, 408–9, 432; cognitive,
　　363–64, 384, 401; confirmation, 167,
　　217, 400; deductive, 406, 408
Bible, 65, 88, 123, 148, 155, 159, 206–7, 250,
　　278, 385
Bird, Alexander, 57
Blackmore, Susan, 158
Blancke, Stefaan, 5, 382
Blumenbach, Johann Friedrich, 446–47
boldness (empirical), 80, 87–89, 95
Boudry, Maarten, 4, 267, 394
bounded rationality, 365, 382–83, 394, 399–401
Boyer, Pascal, 272, 274–76, 278, 280,
　　383–84, 445
Boyle, Robert, 51, 172–74, 178
Braeckman, Johan, 267
Brewster, David, 234–35
Broad, C. D., 256
bullshit, 4, 45, 52–53, 55, 57–59
burden of proof, 173, 218, 297, 354
Burke, Edmund, 412
Burr, Charles, 324, 327
Bush, George W., 137, 213
Butler, Joseph, 398
Byrne, Rhonda, 157

Campbell, George, 122, 130
Carey, Stephen S., 167
Carnap, Rudolf, 54, 103, 184, 453
Carter, James, 220–21
Catholic Church, 151
causal network, 184
Chambers, Robert, 232–33
charlatan, 3, 122, 126–27, 166, 179, 279
chemistry, 13, 51, 63, 66–67, 115, 126, 152,
　　156, 160, 166, 172, 174, 198, 228, 239,
　　241, 354

cherry-picking, 91, 217, 226
chiropractic, 206, 226–27
Chopra, Deepak, 390
Christianity, 237, 281
Christian Science, 128, 205
chrysopoeia, 172–74
Church of England, 234, 237
Cioffi, Frank, 91, 441
circularity, 9, 85, 113, 229, 432
clairvoyance, 31, 35, 85, 146, 154–55, 205,
　　314, 413
classification. See taxonomy
Cleland, Carol E., 5, 191, 200–201
climate change, 3, 5, 21, 183, 194, 342–43,
　　347, 350, 354–56, 398
cluster. See family resemblance
cognitive science of religion, 264, 267–68,
　　276, 385, 394
cold fusion, 16, 93, 176–77, 201
cold reading, 153
community (research), 21, 37–38, 42, 56,
　　63–65, 70, 110, 112, 125, 126, 135,
　　147–50, 158, 165–79, 215, 221–22, 291,
　　298, 300, 345, 348, 356–57, 374, 408–9,
　　425
Comte, Auguste, 106, 145
Condon, Edward, 123, 134–35
conservation of energy, 54, 56, 256–57, 260
conservatism, 3, 49, 131, 212–14, 220, 230,
　　354, 356, 397–98, 408–11, 413, 447
consistency, 32, 50, 167, 174, 311, 345
conspiracy theory, 3, 89, 92, 96, 167–68, 409,
　　445, 456; HIV, 3, 169, 175; 9/11, 166,
　　169, 409
Constitution, 137, 209–10, 216, 278
continental drift, 146
Conway, Erik, 345–46, 349–50
Copernicus, Nicolaus, 12, 151
corroboration, 83, 107, 170, 187, 347, 352
Cosmides, Leda, 362
cosmology, 39, 46, 191, 206, 219; inflation-
　　ary, 207
cover-up. See conspiracy theory
crank, 20, 149–51, 167, 221
creationism, 3, 11, 15–16, 24, 29–30, 35,
　　41–42, 48, 61, 83–84, 86, 88–91, 94, 102,
　　111–12, 117–18, 137, 148, 155, 158–59,
　　204, 206–15, 217, 221, 263–81, 342–45,
　　353–57, 365–74, 385, 403–4, 408–9
Crews, Frederick, 321–22, 441

crop circle, 300
CSI (Committee for Skeptical Inquiry), 135–36, 138
cult, 3, 54, 160. *See also* sect
Cuvier, Georges, 231–32, 413

Darwin, Erasmus, 229–31, 241
Davies, Paul Sheldon, 443, 446–48
Davis, Percival, 213, 372
Dawkins, Richard, 446, 451- 52
deduction, 9, 25, 184–85, 188, 195, 347, 397, 408, 411–12
deductivism, 397–98, 409, 411–12
delusion, 127, 150, 401, 419
demarcation problem: ballpark, 2, 84, 265, 269; criteria of, 9, 11–13, 16–20, 22, 25, 30–41, 49–50, 54–56, 62, 65–75, 80–83, 87, 92, 101, 103–17, 136, 210, 222, 247, 264–65, 269, 322, 334, 351, 412; essentialism, 2, 4, 22, 39–40, 114, 123, 134, 209, 412; fuzzy, 72–73 (*see also* fuzziness); history of, 4, 10, 12–14, 83, 101–18, 121–40, 167–68, 170–74, 221, 225, 227–28, 241, 259, 267; normative, 37–40, 80–84, 90, 92–93, 95; spectrum, 167, 301, 398; territorial, 79–86, 92–93, 95–96
Dembski, William, 267, 279, 368–69
democracy, 35, 105, 115, 117
demographics, 5, 204
demon, 205, 253, 453
denialism, 342–49, 354; climate change, 3, 5, 342–43, 346, 355; HIV, 17, 24, 175; Holocaust, 93, 344
Dennett, Daniel, 2, 81, 272, 370–71, 442–43, 446, 450, 454
Descartes, René, 51, 104–6, 255, 273, 446, 449
De Smedt, Johan, 5, 382
deviance (theories), 4, 68–72, 145, 154, 158–60, 162
devil, 204. *See also* demon
Dewey, John, 132–33
Diderot, Denis, 336
Discovery Institute, 178, 266, 267, 342, 423
Dobzhansky, Theodosius, 239
dogmatism, 39, 109, 176, 351, 406
dualism, Cartesian, 255
dual-process reasoning, 363–64
Du Bois, W. E. B., 133
Duhem, Pierre, 1, 12, 26, 50–51, 87, 91, 188, 253

Duhem-Quine thesis, 26, 91
Dummett, Michael, 195–96, 198, 201
Dunning, Brian, 168
Dupré, John, 10, 15, 19, 21, 31, 402
Durkheim, Émile, 103

economics, 63, 65, 130, 133, 152, 159, 206, 401, 419, 429; behavioral, 167
Eddington, Arthur, 11
Eddy, Mary Baker, 128
education. *See* science: education
Edwards v. Aguilard, 148
Einstein, Albert, 10–11, 80, 108, 150, 189, 400, 402
electrodynamics, 54
Empedocles, 225
Engels, Frederick, 47, 402
Enlightenment, 105, 225, 228
epidemiology, 191, 197, 329
epistemic values, 50, 179, 228, 417, 421–24, 428–35
epistemology, 13–14, 33–34, 57, 62, 82, 104–5, 114, 226–27, 263, 265–68, 270–71, 384, 425–26
equivocation, 86, 88, 93, 324, 327, 431
error, Type I and Type II, 57. *See also* false positive/negative
Erwin, Edward, 90, 321–22, 328, 330
ESP. *See* extrasensory perception
essentialism, psychological, 365–66, 368, 372, 406, 408, 410. *See also* demarcation problem
ethics, 4–6, 65, 103, 130, 287–88, 295, 301–2, 420–26, 229–35
evidentialism, 421–29, 433–35
Evo-Devo, 117–18
evolution (biology), 23, 47–48, 63, 84, 88, 111, 114, 117–18, 130–31, 136, 147–48, 158, 204, 208–15, 217, 225–43, 258, 272, 342, 354–57, 362–70, 383–85, 390, 403–4, 443–44, 448, 452–54
evolutionary psychology, 15, 17, 24, 31, 274, 362, 444
exaptation, 272, 274
expertise, 110–11, 117, 147, 228, 341–42, 349, 354–57
explanatory power, 39, 86, 104, 108, 162, 258–59, 370, 399, 449
extrasensory perception, 85, 134, 154, 206, 361

fallacy, 167, 294–99, 364; ad hominem, 90; *ad ignorantiam*, 297; *ad verecundiam*, 295–96, 299; affirming the consequent, 106; conjunction, 364; false dilemma, 345; genetic, 373; homunculus, 448–49; Loki's wager, 95; *post hoc ergo propter hoc*, 167

fallibilism, 13–14, 39, 106, 113, 389, 399

false positive/negative, 57, 189–91, 193, 195–96, 369, 444, 450. *See also* error, Type I and Type II

falsifiability, 1, 11, 30, 32, 37, 54–55, 71, 73–74, 80, 87, 90, 101, 108, 110–11, 136, 327, 351

family resemblance (Wittgenstein), 4, 19, 21, 25, 51, 406

fertility. *See* fruitfulness

Feyerabend, Paul, 118

Feynman, Richard, 207, 220, 343, 451

Fisher, R. A., 239

Fleischmann, Martin, 177, 211

Fliess, Wilhelm, 441

Fodor, Jerry, 432

Force, James E., 266, 270

Forrest, Barbara, 5, 213, 266–67

Fort, Charles, 138–40

fossil record, 24, 195, 200, 230–31, 235, 237

Frankfurt, Harry, 4, 45, 52–53, 55, 57

Franklin, Benjamin, 228–29

fraud, 4, 31, 35, 45–53, 57–59, 68–69, 102, 117, 130, 157, 173, 216, 301, 317–18, 357

Freud, Sigmund, 1, 10–11, 16, 47, 59, 73, 80, 89–92, 96, 108, 133, 206, 321–38, 362, 402, 440–43, 445, 450–55

fringe science, 65, 101, 138, 172, 175–77, 203, 220, 226–27, 242, 348, 353

Fromm, Erich, 332

fruitfulness (fertility), 12, 32, 39, 66–67, 103, 110, 116–17, 150, 242

fundamentalism, 404

fuzziness: demarcation, 13, 38, 42, 93, 126–27; logic, 25, 42

Galilei, Galileo, 13, 34, 104, 150, 170–71, 298–99

Gall, Franz Joseph, 147

Gardner, Martin, 68, 138, 150

Gee, Henry, 200

Geller, Uri, 135, 161

genetic modification, 227, 243, 385, 410

genetics, 147, 226, 239–40, 363; Mendelian, 225, 239–40, 242

geology, 16, 49, 126, 146, 159, 191–92, 194–95, 234, 352, 408

gerrymandering. *See* ad hoc reasoning

ghost, 38, 145, 204, 386–87. *See also* mind, disembodied

Gigerenzer, Gerd, 363–64, 375, 382, 399, 401

Gish, Duane, 357, 367

global warming, 35, 137, 198, 343, 346, 348–49, 354–55, 403, 406, 409

God, 16, 41, 84, 94–96, 104–5, 122, 131, 135, 148, 157–58, 166, 204, 207, 233, 249–61, 266–69, 271–73, 275, 279, 282, 292, 304, 330, 351, 361, 365, 369, 371, 385–86, 453; as creator, 16, 41, 94, 158, 209, 266–67, 366, 447; interventionist, 94, 228–29, 252, 255, 258, 261; Judeo-Christian, 104; Supreme Being, 215, 270

Godfrey-Smith, Peter, 369, 446, 452–53

Goodman, Alvin, 187

Gordin, Michael D., 139, 221, 227

Gould, Stephen Jay, 226, 241, 334, 402

Gray, Asa, 236, 301

Grice, H. P., 303

grueness, 187. *See also* induction

Grünbaum, Adolf, 90, 321–22, 330, 333, 337

Guthrie, Stewart, 272, 369, 384

Habermas, Jürgen, 301–2

Hacking, Ian, 200

Haeckel, Ernst, 238

Haldane, J. B. S., 239

Hansson, Sven Ove, 31, 36, 86–87

Heisenberg, Werner, 115, 256

Hempel, Carl, 108, 171, 184, 382

hermeneutics, 6, 92, 439, 441, 455

heuristic, 5, 109–10, 112, 116–17, 206, 303, 363–66, 374–75, 399, 401, 413, 443, 448–49; fast and frugal, 363, 365–66, 368, 374

Hines, Terence, 41, 139, 154, 158, 321, 402

holism, 110, 227, 291, 321–13

Holmes, Oliver Wendell, 121–22, 127

Holocaust. *See* denialism

homeopathy, 3, 30, 35, 48–49, 53, 69, 125, 127, 178, 183–84, 197, 307, 310, 315, 361, 372, 402

homunculus, 448–49

Hooke, Robert, 51

Hooker, Joseph, 236

Hovind, Kent, 356–57
Hubbard, Ron L., 387
Hull, David, 22, 170, 402
humanities, 31, 63–65, 74, 105, 169
Hume, David, 5, 9, 106, 250, 255, 257, 263–82, 405, 411, 446
Huxley, Aldous, 343
Huxley, Thomas, 122, 130–31, 236–38, 241, 352, 357
Hyman, Ray, 153
hypothesis: ad hoc, 12, 41; ancillary, 12, 409; auxiliary, 32, 87–88, 188–91, 193, 195, 197, 227, 311

ideology, 15–16, 21, 35, 38, 58, 179, 217, 225, 227–28, 230, 234, 236, 240–41, 331, 343, 345, 418–19, 429, 433
illusion, 6, 263, 293, 401, 435, 439–58
immaterial (beings), 248, 255–58, 260
immunizing strategy, 5, 91
incoherence, 19
induction, 9–10, 14, 80, 106, 126, 186–87, 195, 198–99, 237, 411; new riddle of, 187; problem of, 9, 80, 186–87, 199. *See also* grueness
inference to the best explanation, 14, 411
Inquisition, 298
Intelligent Design, 17, 24, 48, 84, 117, 137–38, 148, 178, 212–16, 258–59, 261, 263, 275, 279, 365, 369, 372, 402, 446, 452
intentional stance, 272, 369–71, 374, 387, 442, 451, 454, 456
interdependence, 82, 92
intuition, 15, 17–18, 91, 277, 291, 365–66, 368–69, 371–74, 388, 391–92, 398, 446; intuitive ontology, 274, 276–77, 386
irrationality, 5, 109, 178, 196, 272, 279, 361–75, 383, 399–400, 428–29, 432, 434
Irving, David, 101, 344

James, William, 327, 334, 455
Jenner, Edward, 409
Johnson, Samuel, 122, 281
Jones, Ernest, 324–25, 327, 332, 337
Jones, John (judge), 112, 114, 212–15
Jung, Carl, 324–25
junk science, 101–2, 113, 140, 203, 348

Kahneman, Daniel, 167, 293, 364
Karén, Michelle, 152

Keeley, Brian, 89, 445, 456
Kelemen, Deborah, 367, 371, 443–44
Kelvin, Lord, 115
Kenyon, Dean, 213
Kepler, Johannes, 12, 51
Kirlian photography, 161
Kitcher, Philip, 32, 42, 72, 87–88, 91, 116, 266, 280–81, 374
Kitzmiller v. Dover Area School District, 111–12, 258, 261
Kuhn, Thomas, 2, 10, 12, 32, 50–51, 62, 72–73, 91, 104, 108–11, 114, 116, 118, 207, 227, 428, 432, 436
Kukla, André, 24, 93
Kurtz, Paul, 135–36

Lakatos, Imre, 32, 37, 48–49, 72, 87, 109, 116, 136, 178; "research programmes," 87, 109
Lamarck, Jean-Baptise de, 231–32, 243, 390, 413
Langmuir, Irving, 72, 101
Larmer, Robert, 256–57
Laudan, Larry, 2, 4, 9–10, 12–22, 25–26, 29, 33–34, 36, 49–50, 54, 79–96, 106, 110–12, 115–16, 118, 136–37, 139, 227, 432
Lavoisier, Antoine, 228
Lear, Jonathan, 336
LeConte, Joseph, 129
libido, 324–27
literary criticism, 46
Loch Ness, 175, 178
Locke, John, 104, 249, 273
logic, 5, 9, 12, 14, 16, 19, 22, 30, 49, 80–82, 87, 90–92, 101, 106–8, 111, 113, 115, 174, 183–91, 195–96, 199, 260, 269, 278, 287–304, 322, 327, 334, 347, 364–65, 368, 375, 383, 393, 401, 441; Bayesian, 187; as discipline, 80, 101, 103, 288; first-order, 186; fuzzy, 25, 42; holistic, 312–13; laws of, 26; and methodology, 32, 37–38, 104; nonclassical, 25, 290–91; of science, 107, 111
logical positivism (empiricism) 14, 71, 80, 103, 107–8, 185
Lombroso, Cesare, 147, 334
Lysenkoism, 38, 138

Mack, John, 159
Malinowski, Bronislaw, 132

Marwen, Jan, 177

Marx, Karl, 10–11, 15–16, 47, 55, 80, 108, 145, 151

materialism, 131, 234, 423

Mayr, Ernst, 22, 240, 366

McCarthy, Jenny, 355

McCauley, Robert, 381, 384–89, 391, 393–94

McClenon, James, 160–61

McDougall, William, 47

McDowell, John, 449

McGinn, Colin, 449

McLean v. Arkansas, 111

meaningfulness, criterion of, 71, 107

memetics, 21, 392, 453

Mencken, Herbert L., 133–34

Mendel, Gregor, 225, 239–40, 242

Mercier, Hugo, 373

mesmerism, 139, 228, 239

metaphor, 88, 173, 289, 291, 294–95, 303, 372, 443, 448, 452, 455

metaphysics, 10, 36, 71–72, 79–81, 84–85, 101, 104, 107, 115, 173, 238, 263–65, 267, 271, 281, 444

method, hypothetico-deductive, 104, 106, 171, 184, 208

methodology, 32, 38, 64, 75, 109, 132–33, 136, 185–86, 197, 251, 307, 432, 434

Michelson-Morley experiment, 59

Mill, John Stuart, 14, 250, 259, 411

Milton, John, 268

mind, disembodied, 248–49, 251

miracle, 85, 193, 204, 250, 252–53, 255, 260, 267, 270–71, 276, 279, 347

modus ponens, 260, 288, 291

Morris, Henry, 148, 213, 368–69, 371–72

multidimensionality, 25, 167–68, 309

multiple endpoints, 88

National Academy of Sciences, 301

naturalism, 34, 82, 84–86, 210–11, 249, 251–53, 259, 264, 266–67, 351, 389, 391, 447; methodological, 84–86, 95, 249, 251–53, 267, 389; ontological, 249

NCSE (National Center for Science Education), 214

necessary and sufficient conditions, 2, 16–17, 19, 22, 25, 30, 33–34, 39–40, 52, 71, 73, 80, 86, 104, 425

Neptune, 12, 188

neuroscience, 81, 207

Newman, John Henry, 235

Newton, Isaac, 13, 34, 46, 51, 104–5, 108, 150–51, 172, 174, 188–89, 278–79, 446–47

New York Times, 129, 159, 220, 226, 346

Nietzsche, Friedrich, 410

9/11 truth movement, 166, 169, 409

nonscience, 1, 29, 31, 45–51, 62, 68, 79, 85–86, 101, 103, 105, 109, 114, 116, 146, 205, 207, 382, 402, 412

null hypothesis, 218

Oedipus complex, 335–36

ontology, 38, 249, 252, 270. *See also* intuition

Oreskes, Naomi, 346, 349–50

origin of life, 213, 249

osteopathy, 373

overdetermination, 185, 191–92, 194–201

Overton, William (judge), 111–12, 114, 209

Paine, Thomas, 95

paleocontact, 385

paleontology, 191, 200, 217, 235–36, 281, 354, 357

Paley, William, 447

paradox, 31, 138, 186–87, 200, 240, 242, 265, 271, 276, 404; of confirmation, 186–87; Sorites, 25

paranoia, 150, 167, 322, 334, 450, 453

paranormal, 3–4, 146–49, 154–57, 159–62, 175, 204, 216, 251, 261–65, 296–97, 361

parapsychology, 16, 39, 85, 90–92, 94, 139, 148–49, 154–62, 296–98

Park, Robert, 102

Parker, Gary, 213

Parmenides, 12

Paul, apostle, 122, 259, 325

Pennock, Robert, 33–34, 84–85, 213, 252–54, 259–60, 264–65

periodic table, 56

Philosopher's Stone, 172–74

philosophy, natural, 104, 122, 124–25, 172, 174, 269–70

phlogiston, 46, 109, 220–21

phrenology, 121–22, 125, 129, 134, 139, 147, 232, 234, 239

physics, laws of, 16, 160

Piaget, Jean, 443

Pigliucci, Massimo, 4, 41, 93, 95, 351, 356

placebo effect, 58

Plait, Phil, 343
Plantinga, Alvin, 253
Plato, 17, 104, 325
Polanyi, Michael, 110, 428
politics, 3, 16, 20–21, 26, 35, 45, 48–49, 102,
    105, 111–13, 115, 137, 173, 217, 228–32,
    238, 263, 342, 345–46, 357, 406, 410,
    432; political science, 167, 408
Pons, Stanley, 201
Popper, Karl, 1, 2, 9–16, 30–32, 46–47,
    54–55, 71–74, 80–83, 87–88, 90–91, 94,
    101, 106–14, 136, 166, 170, 184, 187–91,
    242, 333, 351, 397, 408, 411–12, 432
positivism. See logical positivism
postmodernism, 4, 93, 102
Pratkanis, Anthony, 300
precognition, 154, 176
prediction, 11, 16, 39, 54–56, 63, 69, 86,
    88–89, 104, 106, 109–10, 150–51,
    153–55, 159, 162, 178, 186–93, 195, 200,
    208, 210, 226–28, 240, 253–55, 260, 272,
    275, 288, 293–94, 304, 351, 364, 370,
    387, 413, 442, 444–45
Premack, David, 444
Prince, Morton, 325
problem solving, 32, 88, 108, 116
progress (in philosophy), 12, 14, 21
Protestantism, 131, 235, 264
protoscience, 13, 15, 17, 23, 31, 149, 454
pseudohermeneutics, 6, 92, 441
pseudohistory, 92, 94, 385
pseudophilosophy, 92, 94
pseudoproblem, 2, 33–34
psi, 90–91, 149, 154–58, 160, 401; negative
    psi, 90–91; psychic powers, 149, 154,
    156, 158
psychiatry, 58, 133, 135, 138, 159, 326, 409
psychic, 88, 90, 138–39, 148–49, 152–55,
    158–59, 175–76, 205, 300, 331, 440, 451
psychoanalysis, 1, 3, 5, 10–11, 15, 55, 61,
    73, 80, 89, 91–92, 96, 206–7, 321–22,
    325, 328–33, 335–37, 362, 439–41, 445,
    450–51, 455
psychokinesis, 30, 138, 154, 175, 178
psychology, evolutionary. See evolutionary
    psychology
Ptolemaic. See astronomy
Putnam, Hilary, 291
puzzle, 4, 72–73, 108–9, 337, 431, 451
Pythagoras, 57

quackery, 35, 51, 122, 126–27, 129, 147, 348,
    356
Quine, W. V. O., 1, 24, 26, 91, 95, 108, 291,
    430
Quinn, Philip, 90

race, 58, 105, 133, 217, 233, 236
Radin, Dean, 157–58, 160
Ramsey, Frank P., 448
Ramsey, William M., 449, 458
Randi, James, 300, 304
randomization, 74, 261, 316–17
randomized clinical trial. See RCT
rationality, 39, 109, 184, 363, 368, 375, 424,
    426–32, 434; bounded, 365, 382–83,
    394, 399–401; ecological (adaptive),
    368, 374
RCT, 306–7
recovered memory, 207
reductionism, 31, 41, 103, 220, 227, 248, 449
Reese, David Meredith, 126
regress, logical, 113
Reichenbach, Hans, 411
Reisch, George, 72–73
relativism, 14
relativity, theory of, 11, 146, 150; general, 1,
    10, 51, 54, 80, 189; special, 49
reliability, 14, 33–34, 37, 39, 41–42, 45,
    56–57, 59, 63, 65–71, 73–74, 101, 104–5,
    111, 113, 116–17, 125, 168, 170, 197,
    210–11, 216, 252–53, 269, 308, 316, 332,
    347, 365, 373–74, 388, 404, 413, 419,
    421–22
religion, 41, 83–85, 95–96, 102–3, 115, 122,
    125–26, 128–31, 145, 148–49, 151, 159,
    162, 205, 228, 234, 237–38, 241, 298,
    314, 342, 350–53, 356–57, 371, 374, 381,
    384–93, 408; cognitive science of, 276,
    385; demarcation, 65, 105; popular, 381,
    386, 388–93; religious pseudoscience,
    5, 48, 83, 102, 148, 209–10, 214–15, 261,
    263–81, 387; science and, 84, 128, 130,
    136–37, 211, 265, 388; secular, 237, 241
repeatability, 72–74, 157, 249–50, 252
revolution, scientific. See Scientific
    Revolution
rhetoric, 2, 26, 105, 115, 123–24, 126–28,
    130, 132, 134, 140, 177, 295, 307, 368,
    372, 402, 431, 449
Rhine, Joseph Banks, 102, 149

risk (empirical). *See* boldness
Rousseau, Jean-Jacques, 338, 399, 413
Royal Society of London, 105
Ruse, Michael, 5, 72, 111, 136–37, 448, 455

Sagan, Carl, 135–36, 168, 353, 355
Salem, Bruce, 403
science: borderlines (boundaries) of, 2, 4,
     31, 41, 79, 85, 103, 123, 126, 130–37,
     140, 184, 205, 216, 280; education,
     3, 29, 34–36, 42, 48, 134, 145, 166,
     168, 174, 211, 214, 221, 280–81, 342,
     367; institutions of, 58, 82, 165, 168,
     170–71, 174–78, 219, 266, 301, 391;
     mature, 13, 108, 167, 178, 228, 235, 242;
     natural, 63–65, 74, 95, 107, 121, 157,
     160, 175, 184, 191, 196, 270; nature of,
     2–3, 34–36, 114, 209, 215, 347, 390,
     423; pathological, 101, 203; popular,
     5, 128–30, 225–26, 228, 235, 238–39,
     241–42; power structure of, 3; social,
     51, 55, 63–64, 130, 132–33, 135, 154,
     156, 175, 248, 362, 423, 429, 432; social
     organization of (pseudo), 4, 45, 165,
     175; soft, 5, 17, 23; unity of, 31–32, 56,
     73, 75, 107, 114
scientia, 64, 95, 104
Scientific Revolution, 151, 165, 172
scientism, 103, 114
Scientology, 3, 205, 387–89
Searle, John, 292
sect, 3, 131. *See also* cult
Sedgwick, Adam, 234–35
Seethaler, Sherry, 168
seismology, 197
self-deception (self-delusion), 48, 52, 419
selfish gene, 446, 452–53
Sellars, Wilfrid, 456
Semmelweis, Ignaz, 171
SETI (Search for Extraterrestrial Intelli-
     gence), 24, 135, 206
Seventh-Day Adventism, 123
Shackel, Nicholas, 6, 93
Shakespeare, William, 65, 83, 337
Shermer, Michael, 5, 65, 138–39, 168, 211,
     282, 344, 397, 402
Sidgwick, Henry, 421
Simon, Herbert, 382–83, 401
simplicity, 50–51, 302
Simpson, George Gaylord, 240

sindonology, 65
Singleton, Stephanie, 169
skepticism, 2, 5, 58, 86, 91, 113, 125, 131,
     135, 136, 138, 146, 160, 167–68, 173–74,
     183–84, 197, 218, 221–22, 263, 270, 278,
     293–94, 300, 338, 343, 353; philosophi-
     cal, 253, 449
smoking and cancer, 112, 346–50, 354
Snelling, Andrew, 408
Snow, Charles Percy, 63
Sober, Elliott, 84, 260
social constructivism, 4, 93
Society for Scientific Exploration, 175
sociology, 4, 36, 64, 81, 91, 130, 132, 139,
     157, 174
Socrates, 1–2, 405
soothsaying, 16
species: biological, 22, 39–40, 42, 59, 148,
     233, 352, 366, 371, 392, 447; con-
     cept, 4, 22, 39–40, 42. *See also* family
     resemblance
Spencer, Herbert, 145, 238–39
Sperber, Dan, 89, 363, 373–74
spiritualism, 126, 157, 168, 234
Steiner, Rudolf, 227, 242
Stove, David, 88, 417–18, 435
Strawson, Galen, 445, 450
strong programme (Edinburgh School),
     64, 139
subterfuge. *See* immunizing strategy
Sulloway, Frank, 330
supernatural, 5, 24, 30, 38, 58, 83–86, 89–90,
     94, 96, 131, 145–46, 209, 211, 247–61,
     263–82, 351, 369, 371, 381–94
superstition, 35, 145, 265, 279, 281, 384, 392
synthesis, extended (biology), 118

Tart, Charles T., 154, 156
taxonomy, 4, 20–21, 39, 45, 49, 51, 230, 231,
     235, 402, 408
Taylor, Charles, 454
teleology, 365, 367–69, 371–72, 374, 405,
     443–44, 452
Tennyson, Alfred, 233
testability, 5, 15, 24, 32, 39, 54, 56, 85–86,
     101, 103, 106, 114, 116, 321–22, 325,
     328, 333–34, 336–37, 351
theism, 34, 85, 89, 159, 229, 249, 251,
     259–60, 266, 423; intuitive, 371, 444
theology, 34, 36, 41, 65, 122, 125, 130, 162,

166, 209, 253–54, 277, 280, 365, 381, 386, 388–93; natural, 254; neurotheology, 401; theological correctness, 275, 390

theory of mind, 370, 385, 444

thermodynamics, 51, 56, 192, 198

think tank, 341, 354, 356

TMLC (Thomas More Law Center), 212–13

Tooby, John, 362–63

Toumey, Christopher, 102

transubstantiation, 277

Tremlin, Todd, 268–69, 271–76, 279, 281

Trout, J. D., 187, 391, 441–42, 452–53

truth, 14, 19, 37, 45, 48, 52–53, 57, 59, 105–7, 112–13, 122, 136, 151, 160, 218, 249, 301, 336, 350–53, 419–25, 433–36, 443; probable, 25, 186

Truzzi, Marcello, 157–58

Tyndall, John, 122, 131

ufology, 31, 85, 134, 136–39, 149, 155, 159, 175, 178, 204, 206, 217, 218, 356

uncertainty principle, 256

unconscious, 89, 96, 329, 336, 363, 445, 450, 453, 456

underdetermination, 87, 106, 192–93, 195, 253

unification, 32, 65, 86, 178, 212

Uranus, 12, 188–89

Urbach, Peter, 109

vaccine scare, 21, 183, 356–57, 398, 409

vagueness, 16, 89–90, 153, 304, 351

validity, 149, 151, 153–54, 156, 188, 287, 305

van Frassen, Bas, 423

Velikovsky, Immanuel, 134–35, 175, 385–86, 389, 392–94

Vesalius, Andreas, 151

Vienna Circle, 103

vitalism, 373

von Däniken, Erich, 35, 42, 135, 385

Vulcan, 189

Waelder, Robert, 326–27

Wakefield, Andrew, 356–57

Wallace, Alfred Russel, 168, 233–34, 409

Wason, Peter C., 293

Wegener, Alfred, 146

well-testedness, 83, 212

Whewell, William, 14, 95, 106, 237, 406, 410

White, Ellen G., 123

Wilkins, John, 5, 382

Wilson, David, 256

Wilson, Edward O., 32, 411

Winter, Alison, 125, 139

Wissenschaft, 64, 95

witchcraft, 138, 204, 279

Wittgenstein, 4, 19–22, 25, 42, 103, 107

Wolfram, Stephen, 220–21

Woodruff, Guy, 444

World War II, 93, 134

Worrall, John, 109

Wright, Sewall, 239

Zahar, Elie, 110